71

Advances in Biochemical Engineering/Biotechnology

Managing Editor: T. Scheper

Springer-Verlag Berlin Heidelberg GmbH

Biopolyesters

Volume Editors: W. Babel, A. Steinbüchel

With contributions by
J.-U. Ackermann, W. Babel, U. Breuer, J. Choi,
G. Eggink, S. Hein, D. Jendrossek, B. Kessler,
Y. B. Kim, S. Kobayashi, P. E. Kolattukudy,
S. Y. Lee, R. W. Lenz, Y. Poirier, G. J. M. de Koning,
A. Steinbüchel, H. Uyama, G. A .M. van der Walle,
R. A. Weusthuis, B. Witholt

Springer

Advances in Biochemical Engineering/Biotechnology reviews actual trends in modern biotechnology. Its aim is to cover all aspects of this interdisciplinary technology where knowledge, methods and expertise are required for chemistry, biochemistry, microbiology, genetics, chemical engineering and computer science. Special volumes are dedicated to selected topics which focus on new biotechnological products and new processes for their synthesis and purification. They give the state-of-the-art of a topic in a comprehensive way thus being a valuable source for the next 3–5 years. It also discusses new discoveries and applications.

In general, special volumes are edited by well known guest editors. The managing editor and publisher will however always be pleased to receive suggestions and supplementary information. Manuscripts are accepted in English.

In references Advances in Biochemical Engineering/Biotechnology is abbreviated as Adv. Biochem. Engin./Biotechnol. as a journal.

ISSN 0724-6145

ISBN 978-3-662-14688-0 ISBN 978-3-540-40021-9 (eBook)
DOI 10.1007/978-3-540-40021-9

Library of Congress Catalog Card Number 72-152360

Originally published by Springer-Verlag Berlin Heidelberg New York in 2001.
Softcover reprint of the hardcover 1st edition 2001

Typesetting: Fotosatz-Service Köhler GmbH, Würzburg
Cover: E. Kirchner, Heidelberg
Printed on acid-free paper SPIN: 10719473 02/3020hu – 5 4 3 2 1 0

Managing Editor

Professor Dr. T. Scheper
Institute of Technical Chemistry
University of Hannover
Callinstraße 3
D-30167 Hannover/FRG
E-mail: scheper@mbox.iftc.uni-hannover.de

Volume Editors

Prof. Dr. Wolfgang Babel
UFZ-Umweltforschungszentrum
Leipzig-Halle GmbH
Sektion Umweltmikrobiologie
Permoserstr. 15
04318 Leipzig/FRG
E-mail: babel@umb.ufz.de

Prof. Dr. Alexander Steinbüchel
Westf. Wilhelms-Universität
Münster
Institut für Mikrobiologie
Correnssstr. 3
48149 Münster/FRG
E-mail: steinbu@uni-muenster.de

Editorial Board

Preface

Living systems synthesize seven classes of polymers. Some of them, for instance water insoluble polyesters, have become commercially attractive. Water insoluble polyesters are synthesized by a wide range of different prokaryotic microorganisms including eubacteria and archaea mostly as intracellular storage compounds for energy and carbon. They represent a rather complex class consisting of a large number of different hydroxyalkanoic acids and are generally referred to as polyhydroxyalkanoates (PHA). Water insoluble polyesters are also synthesized by plants as structural components of the cuticle that covers the aerial parts of plants. Eukaryotic microorganisms and animals are not capable of synthesizing water insoluble polyesters; only some eukaryotic microorganisms have been known which can synthesize the water soluble polyester polymalic acid.

The water insoluble polyesters possess interesting properties. They are biodegradable and biocompatible and exhibit physical and material properties making them suitable for various technical applications in industry, agriculture, medicine, pharmacy and some other areas. The microbial polyesters can be produced easily by means of well-known fermentation processes from renewable and fossil resources and even from potentially toxic waste products. However, the price of PHAs is rather high compared with conventional synthetic polymers. If we want to use these biopolymers, it is necessary to improve the economic viability of production process. Therefore, a lot of research work has been done. During the last decade significant progress has been made in elucidating the physiological, biochemical and genetic basis for the biosynthesis and biodegradation of these polyesters and also in developing effective process regimes. Novel applications have been found. The synthesis and intracellular as well as extracellular depolymerization of these polyesters are now understood quite well. The genes encoding the enzymes of the pathways or structural proteins attached to the PHA granules in bacteria have been cloned and characterized from many bacteria. The availability of this knowledge has contributed significantly to establishing new processes for the production of PHAs by means of recombinant bacteria and to tailoring the properties of these polyesters for instance by modifying the synthesis. Meanwhile production of PHAs by transgenic plants has come about, too, and in addition to the in vivo synthesis, purified enzymes are used to prepare this type of polyester in vitro.

This issue of Advances in Biochemical Engineering/Biotechnology presents 10 chapters dealing with different aspects of polyesters from microorganisms and plants, the biochemistry and molecular biology of the synthesis and degradation as well as the technical production and applications of these polyesters. It provides the state-of-the-art knowlegde in this rather rapidly developing, exciting and promising area.

The volume editors are indebted to the authors for their excellent contributions and cooperation in assembling this special volume.

November, 2000 Wolfgang Babel, Alexander Steinbüchel

Contents

Polyesters in Higher Plants

Pappachan E. Kolattukudy

The Ohio State University, 206 Rightmire Hall, 1060 Carmack Rd, Columbus OH 43210, USA
E-mail: Kolattukudy.2@osu.edu

Polyesters occur in higher plants as the structural component of the cuticle that covers the aerial parts of plants. This insoluble polymer, called cutin, attached to the epidermal cell walls is composed of interesterified hydroxy and hydroxy epoxy fatty acids. The most common chief monomers are 10,16-dihydroxy C_{16} acid, 18-hydroxy-9,10 epoxy C_{18} acid, and 9,10,18-trihydroxy C_{18} acid. These monomers are produced in the epidermal cells by ω hydroxylation, in-chain hydroxylation, epoxidation catalyzed by P_{450}-type mixed function oxidase, and epoxide hydration. The monomer acyl groups are transferred to hydroxyl groups in the growing polymer at the extracellular location. The other type of polyester found in the plants is suberin, a polymeric material deposited in the cell walls of a layer or two of cells when a plant needs to erect a barrier as a result of physical or biological stress from the environment, or during development. Suberin is composed of aromatic domains derived from cinnamic acid, and aliphatic polyester domains derived from C_{16} and C_{18} cellular fatty acids and their elongation products. The polyesters can be hydrolyzed by pancreatic lipase and cutinase, a polyesterase produced by bacteria and fungi. Catalysis by cutinase involves the active serine catalytic triad. The major function of the polyester in plants is as a protective barrier against physical, chemical, and biological factors in the environment, including pathogens. Transcriptional regulation of cutinase gene in fungal pathogens is being elucidated at a molecular level. The polyesters present in agricultural waste may be used to produce high value polymers, and genetic engineering might be used to produce large quantities of such polymers in plants.

Keywords. Cutin, Suberin, Hydroxy fatty acid, Epoxy fatty acid, Dicarboxylic acid

Advances in Biochemical Engineering/
Biotechnology, Vol. 71
Managing Editor: Th. Scheper

List of Abbreviations

CAT	chloramphenicol acetyl transferase
CD	circular dichroism
CMC	critical micellar concentration
CPMAS	cross polarization-magic angle spinning
CRE	cutin response element
CTF	cutinase transcription factor
DTE	dithioerythritol
GAL4	β-galactosidase reporter gene
GC-MS	gas chromatography-mass spectrometry
LSIMS	liquid secondary-ion mass spectrometry
NMR	nuclear magnetic resonance
PBP	palindrome binding protein
SDS	sodium dodecyl sulfate
TLC	thin layer chromatography
TMSiI	trimethylsilyl iodide

1 Occurrence

Plants were probably the first to have polyester outerwear, as the aerial parts of higher plants are covered with a cuticle whose structural component is a polyester called cutin. Even plants that live under water in the oceans, such as *Zoestra marina,* are covered with cutin. This lipid-derived polyester covering is unique to plants, as animals use carbohydrate or protein polymers as their outer covering. Cutin, the insoluble cuticular polymer of plants, is composed of interesterified hydroxy and hydroxy epoxy fatty acids derived from the common cellular fatty acids and is attached to the outer epidermal layer of cells by a pectinaceous layer (Fig. 1). The insoluble polymer is embedded in a complex mixture of soluble lipids collectively called waxes [1]. Electron microscopic examination of the cuticle usually shows an amorphous appearance but in some plants the cuticle has a lamellar appearance (Fig. 2).

The periderm, the outer barrier that covers barks and the underground organs such as tubers and roots, is formed by depositing on the walls of the outer one or two cells a polymeric material called suberin, composed of aromatic and aliphatic domains (Fig. 1). Suberized walls are also found in a variety of other anatomical regions within plants such as epidermis and hypodermis of roots, endodermis (casparian bands), the bundle sheaths of grasses, the sheaths around idioblasts, the boundary between the plant and its secretory organs such as glands and trichomes, the pigment strands of grains, the chalazal region connecting seed coats and vascular tissue, and certain cotton fibers [2–4]. The aromatic domains of suberin are derived mainly from cinnamic acid and the esterified aliphatic components are derived from the common cellular fatty acids. These insoluble cell wall adcrustations have soluble waxes associated with them, probably generating the lamellar appearance (Fig. 2).

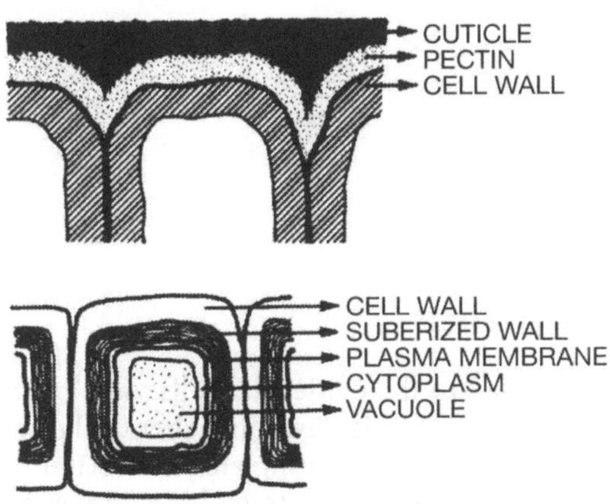

Fig. 1. Schematic representation of the cuticle (*top*) and suberized cell wall (*bottom*)

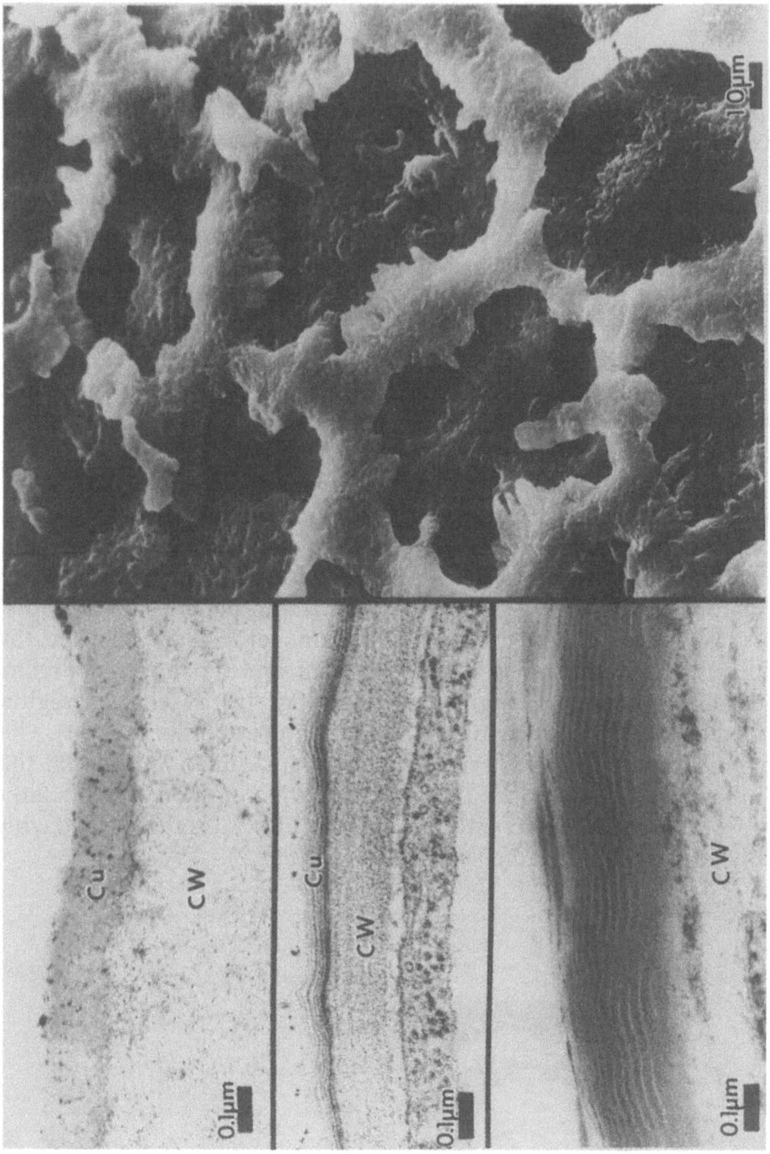

Fig. 2. Electron micrographs illustrating amorphous (*left top*, *Tropaeolum majus*) and lamellar (*left middle*, *Atriplex semibaccata*) cuticle, lamellar structure of potato suberin (*left bottom*), and scanning electron micrograph (*right*) of the underside of tomato fruit cutin showing the protrusions that help to anchor the polymer to the fruit by fitting into the intercellular grooves. Cu = cuticle; CW = cell wall

2 Isolation of Plant Polyesters

The cuticle, being attached to the epidermal cells via a pectinaceous layer, can be released by disruption of this layer by chemicals such as ammonium oxalate/oxalic acid or by pectin-degrading enzymes. After treatment of the recovered cuticular layer with carbohydrate-hydrolyzing enzymes to remove the remaining attached carbohydrates, the soluble waxes can be removed by exhaustive extraction with organic solvents such as chloroform. Scanning electron microscopy of the inside surface of the polymer shows cell-shaped ridges indicating that it is deposited into the intercellular boundaries (Fig. 2). The cutin sheets thus obtained can be powdered and subjected to chemical and/or enzymatic depolymerization [5, 6].

Suberin, being an adcrustation on the cell wall, cannot be separated from cell walls. Instead, suberin-enriched wall preparations can be obtained by digesting away as much carbohydrate polymers as possible using pectinases and cellulases [3, 7]. Depending on the source of the suberized cell wall preparation, the polyester part may constitute a few percent to 30% of the total mass.

3
Depolymerization

Cutin can be depolymerized by cleavage of the ester bonds either by alkaline hydrolysis, transesterification with methanol containing boron trifluoride or sodium methoxide, reductive cleavage by exhaustive treatment with $LiAlH_4$ in tetrahydrofuran, or with trimethylsilyl iodide (TMSiI) in organic solvents [5, 6, 8]. Enzymatic depolymerization can be done with lipases such as pancreatic lipase or cutinases. The chemical methods yield monomers and/or their derivatives depending on the reagent used (Fig. 3). When the polymer contains functional groups such as epoxides and aldehydes, which are not stable to the depolymerization techniques, derivatives useful for identification of the original structure can be generated during the depolymerization process. For example, $LiAlD_4$ would introduce deuterium (D) at the carbon atom carrying the epoxide or aldehyde in such a way that mass spectrometry of the products would reveal the presence of such functional groups in the original polymer [9, 10]. Methanolysis of the oxirane function would give rise to a methoxy group adjacent to a carbinol, diagnostic of the epoxide [11, 12]. Enzymatic depolymerization can give oligomers, as shown when cutinase was first purified [13]. Polyester domains that may also contain non-ester cross-links such as interchain ether bonds or C-C bonds remain as a non-depolymerizable core after such treatments [10, 14]. The monomers can be subjected to standard analytical procedures such as thin-layer chromatography (TLC) and gas-chromatography-mass spectrometry (GC-MS). The monomers are derivatized before gas chromatographic analysis and the most convenient derivative which can be subjected to GC-MS is the trimethylsilyl derivative [5, 6, 10] (Fig. 3). The highly preferred α-cleavage on either side of the mid-chain substituent assists in the identification of cutin monomers by their mass spectra. The enzymatically generated oligomers can also be subjected to structural studies by electron impact and liquid secondary ionization mass spectrometry (LSIMS) and one- or multidimensional NMR spectroscopy [8, 15].

The polyester domains of suberized walls can also be depolymerized using chemical and/or enzymatic approaches similar to those used for cutin. The aromatic domains are far more difficult to depolymerize as C-C and C-O-C cross-links are probably present in such domains. Therefore, more drastic degradation procedures such as nitrobenzene, CuO oxidation, or thioglycolic

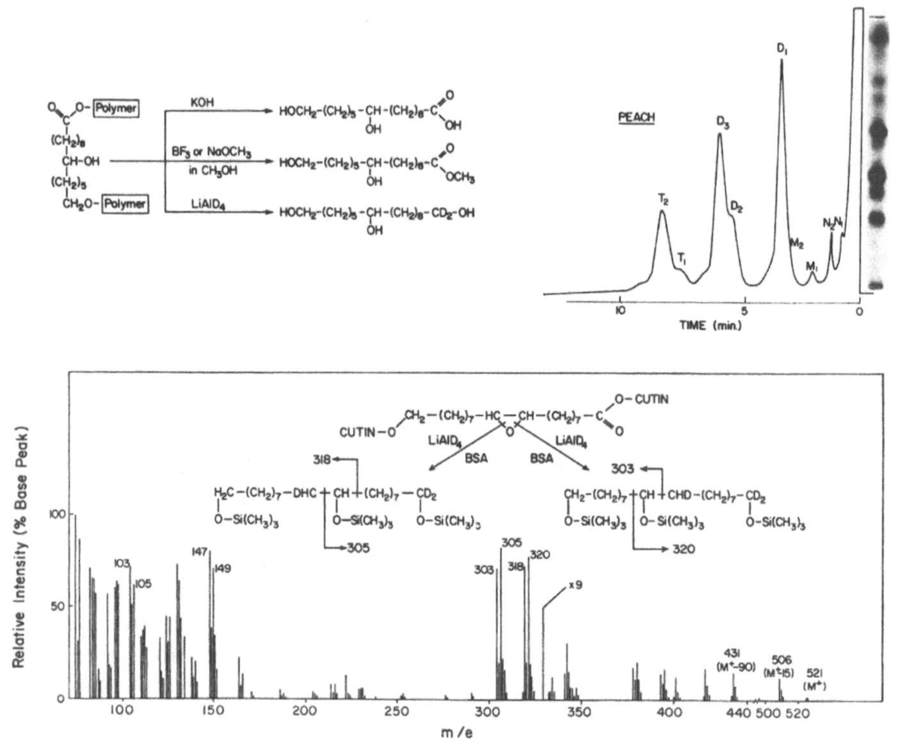

Fig. 3. (*Top left*) Chemical methods used to depolymerize the polyesters. (*Top right*) Thin-layer and gas-liquid chromatograms (as trimethylsilyl derivatives) of the monomer mixture obtained from the cutin of peach fruits by $LiA1D_4$ treatment. In the thin-layer chromatogram the five major spots are, from the bottom, C_{18} tetraol, C_{16} triol, and C_{18} triol (unresolved), diols, and primary alcohol. $N_1 = C_{16}$ alcohol; $N_2 = C_{18}$ alcohol; $M_1 = C_{16}$ diol; $M_2 = C_{18}$ diol; $D_1 = C_{16}$ triol; D_2 and D_3 = unsaturated and saturated C_{18} triol, respectively, T_1 and T_2, unsaturated and saturated C_{18} tetraol, respectively. (*Bottom*) Mass spectrum of component D_3 in the gas chromatogram. BSA = bis-*N,O*-trimethylsilyl acetamide

acid/HCl treatment are used to release aromatic fragments [3, 7, 16, 17]. Since such domains probably do not constitute polyesters, the details of the structures of the nonhydrolyzable aromatic core of suberin are not discussed here.

4 Composition of Cutin

The most common major components of cutin are derivatives of saturated C_{16} (palmitic) acid and unsaturated C_{18} acids (Fig. 4). The major component of the C_{16} family of acids is 9- or 10,16-dihydroxyhexadecanoic acid (and some mid-chain positional isomers), with less 16-hydroxyhexadecanoic acid and much smaller amounts of hexadecanoic acid. In some cases other derivatives are significant constituents. For example, in citrus cutin 16-hydroxy-10-oxo-C_{16} acid, and in young *Vicia faba* leaves 16-oxo-9 or 10-hydroxy C_{16} acid are significant

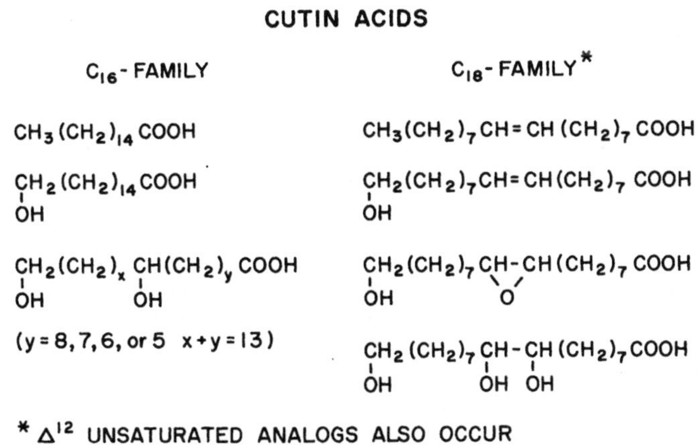

Fig. 4. Structure of the most common major monomers of cutin

components [18–20]. Other oxidation and reduction products of the dihydroxy acids are found as minor components in some plants [21, 22]. Trace amounts of C_{16} dicarboxylic acid are also found. The major components of the C_{18} family of monomers are 18-hydroxy-9,10-epoxy C_{18} acid and 9,10,18-trihydroxy C_{18} acid together with their monounsaturated homologues. Lower amounts of 18-hydroxy C_{18} saturated, mono-, and diunsaturated fatty acids and still lower amounts of their unhydroxylated homologues are found. Fatty acids longer than C_{18}, their ω-hydroxylated derivatives, and the corresponding dicarboxylic acids are minor components of cutin. A list of significant components of cutin is contained in Table 1.

Table 1. Fatty acids with one or more additional functional groups that have been reported as components of cutin or suberin[a]. Adapted from [16]

Monomer		Source	Percentage of total aliphatics
Monohydroxy acids			
8-Hydroxy C_8		*Psilotum nudum* stem	0.7
9-Hydroxy C_9		*Solanum tuberosum* leaf	0.5
12-Hydroxy C_{12}		*Pinus sylvestris* leaf	9
9-Hydroxy $C_{14:1}$		*Coffea arabica* leaf	4.5
14-Hydroxy C_{14}		*Encephalartos altensteinii* leaf	4
9-Hydroxy C_{15}[b]		*Coffea arabica* leaf	1
2-Hydroxy C_{16}		*Conocephalum conicum* leaf	4.8
15-Hydroxy C_{16}		*Astarella lindenbergiana* leaf	72
16-Hydroxy C_{16}	S	*Populus tremula* bark	22
2-Hydroxy C_{18}		*Conocephalum conicum* leaf	3.3
10-Hydroxy C_{18}[b]		*Rosmarinus officinalis* leaf	1.3
12-Hydroxy $C_{18:1}$		*Rosmarinus officinalis* leaf	2.3
18-Hydroxy C_{18}	S	*Cupressus leylandi* bark	8
18-Hydroxy $C_{18:1}$	S	*Solanum tuberosum* storage organ	33

Table 1 (continued)

Monomer		Source	Percentage of total aliphatics
18-Hydroxy $C_{18:2}$		*Spinacia oleracea* leaf	0.1
20-Hydroxy C_{20}	S	*Beta vulgaris* tuber	2.9
22-Hydroxy C_{22}	S	*Gossypium hirsutum* green fiber	70
20-Hydroxy C_{23}		*Conocephalum conicum* leaf	2
20-Hydroxy C_{24}		*Conocephalum conicum* leaf	10
24-Hydroxy C_{24}	S	*Euonymus alatus* "cork wings"	14
26-Hydroxy C_{26}	S	*Quercus ilex* bark	2
28-Hydroxy C_{28}	S	*Fraxinus excelsior* bark	0.9
Dihydroxy acids			
9,15-Dihydroxy C_{15}		*Araucaria imbricate* leaf	1.7
10,15-Dihydroxy C_{16}		*Astarella lindenbergiana* leaf	3.9
7,16-Dihydroxy C_{16}		*Pisum sativum* seed coat	4.1
8,16-Dihydroxy C_{16}		*Hordeum vulgare* leaf	8
9,16-Dihydroxy C_{16}		*Malabar papaiarnarum* fruit	73
10,16-Dihydroxy C_{16}		*Ribes grossularia* fruit	83
10,17-Dihydroxy C_{17}		*Pinus radiata* stem	0.1
10,18-Dihydroxy C_{18}		*Pinus sylvestris* leaf	1.0
10,18-Dihydroxy $C_{18:1}$		*Vaccinium macrocarpon* fruit	1.1
Tri- and pentahydroxy acids			
6,7,16-Trihydroxy C_{16}		*Rosmarinus officinalis* leaf	17
9,10,16-Trihydroxy C_{16}		*Citrus paradisi* fruit	1.9
9,10,17-Trihydroxy C_{17}		*Rosmarinus officinalis* leaf	2.9
9,10,17-Trihydroxy $C_{17:1}$		*Rosmarinus officinalis* leaf	3.0
9,10,18-Trihydroxy $C_{18:1}$		*Citrus paradisi* seed coat	23
9,10,12,13,18-Pentahydroxy C_{18}		*Rosmarinus officinalis* leaf	3.2
Epoxy and oxo acids			
16-Hydroxy-10-oxo C_{16}		*Citrus limon* fruit	34
9-Hydroxy-16-oxo C_{16}[b]		*Vicia faba* embryonic stem	32
9,16-Dihydroxy-10-oxo C_{16}		*Citrus paradisi* fruit	4.2
9,10-Epoxy-18-hydroxy C_{18}		*Citrus paradisi* seed coat	37
9,10-Epoxy-18-hydroxy $C_{18:1}$		*Vitis vinifera* fruit	30
9,10-Epoxy-18-oxo C_{18}		*Malus pumila* young fruit	–
Dicarboxylic acids			
C_9 Diacid		*Solanum tuberosum* leaf	1.7
C_{14} Diacid		*Pinus radiata* stem	0.5
C_{15} Diacid		*Pinus radiata* stem	0.7
6-Hydroxy C15 diacid		*Gnetum gnemom* leaf	7
7-Hydroxy C15 diacid		*Sapindus saponaria* leaf	1.3
8-Hydroxy C15 diacid		*Sphagnum cuspidatum* leaf	0.6
C_{16} Diacid	S	*Citrus paradisi* seed coat	13
$C_{16:1}$ Diacid		*Vaccinium macrocarpon* fruit	0.1
7-Hydroxy C_{16} diacid		*Welwitschia mirabilis* leaf	15
8-Hydroxy C_{16} diacid		*Sphagnum cuspidatum*	7
C_{17} Diacid		*Pinus radiata* stem	5.2
8,9-Dihydroxy C_{17} diacid		*Vaccinium macrocarpon* fruit	0.2
C_{18} Diacid	S	*Ribes nigrum* bark	2.8
$C_{18:1}$ Diacid	S	*Solanum tuberosum* tuber	31

Table 1 (continued)

Monomer		Source	Percentage of total aliphatics
$C_{18:2}$ Diacid		*Vaccinium macrocarpon* fruit	0.02
9,10-Dihydroxy C_{18} diacid	S	*Acer griseum* bark	17
9,10-Epoxy C_{18} diacid	S	*Quercus suber* bark	16
$C_{19:1}$ Diacid		*Pinus radiata* stem	8
C_{20} Diacid	S	*Cupressus leylandi* bark	3.0
C_{22} Diacid	S	*Gossypium hirsutum* green fiber	25
C_{24} Diacid	S	*Citrus paradisi* seed coat	4.8
C_{26} Diacid	S	*Euonymus alatus* "cork wings"	0.1

[a] Monomers from suberin are indicated by S.
[b] Positional isomers also found.

The composition of cutin shows species specificity although cutin from most plants contains different types of mixtures of the C_{16} and C_{18} family of acids. Composition of cutin can vary with the anatomical location. For example, cutin preparations from fruit, leaf, stigma, and flower petal of *Malus pumila* contain 73%, 35%, 14%, and 12%, respectively, of hydroxy and hydroxy-epoxy C_{18} monomers [23]. In general, fast-growing plant organs have higher content of C_{16} family of monomers.

5 Structure of the Polymer Cutin

Cutin is held together mainly by ester bonds' although other types of linkages are also probably present in most plants. The precise nature of the linkages involved in cutin remains unclear. Early studies to elucidate the nature of the linkages present in the amorphous polymer involved indirect chemical modification of free functional groups present in the polymer followed by depolymerization and analysis of the released monomers containing the modifications. One such approach involved oxidation of free hydroxyl groups with CrO_3-pyridine complex followed by depolymerization with sodium methoxide in anhydrous methanol [24]. Another method involved treatment of cutin with methane sulphonyl chloride followed by depolymerization with $LiAlD_4$ that replaces each free hydroxyl group with a deuterium [25]. Combined GC-MS of the resulting mixture of monomers allows quantitation of the products as well as localization of the deuterium indicating the presence of the free hydroxyl group in the original polymer. These methods were applied only to cutins containing the C_{16} family of monomers. The conclusions from both approaches were quite similar; the in-chain hydroxyl group of dihydroxy C_{16} acid accounts for the bulk of the free hydroxyl groups present in the cutin, showing that the primary hydroxyl groups present in the polymer are all esterified. About one-half of the secondary hydroxyl groups were also found to be esterified. For example, the

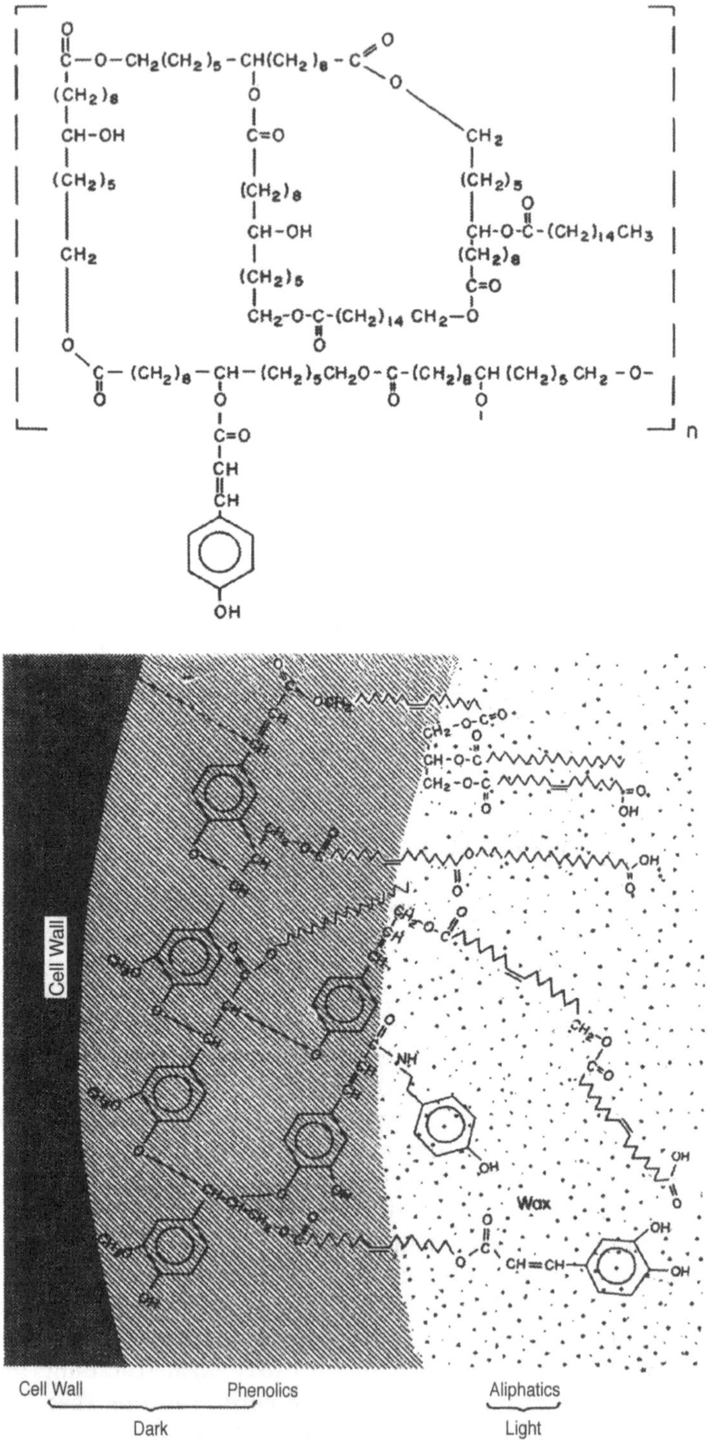

Fig. 5. Models showing the type of structures present in the polymers cutin (*top*) and suberin (*bottom*)

mesylation technique showed that tomato fruit cutin contained approximately 0.4 free hydroxyl groups per monomer, a value similar to that obtained by measurement of the label incorporated into the polymer acetylated by radioactive acetylating agents. The CrO_3 oxidation technique indicated a slightly higher number of free in-chain hydroxyl groups. More recently, NMR approaches have been used to examine the structural features of the polymer. Solid-state NMR analysis (CPMAS NMR) indicated that cutin is a moderately flexible netting with motional constraints at cross-link sites [26]. More than half of the methylenes were found to be in the rigid category, with about 36% in the mobile category. Since this result was obtained with citrus cutin that contains mid-chain carbonyl groups that would give little ability to form cross-links when compared to the corresponding mid-chain hydroxylated monomers, the flexibility observed in the citrus cutin might be slightly more than that present in other cutins. Based on the monomer composition and the number of free primary and secondary hydroxyl groups, a general hypothesis concerning the structure of cutin was proposed (Fig. 5). More recently, the postulated types of linkages were observed in oligomers generated by enzymes [15]. Pancreatic lipase and fungal cutinase are two enzymes that can hydrolyze preferentially the primary alcohol ester linkages in cutin to generate oligomers as observed when the fungal cutinase was first purified [13]. Such oligomers were recently isolated and subjected to structural studies using NMR and secondary-ion mass spectrometry (LSIMS). These results demonstrated the presence of secondary alcohol esters formed at the 10-hydroxy group of the dihydroxy C_{16} acid (Fig. 6). A chemical depolymerization using trimethylsilyl iodide, that preferentially cleaves sterically hindered ester bonds, generated several oligomers which were separated and subjected to structural studies by LSIMS and multi-dimensional NMR [8]. The structures of these oligomers (Fig. 6) also confirmed the general structural features deduced from indirect chemical studies on the polymer. Although these oligomers illustrate the type of structures present in the polymer, the quantitative distribution of such linkages present in the polymer cannot be deduced from such approaches. However, what is clear is that the polymer is held together mostly by primary alcohol ester linkages with about half of the secondary hydroxyl groups being involved in ester cross-links and/or branching.

Exhaustive treatments of cutin which cleave ester bonds, such as hydrolysis, hydrogenolysis, or transesterification, leave behind insoluble residues from virtually all cutin samples [10, 14]. This depolymerization-resistant residue is thought to represent cutin monomers held together by non-ester bonds. Treatment of such residues with HI generates soluble materials indicating the presence of ether bonds. NMR studies on the insoluble material remaining after exhaustive hydrogenolysis with $LiAlH_4$ of cutin from the fruits of apple, pepper, and tomato reveal the presence of methylene chains [27]. Similar ^{13}C CP-MAS NMR studies of the residue remaining after treatment of lime fruit cutin with TMSiI showed the presence of polymethylenic functions. The non-ester-bound polymeric materials found in fossilized cuticles has been called "cutan" and was considered to be cutin-derived [28]. Such non-ester-bound polymeric materials have been also found in modern plant cuticles [29]. Such materials

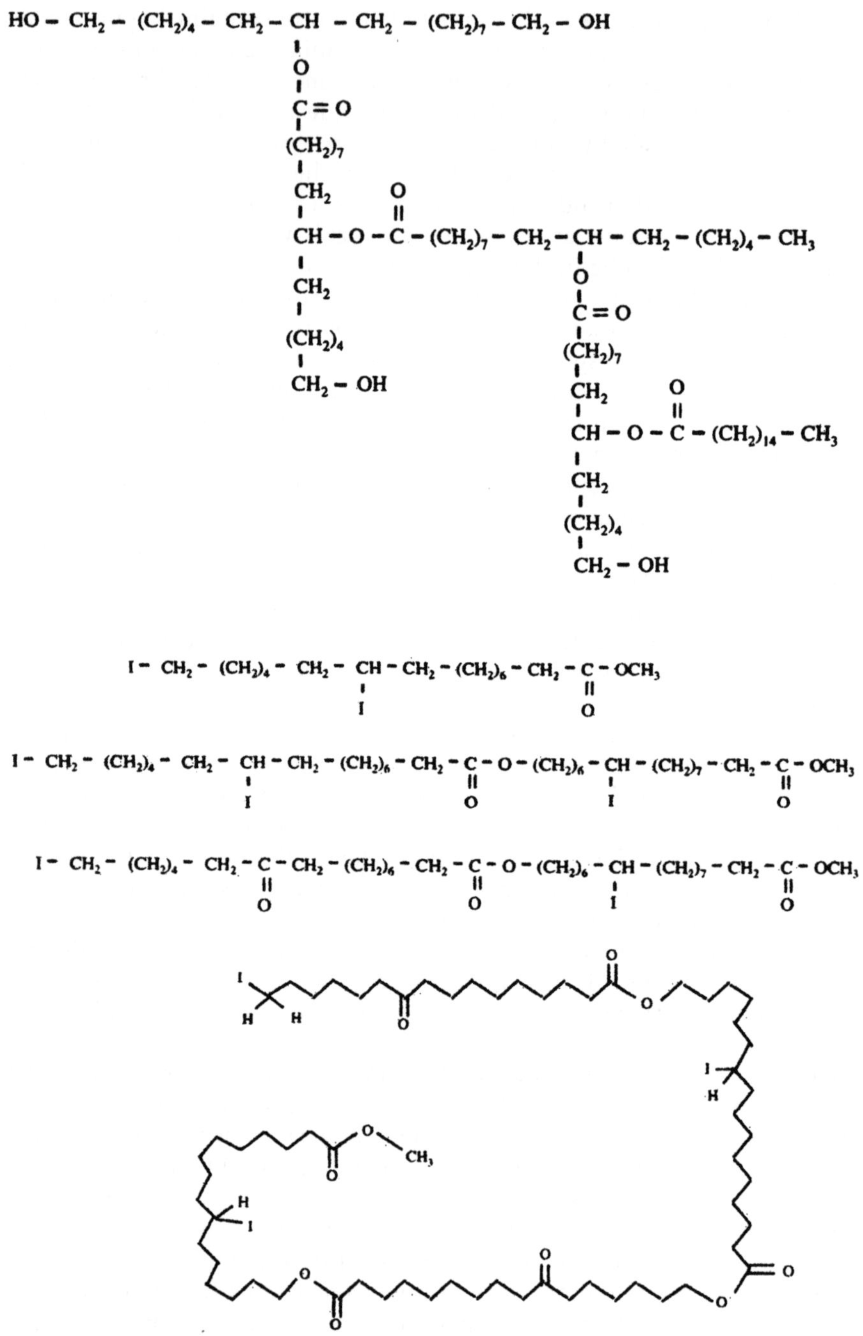

Fig. 6. Proposed chemical structures of isolated soluble products of lime cutin depolymerization with TMSiI (*bottom*) and pancreatic lipase (*top*)

have been subjected to pyrolysis-coupled gas liquid chromatography and mass spectrometry. The products included not only those expected from C_{16} and C_{18} fatty acids but also hydrocarbons in the range of 19–26 carbons. Recently, the depolymerization-resistant fractions of *Clivia miniata* and *Agave americana* were studied by Fourier Transform infrared and ^{13}C NMR spectroscopic analyses, calorimetry, X-ray diffraction, and exhaustive ozonalysis [30]. The results suggested that the polymeric core materials consist of an amorphous three-dimensional network of polymethylenic molecules linked by ether bonds, containing double bonds and free carboxylic acid functions, and part of this core was ether linked as HI-treatment released part of the label. The biosynthetic evidence that polyunsaturated fatty acids are preferentially incorporated into the depolymerization-resistant core [30, 31] is consistent with the chemical evidence. The relative content of the depolymerization-resistant cutin varies a great deal from plant to plant. This core may contain, in addition to the polymethylenic structure, some phenolics and possibly some carbohydrates. Phenolics have been found to be associated with the cuticular structure [32, 33] and peroxidases are expressed in epidermal cells [34]. Therefore, it is probable that some cuticular components including the phenolic materials are peroxidatively coupled, generating C-C bonds and C-O-C bonds. It is likely that the major part of the non-depolymerizable portion of cutin is composed of polymethylenic components.

6 Suberin Composition

The aliphatic monomers of suberin constitute 5–30% of the suberin-enriched cell wall preparations [16, 35, 36]. The most common aliphatic components are fatty acids, fatty alcohols, ω-hydroxy fatty acids, and dicarboxylic acids. The fatty acid and alcohol portions of suberin are characterized by the presence of very long chain (20–30 carbons) components. In the ω-hydroxy acid and dicarboxylic acid fractions, saturated C_{16} and monounsaturated C_{18} acids are the common major components. Homologues containing more than 20 carbons with an even number of carbon atoms are often significant components of such fractions, unlike those found in cutin. The more polar acids which contain epoxy, hydroxy, and dihydroxy functions similar to those found in cutin are usually minor components in suberin, although in some bark suberin samples they can be significant components. The compositional distinction between cutin and suberin (Table 2) originally formulated in 1974 [11] based on a limited number of analyses has been essentially confirmed by the results obtained by the more recent extensive analyses of such polymers from a large number of plant species [16, 37]. The more characteristic feature of the aliphatic components of suberin is the presence of very long chain ($>C_{18}$) components and dicarboxylic acids, mostly unsubstituted dicarboxylic acids with small amounts mid-chain hydroxy or epoxy acids. The major polyfunctional aliphatic components found in suberin are listed in Table 1. The presence of a large number of carboxyl groups in excess of the number of hydroxyl groups present in the monomer would suggest that these carboxyl groups may be esterified to other hydroxyl-containing cell wall components such as phenolics and carbohydrates.

Table 2. Compositional difference between cutin and suberin

Monomer	Cutin	Suberin
Dicarboxylic acids	Minor	Major
In-chain-substituted acids	Major	Minor [a]
Phenolics	Low	High
Very long-chain (C_{20}–C_{26}) acids	Rare and minor	Common and substantial
Very long-chain alcohols	Rare and minor	Common and substantial

[a] In some cases substantial.

The phenolics may be rich in unreduced phenylpropanoic acids [38] and some of those acids are in amide linkage with tyramine [39]. Some of the carboxyl groups may be esterified to glycerol in suberin [40, 41]. The green cotton fibers that were shown to be suberized contain caffeoyl-fatty acid-glycerol esters in their wax fraction. The insoluble suberin fraction was also shown to contain glycerol. A recent analysis of the insoluble suberin material that had been thoroughly extracted with SDS showed the presence of glycerol in the suberin polymer of not only cotton fiber but also potato periderm [41], and during the purifications of potato periderm suberin the glycerol content and the dicarboxylic acid content increased in a similar manner, suggesting that glycerol was an integral part of suberin.

7
Structure of Suberin

Since suberization involves deposition of phenolic and aliphatic materials on the plant cell wall, the isolated material enriched in suberin is composed of complex polymers including cell wall components, phenolic polymeric material, and the polyester domains [3, 7, 16]. How the aliphatic components are linked together is not known. Indirect chemical studies revealed the presence of few, if any, free hydroxyl groups in the aliphatic components. From the composition of the monomers it would appear that a linear polymer composed of ω-hydroxy acids can be made. However, the number of carboxyl groups exceeds the number of hydroxyl groups available in the aliphatic components. The recently reported presence of glycerol would provide hydroxyl sites to esterify some of the carboxyl groups of dicarboxylic acids and help produce a polymer network [40, 41]. However, no oligomeric polymethylenic components have been isolated from suberin and therefore there is no direct evidence concerning the linkages. ^{13}C CPMAS has been used to examine suberized preparations and such studies revealed the presence of polymethylenic polyesters in suberized walls [3, 38, 42–46]. The NMR spectrum of suberin from *Solanum tuberosum* showed the presence of a high proportion of aliphatic CH_2 but also had a large amount of CHOH carbon, probably from contaminating cell wall carbohydrates [3]. How the polyester domain is attached to the cell wall is not known. However, many lines of evidence suggest that the phenolic materials are probably attached to the cell wall and the aliphatic components are attached to the phenolics. A working hypothesis depicting this concept was proposed many years ago [25] and most of the experimental evidence obtained since then is consistent with such a general picture. Such a hypothetical structural organization of suberin that also takes into account some of the more recent results indicated above is shown in Fig. 5. There is no direct proof for the structural details. This working model incorporates the known structural features, explains the observed acidic character of the polymer [47], and shows the types of structures that may be present in suberin and the general organization of the suberized wall. The following observations support the overall hypothesis about the organization of suberin [3, 7]:

1. Depolymerization techniques that cleave ester bonds release the indicated aliphatic monomers and phenolic components from suberin.
2. Treatment of suberin with nitrobenzene generates vanillin, *p*-hydroxy benzaldehyde, but not much syringaldehyde that arises mostly from lignin.
3. Suberized cell walls stain positively for phenolics with indications that suberin contains monohydroxyphenolic rings and has fewer *O*-methoxy groups than lignin.
4. The inability to solubilize aromatic components of suberin-enriched preparations by the methods used for lignin suggests that suberin structure is distinctly different from that of lignin, probably due to the aliphatic cross-linking and the higher degree of condensation present in suberin.
5. Phenolic acids and aliphatic acids are both involved in the biosynthesis of suberin, and phenolic acids are not synthesized in tissue slices that do not undergo suberization.
6. Inhibition of synthesis of the aromatic matrix by inhibitors of phenylalanine: ammonia lyase causes the inhibition of deposition of aliphatic components and prevents development of diffusion resistance. Inhibition of synthesis of peroxidase, the enzyme involved in the deposition of the polymeric phenolic matrix, caused by iron deficiency, prevents deposition of aliphatic components of suberin.
7. The time-course of deposition of aromatic monomers into the polymer laid down by suberizing tissue slices indicates that the phenolic matrix is deposited simultaneously with or slightly before the aliphatic components. The specific anionic peroxidase appeared with a time-course consistent with its involvement in the polymerization and deposition of the phenolic matrix of the suberin. Increase or decrease in suberin content involves similar changes in both the aliphatic and aromatic components and such changes are associated with the expected increase or decrease in the anionic peroxidase activity caused by physical or biological stress.

Removal of the aliphatic materials by hydrogenolysis leaves a residue that contains low amounts of polymethylenic components, suggesting that the suberized material contains some aliphatic components not susceptible to cleavage by such methods [3]. On the other hand, removal of suberin from cork cell wall preparations was examined by CPMAS and the results showed that the aliphatic components were nearly completely removed from this suberin preparation as the spectra showed that the residual material was virtually devoid of methyl and methylene peaks [3, 45]. The spectra of the completely desuberized material from the cork could be accounted for by the presence of phenolic materials and carbohydrates. Existence of an insoluble non-hydrolyzable aliphatic biomacromolecule called "suberan" (in analogy to the term "cutan") in the periderm of tissues of some angiosperm species has been reported [48] and a high molecular weight material containing aliphatic components was recently reported to be present in the suberin preparation from *Quercus suber* [35]. It is possible that the amount of aliphatic materials that cannot be removed by the ester cleaving reactions would depend on the origin of the suberized material and may not be a general feature of suberin. Much more work will be required to elucidate the precise nature of the linkages involved in this extremely complex polymeric material.

8 Biosynthesis of Cutin

8.1 Cutin Monomers

Early attempts to study the biosynthesis of cutin involved measurements of fatty acid levels in wounded tissue and oxygen uptake in cell free preparations caused by addition of fatty acids that were tested as potential substrates for cutin synthesis [49]. Systematic biochemical studies on cutin synthesis started when it was found that rapidly expanding leaves of *V. faba* incorporated radioactive precursors into an insoluble polymer [50, 51]. When the insoluble polymer was subjected to depolymerization by $LiAlH_4$ hydrogenolysis, the ether-soluble extracts containing the cutin monomers were found to be radioactive and these products could then be subjected to standard analytical methods such as TLC and radio gas chromatography. Using such an approach it was found that the most rapidly expanding tissues synthesized cutin most rapidly. The epidermis was demonstrated to be the site of cutin biosynthesis. For example, excised epidermis of leaves from *V. faba*, *Senecio odoris* (*Kleinia odora*), and pea incorporated labeled acetate and palmitic acid into cutin monomers. In developing fruits of apple, only the skin and not the internal tissue incorporated exogenous labeled fatty acids into cutin monomers [31]. In both leaves and fruit incorporation of exogenous labeled precursors into cutin increased in proportion to the rate of expansion of the organ and the rate drastically decreased as the tissue expansion slowed down. Thus, most rapidly expanding tissues were found to be appropriate for studying cutin biosynthesis [27].

8.1.1 *Biosynthesis of the C_{16} Family of Cutin Acids*

Leaf discs from rapidly expanding *V. faba* leaves incorporated ^{14}C-labeled palmitic acid into cutin. After removal of the soluble lipids and other materials, the insoluble residue was subjected to $LiAlH_4$ hydrogenolysis and the labeled reduction products of cutin monomers were identified by chromatography as hexadecane-

1,16-diol and hexadecane-1,7,16-triol, obviously derived from ω-hydroxypalmitic acid and 10,16-dihydroxypalmitic acid of cutin [52]. A similar labeling pattern was observed when [1-^{14}C]palmitic acid was incubated with *S. odoris* leaf disks or apple fruit skin discs [31]. The major radioactive component of the polymer derived from the labeled C_{16} acid was the dihydroxy acid and smaller amounts of label were found in ω-hydroxypalmitic acid and palmitic acid itself. On the other hand labeled stearic acid and oleic acid were poorly incorporated into *V. faba* cutin and the small amount of label that was incorporated was found mainly in non-hydroxy acids with small amounts in ω-hydroxyacids. Thus, the in-chain hydroxylated C_{16} monomer was found to be derived mainly from palmitic acid. The time-course of incorporation of palmitic acid showed that hydroxy acids derived from it did not accumulate in the soluble lipids although they could be detected by autoradiography, indicating that the cutin monomers were incorporated into the insoluble polymer as soon as they were made. Exogenous 16-hydroxypalmitic acid was incorporated into cutin in *V. faba* leaf disks and the major part of the radioactivity from this monomer was found in the dihydroxypalmitic acid of the polymer, the rest being in ω-hydroxypalmitic acid. This result suggested that ω-hydroxy acid is the precursor of the dihydroxy acid. The mid-chain hydroxylated acid containing no hydroxyl group at the ω position was never found in any of these studies, suggesting that the biosynthesis involved ω-hydroxylation followed by mid-chain hydroxylation and subsequent incorporation into the polymer.

A microsomal preparation from the shoots of the *V. faba* seedlings catalyzed ω-hydroxylation of palmitic acid with NADPH and O_2 as required co-factors [53]. This mixed-function oxidase was inhibited by CO, suggesting the involvement of a CytP$_{450}$-type enzyme. However, the inhibition could not be reversed by light. Oleic acid was hydroxylated by this preparation at a comparable rate but stearic acid was a very poor substrate. ω-Hydroxylation was recently demonstrated to be catalyzed by a CytP$_{450}$ induced by clofibrate in *Vicia sativa* seedlings and antibodies raised against NADPH-CytP$_{450}$ reductase inhibited the reaction [54]. This induced hydroxylase could also hydroxylate mid-chain modified acids such as those containing mid-chain epoxide and diols [55], raising the possibility that this clofibrate-induced enzyme may be more like the typical xenobiotic metabolizing enzymes and may not be a truly biosynthetic enzyme. More recently, a CytP$_{450}$-dependent ω-hydroxylase from clofibrate-treated *V. sativa* seedlings was described [56]. This CytP$_{450}$ enzyme hydroxylated the methyl end of saturated and mono-, di-, and triunsaturated C_{18} fatty acids without demonstrating any stereospecificity for the diunsaturated C_{18} acid. The mRNA for this CytP$_{450}$ began to accumulate after 90 min exposure of the seedling to clofibrate. More relevant to the biosynthesis of cutin was the observation that the mRNA level for this CytP$_{450}$ increased during plant development and after wounding of tissues, possibly indicating its role in the ω-hydroxylation involved in the biosynthesis of cutin and suberin monomers. However, the specific localization of this enzyme in the epidermal cells that are involved in cutin biosynthesis or in the periderm cells involved in suberization (wound healing) has not been demonstrated and therefore it remains unclear whether such a xenobiotic-inducible CytP$_{450}$ represents the enzyme involved in the biosynthesis of cutin monomers.

The mechanism of conversion of ω-hydroxypalmitic acid into the dihydroxy acid could either involve the formation of a double bond in a mid-chain position followed by hydration or a direct hydroxylation by a mixed-function oxidase. Neither palmitoleic acid nor palmitelaidic acid was incorporated into 10,16-dihydroxypalmitic acid in cutin, suggesting that hydration of the Δ^9 double bond is probably not involved in the introduction of the mid-chain hydroxyl group involved in cutin synthesis [52]. Double labeling experiments indicated that the introduction of the mid-chain hydroxyl group involved loss of a single hydrogen atom, indicating a direct hydroxylation rather than involvement of a double bond. This conversion of ω-hydroxy acid required molecular oxygen and was inhibited by chelators with the reversal of this inhibition by Fe^{+2}, suggesting that a direct hydroxylation by a mixed-function oxidase is involved in the mid-chain hydroxylation. A cell-free extract from the excised epidermis from *V. faba* leaves catalyzed the conversion of 16-hydroxypalmitic acid into the 10,16-dihydroxy acid [57]. This reaction required NADPH, ATP, and CoA. In such cell-free preparations the exogenous 16-hydroxypalmitic acid also underwent β-oxidation generating 3-hydroxy acids. To eliminate complications caused by such multiple products, an assay was developed in which the positional isomers of the hydroxy acids were resolved by HPLC. Using this assay it was shown that the mid-chain hydroxylation required O_2 and was inhibited by carbon monoxide in a photoreversible manner [58]. All of the results thus suggest that the mid-chain hydroxylation is catalyzed by a mixed-function oxidase involving a $CytP_{450}$. However, such an enzyme has not been purified to demonstrate directly the involvement of such a $CytP_{450}$. The occurrence of mid-chain positional isomers of the dihydroxyfatty acid in a species-specific manner in plants suggest that the positional specificity of the mid-chain hydroxylase may vary in a species-specific way. Developmental changes in the positional isomer composition suggested the possibility that two different hydroxylases with different positional specificity are involved in the synthesis of these positional isomers. The presence of higher 9-hydroxy isomer content in the cutin of etiolated *V. faba* stem and the increase in 10-hydroxy isomer content caused by light exposure of the stem supports the dual hydroxylase hypothesis [59]. Biosynthesis of C_{16} monomers of cutin is summarized in Fig. 7.

8.1.2
Biosynthesis of the C_{18} Family of Cutin Acids

Biosynthesis of this family of monomers was studied using plant tissues that have the C_{18} family of acids as the major cutin monomers. Thus, in expanding grape berry skin slices, exogenous labeled oleic acid was converted mainly into 18-hydroxyoleic acid and 18-hydroxy-9,10-epoxy C_{18} acid whereas in skin slices of rapidly expanding young apple fruit, labeled oleic acid was incorporated into the same hydroxy and epoxy acids and into 9,10,18-trihydroxy C_{18} acid [31]. This incorporation pattern reflected the composition of the C_{18} monomers in the two tissues; in grape berry, the epoxy acid is a major cutin monomer whereas in the apple cutin the trihydroxy acid is a major component. Exogenous stearic acid was not incorporated into any mid-chain hydroxylated monomers,

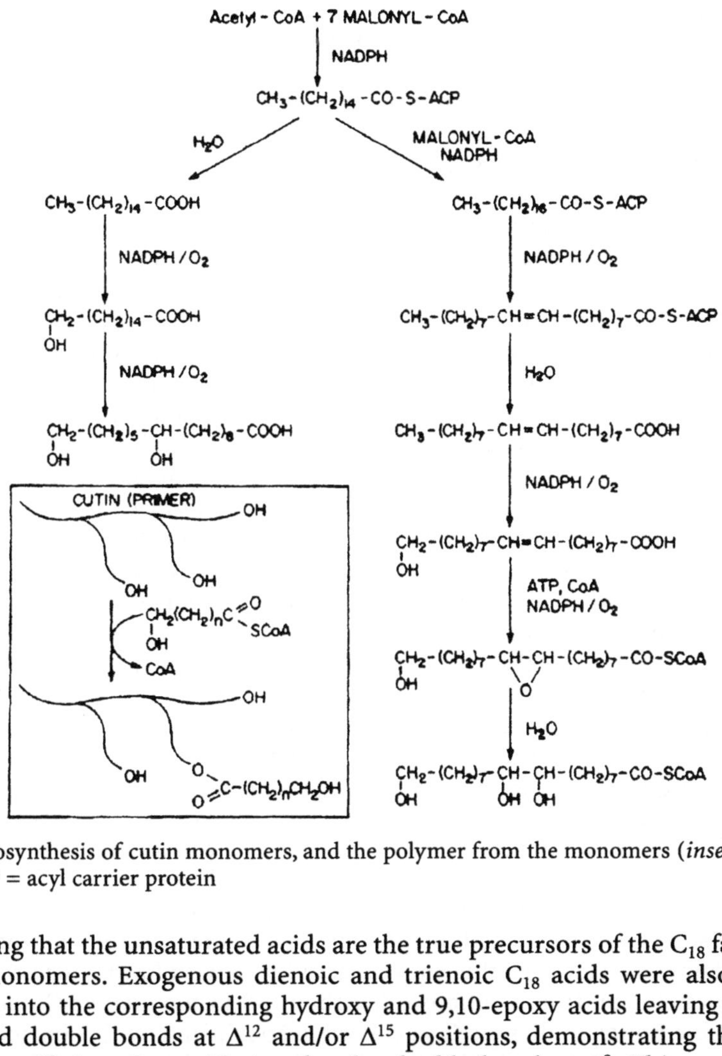

Fig. 7. Biosynthesis of cutin monomers, and the polymer from the monomers (*inset, bottom left*). ACP = acyl carrier protein

indicating that the unsaturated acids are the true precursors of the C_{18} family of cutin monomers. Exogenous dienoic and trienoic C_{18} acids were also incorporated into the corresponding hydroxy and 9,10-epoxy acids leaving the unmodified double bonds at Δ^{12} and/or Δ^{15} positions, demonstrating the positional specificity of epoxidation for the double bond at Δ^9. This specificity, although very common in plants, is not confined to the Δ^9 double bond in all plants. For example, in *Rosemarinus officinalis* both Δ^9 and Δ^{12} double bonds are epoxidized and the epoxides are hydrated to generate 9,10,12,13,18-pentahydroxy C_{18} acid. In *R. officinalis* leaf slices exogenous labeled linoleic acid was incorporated into not only the Δ^9 double bond-modified products indicated above but also into the 9,10,18 trihydroxy-12,13-epoxy C_{18} acid and the pentahydroxy acid [60]. Incorporation of the di- and trienoic C_{18} acid into cutin was reflected in the composition of the cutin in the developing apple fruit. In the younger fruit that are green and contain di- and trienoic C_{18} acids, their 18-hydroxy derivatives, 18-hydroxy-9-epoxy Δ^{12}- and -$\Delta^{12,15}$ acids, and 9,10,18-trihydroxy Δ^{12}- and $\Delta^{12,15}$ C_{18} acids were found as significant components, whereas in the less green and more mature fruit such acids were only minor components

[31]. Based on the composition of the C_{18} family of cutin monomers we postulated that oleic acid would be ω-hydroxylated first, followed by epoxidation of the double bond at C-9 followed by the hydrolytic cleavage of the oxirane to yield 9,10,18-trihydroxy acid. This postulate was experimentally verified by the demonstration of specific incorporation of exogenous 18-hydroxyoleic acid into 18-hydroxy-9,10-epoxy C_{18} acid in grape berry skin slices and apple fruit skin disks, and incorporation of exogenous labeled 18-hydroxy-9,10-epoxy C_{18} acid into 9,10,18-trihydroxy C_{18} acid of cutin in apple fruit skin slices [61].

To test for the occurrence of the postulated biochemical reactions in cutin-synthesizing plant tissues, cell-free preparations were made from tissues that produce the epoxy acid as a major component or from tissues that produce the trihydroxy acid as a major component. A particulate preparation from young spinach leaves, that produce the epoxy acid as the major cutin component, catalyzed epoxidation of 18-hydroxy [18-^{3}H]oleic acid to the corresponding *cis*-epoxy acid [62]. This reaction required ATP and CoA, indicating that the substrate of the epoxidation was the CoA ester. This epoxidation also required NADPH and O_2. It was inhibited by CO and this inhibition was reversed by light at 450 nm, suggesting that a $CytP_{450}$-type enzyme is involved in this epoxidation. This epoxidation was maximal with the natural substrate, namely, ω-hydroxyoleic acid, whereas the *trans*-homologue, 18-hydroxyelaidic acid, was a very poor substrate, as was oleic acid. This high degree of substrate specificity supports the hypothesis that this enzyme is in fact the one that is involved in the biosynthesis of the epoxy cutin monomer. Enzyme preparations capable of epoxidizing the 18-hydroxyoleic acid were also obtained from the skin of rapidly expanding apple fruit and from excised epidermis of *S. odoris* leaves, but not from internal tissues from these organs, again demonstrating the biosynthetic relevance of this enzymatic activity. More recently, two newly-found enzyme activities in soybean seedlings were suggested to be involved in the biosynthesis of cutin monomers [63, 64]. A hydroperoxide-dependent epoxidase found in the microsomes catalyzed epoxidation of oleic acid. The specificity of this activity for *cis* olefin is consistent with its possible involvement in the biosynthesis of cutin monomers. However, this enzyme activity showed low regioselectivity in that *cis* double bonds in positions other than C-9 in monoenoic C_{18} acids and both double bonds in dienoic C_{18} acids were epoxidized by this enzyme. Even more noteworthy was the finding that the ω-hydroxyoleic acid was found to be a poor substrate for this epoxidase, unlike the $CytP_{450}$ enzyme activity obtained from cutin-synthesizing tissues. Whether the hydroperoxide-dependent epoxidase is present in cutin-synthesizing epidermal tissues, as previously noted for the $CytP_{450}$-dependent epoxidase, remains unknown. These observations cast doubt about whether the hydroperoxide-dependent enzyme is in fact involved in the biosynthesis of the epoxy acids found in cutin. In developing seeds of *Euphorbia lagascae*, which produce *cis*-12,13-epoxy-9-hydroxydienoic C_{18} acid (vernolic acid), both a $CytP_{450}$-type epoxidase and a hydroperoxide-dependent epoxidase were found, but in germinating seeds – which do not synthesize the epoxy acid – only the latter epoxidase was found [65], suggesting that the $CytP_{450}$-type epoxidase may be the biosynthetic enzyme whereas the other enzyme activity may be involved in the degradation of lipids during germination.

It is probable that a similar situation exists in cutin biosynthesis in the apple skin slices and excised epidermis of *S. odoris*, where $CytP_{450}$-type epoxidase is the biosynthetic enzyme.

The final step of biosynthesis of the major C_{18} monomer would involve the hydrolytic cleavage of the oxirane by an epoxide hydrase. A particulate fraction prepared from the homogenates of the skin of rapidly expanding young apple fruit catalyzed the hydration of 18-hydroxy-*cis* 9,10-epoxy-C_{18} acid to *threo*-9, 10, 18-trihydroxy C_{18} acid [66]. This epoxide hydration required no co-factors and was localized mainly in a particulate fraction. The internal tissue of apple fruit did not catalyze this epoxide hydration, indicating that this activity was confined to the cells that produced cutin. The biosynthetic relevance of this enzyme was further demonstrated by the substrate specificity of this epoxide hydrase activity. The maximal activity was obtained with 18-hydroxy-*cis*-9,10-epoxy C_{18} acid; *cis*-9,10-epoxystearic acid was a poor substrate as was styrene oxide, the substrate used by mammalian catabolic epoxide hydrases. Such an epoxide hydrase activity was also detected in enzyme preparations from spinach leaves and from excised epidermis of the leaves from *S. odoris*. These observations strongly suggest that this particulate epoxide hydrase is involved in the biosynthesis of the cutin monomer. The cytosol from the actively cutin-synthesizing apple tissue showed no epoxide hydrase activity. A soluble epoxide hydrase cDNA has been cloned from *Arabidopsis thaliana* and potato [67, 68]. The level of their transcripts was elevated by auxin treatment and wounding, and indirect arguments have been presented to suggest that such soluble epoxide hydrases may be involved in cutin and suberin biosynthesis. A soluble epoxide hydrase was also found in soybean seedlings [69, 70]. This enzyme showed a preference for *cis*-epoxide but 18-hydroxy-9,10-epoxy C_{18} acid was a poor substrate, unlike the particulate epoxide hydrase found in the skin slices of the young apple fruit and other cutin-synthesizing tissues indicated above. It is uncertain whether such soluble epoxide hydrases are actually involved in cutin and suberin biosynthesis. The soybean enzymes would epoxidize the double bonds and hydrate the epoxide without requiring an ω-hydroxyl group. If such a specificity is manifested in the cell, the mid-chain modified molecules containing no ω-hydroxy groups might be present and should be incorporated into the polymer. However, such molecules have not been found in cutin. Therefore, the specificity of the soybean enzyme would not be consistent with the known composition of cutin monomers. Until the enzymes are shown to be present specifically in the cells involved in cutin synthesis or some other biological connection between these enzymes and cutin biosynthesis is demonstrated, the relevance of such an activity in cutin biosynthesis remains unclear. Biosynthesis of the C_{18} monomers of cutin is summarized in Fig. 7.

8.2
Synthesis of the Polymer from Monomers

Synthesis of the insoluble cutin polymer that is deposited outside the epidermal cell walls in rapidly expanding plant organs would have to occur at the site of the final deposition of the polymer. A cutin-containing particulate preparation from the excised epidermal tissue of rapidly expanding *V. faba* leaves was found to incorporate labeled C_{16} monomers into an insoluble material with ATP and CoA as required co-factors [71]. That this incorporation represented synthesis of cutin was demonstrated by the fact that only chemical treatments that are known to release esterified monomers could release the incorporated label. Even more significantly, cutinase, but no other hydrolytic enzymes, released the label incorporated into the insoluble material by the particulate preparation. That this enzymatic activity is involved in the biosynthesis of cutin is suggested by the observation that particulate preparations from the epidermal tissue of *V. faba* and *S. odoris*, but not from the mesophyl tissue, catalyzed incorporation of the labeled C_{16} monomer into the insoluble material. Presumably the hydroxyacyl moiety was transferred from the CoA ester to the growing polymer. ω-Hydroxy C_{18} acid and other fatty acids up to C_{18} could also be incorporated into insoluble material by the enzyme preparations. However, the C_{16} family of acids was preferred as expected from the composition of the *V. faba* cutin. Methylation of the carboxyl group, but not acetylation of the ω-hydroxyl group, of the C_{16} monomer prevented incorporation into cutin, suggesting that the carboxyl end of the incoming monomer is transferred to the free hydroxyl of the polymer. Since the particulate preparation contained cutin primer into which the incoming monomers would be incorporated, the nature of the primer involved in this process could not be studied until the enzyme was dissociated from the primer. Mild sonication of the particulate preparation yielded a soluble enzyme preparation that required exogenous purified cutin as a primer. The transferase activity was proportional to the amount of cutin primer added and the system required the same co-factors as the particulate preparation. *V. faba* cutin powder was strongly preferred as a primer although cutin from other plant species could substitute less well. Other polymers such as cellulose were ineffective as acyl acceptors. Acetylation of the cutin primer decreased its priming efficiency, confirming the requirement for free hydroxyl groups in the primer. Cutin prepared from very young *V. faba* leaves was a more efficient primer than cutin from mature fully expanded leaves, suggesting that the enzyme prefers the open structure of the less developed polymer. Opening of the polymer structure by brief treatment with cutinase increased the efficiency of the primer. Chemical treatments that increased the number of hydroxyl groups or opened the polymer matrix also increased priming efficiency. Such enzymatic cutin-synthesizing activities could also be obtained from flowers of *V. faba* and excised epidermis of *S. odoris* leaves. The hydroxyacyl-CoA:cutin transacylase involved in the synthesis of the polymer from monomers has not been purified from any source. The biosynthesis of the polymer is depicted in Fig. 7.

The biosynthetic origin of the depolymerization-resistant core of cutin (cutan) remains to be established. The early observation that linoleic acid and linolenic acid were preferentially incorporated into the non-depolymerizable core of cutin in apple skin slices suggested that the ether-linked or C-C-linked core might arise preferentially from the *cis*-1,4-pentadiene system [31]. The insoluble residue, that contained the label from the incorporated polyunsaturated C_{18} acids, released the label upon treatment with HI, supporting the notion that some of those aliphatic chains were held together by ether bonds. More recently, preferential incorporation of labeled linoleic acid into the non-ester-bound part of cutin in *C. miniata* leaf disks was reported [30]. The preferential incorporation of pentadiene-containing fatty acids into the non-ester-bound part of the polymer suggests the involvement of lipoxygenase- and peroxidase-type reactions in the formation of such materials. The observation that the ether-linked portion can be degraded by ozonolysis [30] indicates that there are double bonds in this non-hydrolyzable core. This observation would be consistent with the biosynthetic origin of this part of the polymers from polyunsaturated acids, probably via the involvement of lipoxygenase. It would be interesting to determine whether the organs that produce cutin containing larger proportion of such non-ester-bound polymeric material contain higher levels of lipoxygenases.

How the polyester is anchored to the epidermal cell wall is not known. There is evidence that the ω-oxo function in the major cutin monomer may be involved in acetal type linkages that anchor the polymer in the young leaves and the further expansion of the polymer would involve ester linkages without needing the oxo derivative. Developmental changes in the monomer composition of expanding *V. faba* leaves suggested this possibility. In the disubstituted C_{16} acid fraction that constitutes the major components of cutin in *V. faba*, the major portion of 9-hydroxo C_{16} acid contained an aldehyde function at the ω-carbon. As the leaves developed the oxo isomer decreased from 50% in the youngest tissue to 10% in the mature leaf [19]. The ω-oxo acid was also found in other plant cutins. In young apple fruit where C_{18} monomers are major components 18-oxo-9,10-epoxy C_{18} acid was found, suggesting the possible involvement of the ω-oxo monomers in anchoring the polyester to the epidermis [72].

9
Biosynthesis of Suberin

9.1
Biosynthesis of the Aliphatic Monomers of Suberin

In suberizing potato tuber disks, labeled oleic acid was incorporated into ω-hydroxyoleic acid and the corresponding dicarboxylic acid, the two major aliphatic components of potato suberin [73]. Exogenous labeled acetate was also incorporated into all of the aliphatic components of suberin, including the very long chain acids and alcohols in the wound-healing potato slices. The time-course of incorporation of the labeled precursors into the suberin components was consistent with the time-course of suberization. The biosynthetic pathway for the major aliphatic components of suberin is shown in Fig. 8a.

The unique suberin components that are not found as significant components of cutin are the very long chain molecules and the dicarboxylic acids. Therefore, chain elongation and conversion of ω-hydroxy acids to the corresponding dicarboxylic acids constitute two unique biochemical processes involved in the synthesis of suberin. Incorporation of labeled acetate into the very long chain components of suberin was demonstrated and this ability developed during suberization in potato tuber disks [73]. The enzymes involved

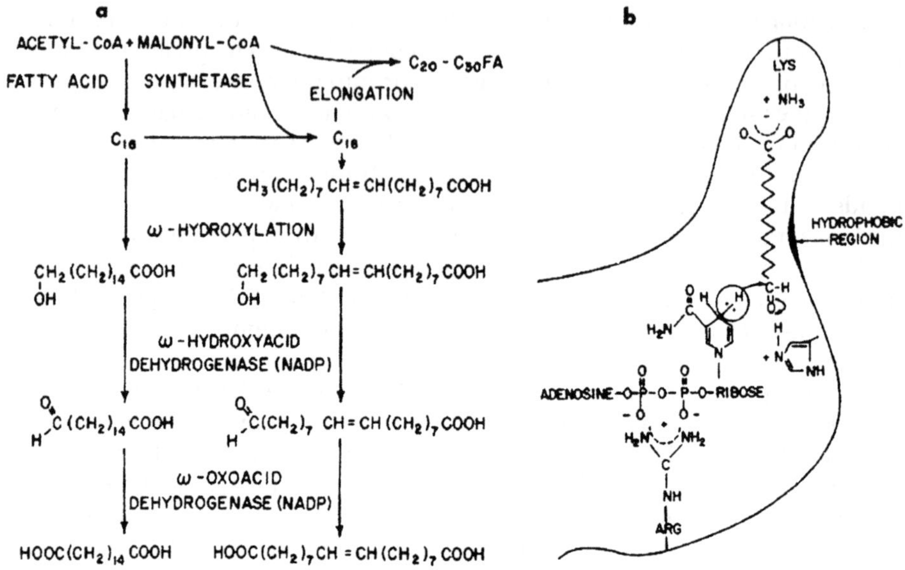

Fig. 8a, b. **a** Biosynthetic pathways for the major aliphatic components of suberin. **b** Representation of the active site of ω-hydroxy acid dehydrogenase involved in the synthesis of the dicarboxylic acids characteristic of suberin. From [74]

in chain elongation are yet to be purified and characterized from any tissue, although this process plays key functions in plants and animals. The enzymology of oxidation of ω-hydroxy acids to the corresponding dicarboxylic acids in suberizing potato tissue has been elucidated. Extracts from suberizing potato tuber slices catalyzed conversion of ω-hydroxypalmitic acid to the corresponding dicarboxylic acid with NADP or NAD as the co-factor, with a slight preference for the former [74]. This dehydrogenase activity, located largely in the soluble supernatant, is different from alcohol dehydrogenase. Conversion of the ω-hydroxy acid to the dicarboxylic acid involves the ω-oxoacid as an intermediate. The enzyme activity involved in the conversion of the hydroxy acid to the oxoacid with NADP as the cofactor could be separated from that which catalyzed conversion of the ω-oxoacid to the dicarboxylic acid. The ω-hydroxy acid dehydrogenase, but not the oxoacid dehydrogenase, was found to be induced by wounding potato tubers in a time-course that strongly suggested its involvement in the deposition of suberin; the oxoacid dehydrogenase was present constitutively. The ω-hydroxy acid dehydrogenase was purified to near homogeneity from wound-healing potato tuber disks. The purified enzyme showed a native molecular weight of 60,000 and a monomer molecular weight of 30,000, indicating that it was a dimer of identical subunits [75]. Surprisingly, this enzyme did not show any stereospecificity for hydride transfer as both pro-R and pro-S hydrogen from NADPH were equally transferred to the 16-oxo acid. Chemical modification studies of this key enzyme involved in suberization showed the presence of an essential arginine, histidine, and lysine [76]. The chain-length specificity of the dehydrogenase was determined using a series of synthetic *n*-alkanals containing 3–20 carbon atoms. As the length of the alkanal increased from C_3 to C_8 the Km for the substrate decreased from 700 µmol/l to 90 µmol/l and further increase in chain length from C_8 to C_{20} resulted in only a small decrease in Km. The V_{max} drastically decreased when the chain length of the aldehyde was increased above C_{18}, indicating that the substrate-binding pocket has a limited depth to accommodate the ω-hydroxy acid. Based on the chemical modification studies and kinetic studies, it was concluded that the substrate is bound into a pocket that has a lysine residue which forms an ionic bridge with the distal carboxyl group of the substrate. The arginine residue is involved in the binding of the pyrophosphoryl group of the nucleotide substrate and the histidine participates in catalysis by protonating the carbonyl when the hydride from the nucleotide is transferred to the carbonyl (Fig. 8b). The chain length specificity of the dehydrogenase can explain the finding that suberin contains dicarboxylic acids in the range of C_{18} but the ω-hydroxy acid contains longer chains, presumably because the dehydrogenase cannot readily accommodate the longer ω-hydroxy acids into its substrate binding pocket.

9.2 Incorporation of the Aliphatic Components into the Polymer

How the aliphatic monomers are incorporated into the suberin polymer is not known. Presumably, activated ω-hydroxy acids and dicarboxylic acids are esterified to the hydroxyl groups as found in cutin biosynthesis. The long chain fatty alcohols might be incorporated into suberin via esterification with phenylpropanoic acids such as ferulic acid, followed by peroxidase-catalyzed polymerization of the phenolic derivative. This suggestion is based on the finding that ferulic acid esters of very long chain fatty alcohols are frequently found in suberin-associated waxes. The recently cloned hydroxycinnamoyl-CoA: tyramine *N*-(hydroxycinnamoyl) transferase [77] may produce a tyramide derivative of the phenolic compound that may then be incorporated into the polymer by a peroxidase. The glycerol triester composed of a fatty acid, caffeic acid and ω-hydroxy acid found in the suberin associated wax [40] may also be incorporated into the polymer by a peroxidase.

9.3 Enzymatic Polymerization of the Aromatic Components of Suberin

The deposition of the aromatic polymer is probably catalyzed by a highly anionic peroxidase. Such an enzyme was found to be induced in wound-healing potato tuber disks. From the external layer of cells from such disks a highly anionic peroxidase (pI, 3.15) was purified to homogeneity and the cDNA for this enzyme was cloned [78]. Immunoblot and RNA blot analyses indicated that the time-course of appearance of the protein and transcript for this enzyme was exactly what was expected from its involvement in suberization. Immunocytochemical localization of this highly anionic peroxidase showed that it was present only in the cell walls of suberizing cells and only during suberization [79]. Transgenic potato that over-expressed this anionic peroxidase gene showed enhanced suberization (B. Sherf and P.E. Kolattukudy, unpublished). Factors that induce suberization such as mechanical injury of fruit, leaves, and tubers [79, 80], Mg^{+2} deficiency in corn roots [81], abscisic acid treatment of potato callus culture [82], and fungal invasion of tomato vasculature [83] elevated the levels of the anionic peroxidase. Suberization inhibition caused by Fe^{+2} deficiency in bean roots [84], and by thorough washing of the wound surface in potato tuber disks [82], decreased the level of anionic peroxidase as expected from its involvement in suberization. Fungal induction of suberization that occurs selectively in resistant tomato lines is associated with selective induction of expression of the anionic peroxidase gene in this line but not in the near isogenic susceptible line [85]. Antisense expression in transgenic tomato plants abolished the appearance of the transcripts of this anionic peroxidase in wound-healing fruits. However, the periderms formed on the wounded fruits were suberized, most probably using alternate peroxidases, and such peroxidases were found in the wound periderms [86]. Promoter analysis using β-glucuronidase reporter gene fusion in transgenic tobacco plants showed that the normal tissue-specific and developmentally regulated expression of this gene requires about 200 bp 5′-flanking sequence and the wound-induced and pathogen-induced high level expression requires an additional 150 bp of 5′ flanking region [34].

10 Cutin Degradation

10.1 Cutin Degradation by Bacteria

Bacteria were isolated from soil using apple cutin sheets buried in an apple orchard as a nutrient trap. The bacteria attached to these sheets, isolated by enrichment culturing methods, were found to be capable of degrading cutin [87]. However, the enzymes involved in the degradation were not isolated. Bacteria isolated from the surface of aerial plant organs were found to degrade cutin. A bacterial culture, thought to be capable of enhancing nitrogen nutrition when sprayed on crops [88], was found to have two bacterial species, one that secreted cutinase to generate nutrients from cutin and the other that was able to grow without requiring fixed nitrogen [89]. From the extracellular fluid of the cutin-grown culture of the former, that was first thought to be *Pseudomonas putida* (but was later identified to be *Pseudomonas mendocina*), cutinase was purified and characterized [90], and the gene that encodes this protein has been cloned and sequenced [91]. This bacterial cutinase is a 30-kDa protein with an amino acid composition distinctly different from that of fungal cutinases, and it does not show an immunological relationship with fungal cutinases [90]. It hydrolyzes *p*-nitrophenyl esters of C_4 to C_{16} fatty acids and short chain triacylglycerols such as tributyrin, although long chain esters are less readily hydrolyzed. This bacterial cutinase uses a catalytic triad involving active serine for catalysis. This enzyme has potential for cleaning applications [92] and as adjuvents for agricultural chemical formulations [93]. Cutinase from *Streptomyces scabies* has also been purified and characterized [94]. More recently other bacteria, especially the thermophylics such as *Thermomonospora*, have been found to produce cutinase activity [95]. Such species are being examined for potential commercial applications.

10.2
Cutin Degradation by Fungi

Since cutin is a major structural component of the aerial organs of plants, fungal ingress into plants must involve penetration through this polymeric barrier. Because of the potential significance of this process in fungal pathogenesis, that claims the greatest loss in the production of food and fiber, this area has been investigated fairly extensively. Early studies detected fungal cutinase activity using assays for fatty acid production but the enzymes were not isolated [96–99]. Cutinase was first purified in the 1970s from *Fusarium solani* f. *pisi* grown on cutin as a sole source of carbon [13, 100]. Two forms of this enzyme were separated by ion exchange chromatography and no real differences in their catalytic capabilities were detected.

10.2.1
Isolation of Fungal Cutinases and their Molecular Properties

To isolate fungal cutinases, fungi are grown on a mineral medium containing cutin powder as the sole source of carbon [101]. The extra-cellular fluid is concentrated to give a dark viscous solution. Gel filtration through a Sephadex G-100 column often gives two esterase fractions. The one at the higher molecular weight region catalyzes the hydrolysis of *p*-nitrophenylesters of fatty acids with 2–18 carbon atoms but does not hydrolyze cutin at significant rates. The second peak that is retarded in the column catalyzes the hydrolysis of cutin as well as *p*-nitrophenyl esters of short chain fatty acids such as butyrate, but not the esters of very long chain fatty acids. This second esterase fraction is subjected to ion exchange chromatography using QAE-Sephadex that retains all of the color and the unretained enzyme emerges as a colorless material. This enzyme preparation is subjected to hydrophobic chromatography using octyl sepharose chromatography and the resulting product is then subjected to a cation exchange chromatography on SP-Sephadex that separates isozymes when present, and yields electrophoretically homogeneous enzyme preparations. This procedure has yielded pure cutinase from a variety of fungi [102].

Molecular weight and subunit composition of most of the fungal cutinases so far purified show that they are single peptides with a molecular weight in the range of 20–25 kDa [102]. Occasionally, a proteolytic nick is observed and the cleavage products appear to be separable by SDS-PAGE [94]. Since the nicked protein is held together by a disulfide bridge, such species co-purify with the unclipped enzyme and they separate only when SDS-PAGE is done in the presence of reducing agents. Fungal cutinases contain very low amounts of carbohydrates, and these carbohydrates are attached by *O*-glycosidic linkages as indicated by the appearance of a chromophore that absorbs at 241 nm upon treatment with alkali (base elimination) [94, 103, 104]. Analysis of the released carbohydrates after reduction with NaB^3H_4 showed that monosaccharides are *O*-glycosidically attached to this protein at serine and threonine residues and in some cases at *β*-hydroxyphenylalanine and *β*-hydroxytyrosine residues. The *O*-glycosidically attached sugars so far identified are mannose, *N*-acetylglucosamine, glucuronic acid, and arabinose [94, 104]. Fungal cutinase constitutes the first case where the novel structure of *O*-glycosidically-attached monosaccharides such as *N*-acetylglucosamine are found in proteins. Some years later such structures were rediscovered in nuclear proteins using the same methodology as those used to discover such structures in cutinases and such structures are thought to be important for the nuclear function of such proteins [105, 106].

Since the original isolation and characterization of fungal cutinases, such enzyme activities have been found to be produced by a large number of phytopathogenic fungi even though in some cases the enzyme was not purified and characterized. Many of the cases, where characterization has been done, show small differences in molecular weight but all of them use the active serine catalytic triad for hydrolysis. Depending possibly on the anatomical region attacked by the fungus, a dichotomy in the optimal pH for hydrolysis was noted; some showed a near neutral pH optimum and not the alkaline pH optimum shown by the others [107–109]. There is also a possibility that in some cases the cutinase gene used for saprophytic growth may be different from that used for pathogenesis [108]. However, the validity of such generalizations remains to be tested. Some cutinases of very different sizes have also been reported [110, 111]. However, some of these that show very low activity may be esterases similar to those found to be produced by cutin-grown *F. solani* f. *pisi* when cutinase was first purified [100] rather than true cutinases. However, the occurrence of membrane-bound constitutive cutinases [110, 112] cannot be ruled out.

When polyester-hydrolyzing activity was isolated using synthetic polyesters such as polycaprolactone, and the enzyme was examined in detail, it was found that it was a cutinase that was responsible for the hydrolysis [113]. Similarly, the polyester domains of suberin were found to be degraded by cutinase. Cutinase is a polyesterase, and similar enzymes may be widely distributed and can degrade a variety of natural and synthetic polyesters. Microbial polyhydroxyalkanoic acids that are attracting increasing attention as biodegradable polyesters can be hydrolyzed by bacterial polyesterases that share some common features with cutinases [114] and this area is covered in another chapter [115].

10.2.2
Catalysis by Cutinase

Fungal cutinase catalyzes hydrolysis of model substrates and in particular *p*-nitrophenyl esters of short chain fatty acids, providing a convenient spectrophotometric assay for this enzyme activity [101, 102, 116]. Hydrolysis of model esters by this cutinase showed the high degree of preference of this enzyme for primary alcohol ester hydrolysis. Wax esters and methyl esters of fatty acids were hydrolyzed at low rates. Alkane-2-ol esters were hydrolyzed much more slowly than wax esters and esters of mid-chain secondary alcohols were not hydrolyzed at significant rates. Triglycerides were hydrolyzed by the purified fungal cutinases at slow rates and this activity was as sensitive as cutinase to active site-directed reagents, showing that both activities involve the same catalytic site. Trioleyl glycerol and tributyryl glycerol were hydrolyzed 5–30 times as rapidly as tripalmitoyl glycerol by several fungal cutinases. Time-course of formation of products from the glycerides also showed that the enzyme had a high degree of preference for hydrolysis of primary alcohol esters. Cinnamoyl esters of alcohols and cholesterol esters were not hydrolyzed at measurable rates, whereas cyclohexyl esters were readily hydrolyzed. With cutin as the substrate fungal cutinase showed both *exo*- and *endo*-esterase activity. Thus, short-term incubation of biosynthetically labeled cutin with purified cutinase released oligomeric and monomeric labeled products that could be separated by gel filtration [13]. Cutinase also catalyzed hydrolysis of the oligomers to monomers.

Cutinase is a serine esterase that catalyzes hydrolysis of ester bonds using the catalytic triad involving histidine, aspartic acid and "active" serine (Fig. 9) [117]. A variety of organic phosphates and other reagents that are known to react with active serine irreversibly inhibited fungal cutinases. Some of them showed 50% inhibition at lower than nmol/l concentrations, strongly suggesting that the enzymes contain extremely reactive serine [102]. Reversible inhibitors such as organic boronic acids also inhibited cutinases. Phenylboronic acid showed a competitive inhibition of cutinase with a Ki of 140 mmol/l. Alkylboronic acids with aliphatic chains containing 4–18 carbon atoms also showed competitive inhibition with much lower Ki values in the range of several µmol/l. These boronic acids protected the enzyme from modification by organic phosphates as expected from the reversible formation of the complex

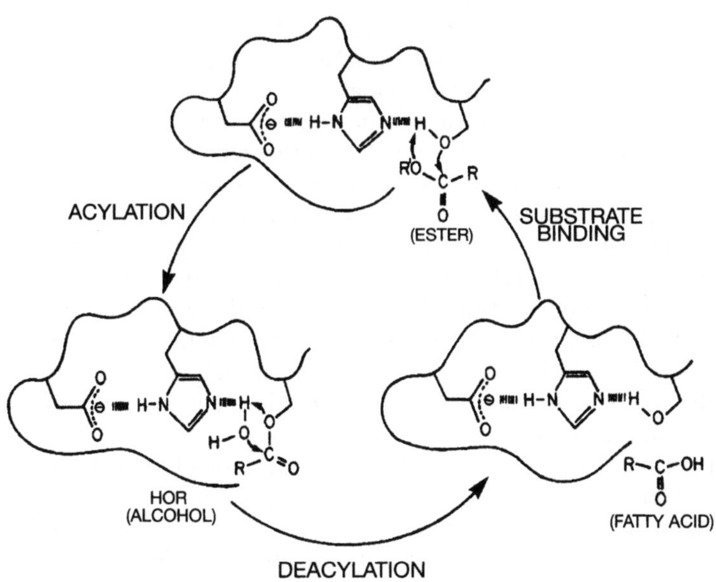

Fig. 9. Mechanism of catalysis by cutinase

with the active serine. Alkyl isocyanates, known to react with active serine-containing enzymes, were found to be potent inhibitors of fungal cutinase [118].

Fungal cutinase is very unusual in that it could not be readily modified by treatment with reagents that are known to react with the other residues of the catalytic triad. However, when it was found that SDS, above the critical micellar concentration (CMC), inactivated the enzyme and that the cutinase activity could be fully and rapidly recovered by the addition of Triton X-100, chemical modification could be achieved in the presence of SDS [117]. Thus, cutinase could be modified by treatment by diethylpyrocarbonate, a histidine-specific reagent, in the presence of increasing concentrations of SDS followed by determination of the residual activity under renaturing conditions. Diethylpyrocarbonate rapidly inactivated the enzyme in the presence of greater than 3 mmol/l SDS. That this inactivation was due to a selective modification of His was suggested by the increase in absorbance at 237 nm expected from *N*-carbethoxyhistidine. Hydroxylamine and alkaline conditions that are expected to remove the carbethoxy group from histine caused reversal of inactivation of cutinase resulting from the diethylpyrocarbonate treatment. Modification of approximately one His residue per molecule of cutinase was required for complete inactivation of the enzyme. Modification of the carboxyl group in the active site by 1-ethyl-3-(3-dimethylaminopropyl)-carbodiimide (EDC) also required the presence of SDS. The labeling of the carboxyl-modified enzyme with radioactive glycine ethyl ester confirmed the modification of the carboxyl group and it was concluded that one essential carboxyl group modification caused inactivation of the enzyme. An acyl enzyme intermediate was detected with cutinase that had been modified by carbethoxylation at the histidine residue because this modification inhibited the deacylation step by a factor of about 10^5. Labeled acetyl cutinase was isolated after incubation of the carbethoxylated cutinase with *p*-nitrophenyl[1-^{14}C] acetate.

Other chemical modification studies also revealed some of the properties of the enzyme that are important for catalyzing cutin hydrolysis [102]. For example, phenylglyoxal treatment of the enzyme resulted in inactivation of cutinase in a pseudo first-order reaction and the degree of phenylglyoxal modification of Arg residues correlated with the degree of inactivation of the enzyme. Two Arg residues were found to be essential for the activity. Hydrolysis of *p*-nitrophenylbutyrate was not affected by this Arg modification whereas cutin hydrolysis was severely inhibited, indicating that the modified arginines were involved in the interaction of the enzyme with cutin. This hypothesis was supported by the experimental verification that phenylglyoxal treatment inhibited the binding of the enzyme to cutin. Presumably the Arg residues interact ionically with the polymer and this interaction is important in the hydrolysis of the insoluble polymer.

Fungal cutinases show no free SH groups but have 4 Cys residues, indicating that they are in disulfide linkage [119]. The reaction of the native enzyme with DTE was extremely slow but in the presence of SDS at its CMC rapid reduction could be observed [102]. Reduction of the disulfide bridge resulted in irreversible inactivation of the enzyme and the protein tended to become insoluble. CD spectra of cutinase in the 205–230 nm region, before and after DTE reduction, suggested that the disulfide bonds impose strong structural constraints on the secondary structure, and upon reduction the CD spectrum in the presence of SDS became that expected from the amino acid composition of cutinase. The conformational tightness of this protein can be dramatically demonstrated by the extreme resistance of the native enzyme to proteolytic degradation by trypsin, chymotrypsin, elastase, proteinase K, V8 proteinase, and clostripain. However, in the presence of SDS, incubation with the proteinases caused rapid hydrolysis and loss of cutinase activity. The insensitivity of the protein to chemical modification in the absence of SDS also indicates that the conformation of the protein is very tight.

Conformational changes brought about by amphipaths in cutinases could also be seen by their effect on the absorbance spectrum of the enzyme [102]. SDS caused a decrease in absorbance at around 285–295 nm and an increase in absorbance at 260–270 nm. The second derivative spectra showed that SDS caused a shift in the Trp band at 290. The fluorescence spectrum of the enzyme in the absence of SDS was that of a protein with little Trp but the excitation spectra was characteristic of a Trp-containing protein. On the other hand, in the presence of increasing concentrations of SDS or sodium dodecanoate, the fluorescence spectrum became increasingly similar to that of a normal Trp-containing protein. Irradiation of the protein at the excitation maximum of Trp for a period of 10 min also caused changes similar to those brought about by SDS in the emission spectrum, causing a 2.6-fold enhancement at the emission maximum and a shift in the emission maximum from 312 nm to 335 nm. The emission spectrum of Trp in the native enzyme was very heavily quenched and irradiation or amphipath treatment released Trp from this quenched state by bringing about conformational changes in the protein. CD spectral studies also indicated that the Trp residue is in a less asymmetric environment when the enzyme is in the presence of SDS. A recent study suggested that the irradiation-induced release of Trp from its quenched state involves cleavage of the disulfide bridge in the enzyme that should cause conformational changes in the protein [120]. Denaturation of proteins by CMC of SDS is a common occurrence; however, cutinase is remarkable in that this inactivation is fully reversed by the presence of TritonX-100.

Monomeric amphipaths caused conformational changes at the active site region as first suggested by the observation that the reactivity of the active serine against organophosphates was increased up to several orders of magnitude by the presence of very low concentrations of SDS [102]. Furthermore, hydrolysis of soluble model esters and diisopropylphosphorylation of the active serine became pH-dependent only in the presence of low concentrations of SDS. Nonionic TritonX-100 did not exhibit such effects but sodium dodecanoate showed the same effect as SDS, indicating that the interaction required an ionic amphipath [102]. The maximum change in the Km and V_{max} occurred prior to reaching the CMC of the amphipaths, suggesting that the amphipath is bound to the enzyme as a monomer. The localized conformational changes occurring at the active site as a result of binding of the monomeric amphipath could be studied by changes in the spectral properties of a fluorescence probe attached to the active site. Pyrenebutylmethanephosphoryl fluoride at 50% molar excess completely inactivated cutinase in 3 min by covalent attachment at the active site. Addition of low concentrations of SDS caused dramatic changes in the absorbance spectrum of this modified cutinase. All absorption bands were enhanced and two new vibrational absorbance components appeared in the presence of SDS. The CD spectrum of the modified enzyme also showed dramatic changes due to the presence of SDS. The changes caused by low concentrations of SDS were limited to the CD bands of the pyrene attached at the active site; the Trp band showed little change. The spectral changes and the catalytic parameters changed concomitantly with the addition of very low concentrations of the amphipath. Direct measurements of the binding of labeled SDS to the protein at different SDS concentrations suggested that 2 mol of the amphipath were bound per mol of the enzyme. The binding of the two molecules correlated with changes in the spectral properties, indicating that the localized conformational changes relevant to the changes in catalytic parameters were due to the binding of these two molecules of SDS. Since two Arg residues were found to be essential for cutinase activity, it appears possible that the essential Arg residues might be the ones involved in SDS binding.

The crystal structure of cutinase from *F. solani* f. *pisi* (Fig. 10) indicated that this fungal cutinase constitutes a separate class of enzyme that may be regarded as a bridge between esterases and lipases in that the free cutinase has a well-de-

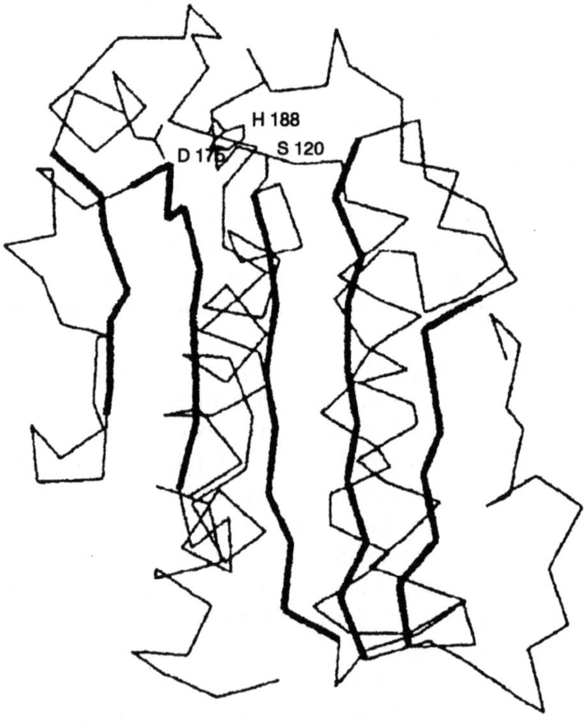

Fig. 10. Structure of *F. solani* f. *pisi* cutinase expressed in *E. coli* [121]. Locations of active site residues are indicated

fined active site and a preformed oxyanion hole and that it does not need any rearrangements to bind its substrate [121, 122]. This conclusion is not totally consistent with the activation of the enzyme observed in solution caused by the amphipaths and the localized conformational changes in the active site region brought about by the low concentrations of the amphipath [102], indicated above. However, these observations are consistent with the results of a recent solution structure study of the enzyme in complex with phosphonate inhibitors by NMR that showed that the crystal structure probably does not represent the actual conformation of the enzyme in solution [123, 124]. The NMR studies indicated that the enzyme adopts its active conformation only upon binding to the inhibitor. While the active site Ser 120 is rigidly attached to the stable α/β core of the protein, the remainder of the binding site is thought to be very flexible in the free enzyme. The other two active site residues, Asp and His, as well as the oxyanion hole residues (Ser Gln), are only restrained into their proper positions upon binding of the substrate-like inhibitor. This conclusion from the NMR studies, that cutinase does need conformational rearrangements to bind its substrate, which may form the rate limiting step in catalysis, is consistent with the need for SDS for achieving chemical modification of His and Asp residues and the localized conformational changes at the active site with the concomitant activation of the enzyme caused by the amphipath observed in the earlier studies [102].

10.3
Cutin Degradation by Animals

Large amounts of cutin are ingested by animals as part of the vegetables and fruits in their diet. To determine whether the polyester can be utilized, radioactive apple cutin was fed to rats and the label was found in all tissues and organs and, thus, metabolism of the radioactive monomers was clearly established [125]. The results suggested that a pancreatic enzyme was probably involved in the hydrolysis of this polyester. Purification of the cutin-hydrolyzing activity from porcine pancreas showed that the enzyme responsible is pancreatic lipase [126]. Bile salts stabilize the pancreatic lipase and co-lipase reversed the inhibition caused by bile salt with cutin as a substrate. Thus the interaction of the enzyme with the cutin surface involving co-lipase and bile salt is similar to that observed for triglycerides. This lipase releases oligomers and monomers just as the fungal cutinase does. Oligomers generated from cutin have been recently isolated and subjected to structural studies [15].

10.4
Cutin Degradation by Plants

Since the insoluble cutin polymer covers expanding plant organs, this polymer may undergo a "make and break" type of expansion during the growth of the organ. However, how the polymer structure adjusts to the expansion of the organ it covers is not known. No enzymatic degradation of cutin in the plant organ has been detected except for pollen cutinase that may be involved in fertilization. Since stigma of plants are exposed to potential contact with pollen from many species, without having the luxury of active mate selection available to animals, genetic fidelity is maintained by controlling compatibility in fertilization. In plants with dry stigma that have intact cuticle at maturity, the breaching of this cuticular barrier by germinating pollen is thought to be a crucial step in determining compatibility [127–129]. A pollen cutinase is thought to be involved in gaining access through the stigmatic cuticle. When cutin was added to germinated pollen, a slight increase in acidity was noted, suggesting hydrolytic release of cutin monomers by pollen enzyme(s) [130, 131]. When labeled apple cutin was incubated with nasturtium pollen, all types of cutin monomers were released [132]. The nasturtium pollen cutinase was purified to homogeneity [133] and this is the only plant cutinase so far purified. This pollen cutinase is a single peptide of 40 kDa containing about 7% of *N*-glycosidically-attached carbohydrates. The amino acid composition of nasturtium pollen was quite different from that of fungal cutinases as it contained a much higher content of acidic amino acids and Cys residues. Antibodies prepared against *F. solani* cutinase did not cross-react with this pollen cutinase.

Catalytic properties of the pollen enzyme were drastically different from those of bacterial and fungal cutinases. The pH optimum for the pollen enzyme was 6.8, similar to that observed for a few fungal cutinases, whereas most of the microbial enzymes usually show much higher alkaline pH optima. In contrast to the fungal enzyme that showed stability at both acidic and basic conditions, the pollen enzyme was unstable except at neutral condition [133]. The catalytic mechanism of this pollen enzyme does not involve the active serine catalytic triad, as this enzyme was totally insensitive to active serine-directed reagents. On the other hand, the pollen enzyme was extremely sensitive to inhibition by thiol-directed reagents that have no effect on fungal cutinases. Thus, this pollen enzyme seems to a thiol polyesterase. In spite of such contrasting molecular and catalytic properties, the substrate specificity of the pollen enzyme resembled that of the microbial cutinase. Pollen cutinase showed a high degree of preference for hydrolysis of primary alcohol esters. It also hydrolyzed *p*-nitrophenyl esters of $C_2 - C_{18}$ fatty acids.

Remarkably, *Brassica napus* pollen was reported to have a 22 kDa cutinase that cross-reacted with antibodies prepared against *F. solani* f. *pisi* cutinase [134]. Although a 22 kDa and a 42 kDa protein that catalyzed hydrolysis of *p*-nitrophenyl butyrate were found in this pollen, only the former catalyzed cutin hydrolysis. Immunofluorescence microscopic examination suggested that the 22 kDa protein was located in the intine. Since the nature of the catalytic mechanism of this enzyme has not been elucidated, it is not clear whether this represents a serine hydrolase indicating that plants may have serine and thiol cutinases. The role of the pollen enzyme in controlling compatibility remains to be established.

11 Suberin Degradation

Suberized cell walls are some of the last components of tree barks that remain after being buried in soil for years [135]. Fungi can penetrate suberized walls as indicated by the ultrastructural evidence summarized elsewhere [3]. Although such evidence suggested that suberin can be degraded by microbes, direct evidence for enzymatic degradation became available only more recently. Ability to degrade suberin has been reported for *Rosellinia desmazieresii* [136], *Armillaria mellea* [137], and *Mycena meliigena* [138], and a variety of fungi were found to grow on potato suberin as the sole source of carbon [3]. *F. solani* f. *pisi* was found to grow more rapidly than the other fungi. The culture filtrate was tested for enzymes that could release labeled components from potato suberin biosynthetically labeled by incorporation of labeled cinnamic acid and labeled oleic acid. The enzyme that released the labeled esterified aliphatic components was purified to homogeneity and characterized [139]. This enzyme was found to be identical to the cutinase produced by *F. solani* f. *pisi* grown on cutin. When the periderm of *Rubus idaeus* was incubated with this purified enzyme, all types of aliphatic monomers of this suberin were released. It is not surprising that the fungus produces the same polyesterase to grow on the polyester-containing suberin as that produced upon growth on cutin. The enzyme(s) that released the aromatic components was separated from this polyesterase during protein fractionation. Only $< 10\%$ of the label derived from radioactive cinnamate could be released by the enzyme preparation and the identification of the released products showed that the extracellular enzyme released only the esterified phenolic components [2]. Phenolic esterases have been purified from fungi [140].The aromatic components held together by the more refractory linkages are probably degraded by enzymes similar to those used in lignin degradation.

12 Function

12.1 Function of Cutin

12.1.1 *Interaction with Physical Environmental Factors*

The major function of cutin is to serve as the structural component of the outer barrier of plants. As the major component of the cuticle it plays a major role in the interaction of the plant with its environment. Development of the cuticle is thought to be responsible for the ability of plants to move onto land where the cuticle limits diffusion of moisture and thus prevents desiccation [141]. The plant cuticle controls the exchange of matter between leaf and atmosphere. The transport properties of the cuticle strongly influences the loss of water and solutes from the leaf interior as well as uptake of nonvolatile chemicals from the atmosphere to the leaf surface. In the absence of stomata the cuticle controls gas exchange. The cuticle as a transport-limiting barrier is important in its physiological and ecological functions. The diffusion across plant cuticle follows basic laws of passive diffusion across lipophylic membranes [142]. Isolated cuticular membranes have been used to study this permeability and the results obtained appear to be valid for intact leaves. The cuticular waxes play a major role in controlling this diffusion and therefore the role of the polyester may be predominantly to serve as the matrix that holds the wax that provides the major diffusion barrier.

Foliar uptake can occur from vapor, liquid, or solid interphases. Most environmental pollutants are taken up as vapors or dissolved in water [143]. The situation is more complicated with agricultural sprays because active ingredients are usually formulated using a variety of additives called adjuvants [144]. Such materials affect the physical properties of the spray liquid and may serve as emulsifiers, wetting agents, spreaders, stickers, antifoaming agents, or buffers making the analysis of the overall process very complex, although adjuvants improve the biological effectiveness of active ingredients. In the case of weak acids such as those found in herbicides or growth regulators, the penetration is maximal when the content of the nonionized species is maximal. The role of the adjuvant in maximizing the penetration has been studied extensively [144]. The cuticle is thought to play a significant role in the interaction between electromagnetic radiation and the plant [145]. In most such interactions the polyester itself may not be playing as major a role as the waxes.

12.1.2
Interaction with Biological Factors in the Environment

The cuticle, often being the first contact point with environmental microbes, probably plays a highly significant role in the interaction of the plant with microorganisms (Fig. 11). It can provide the carbon source for the growth of

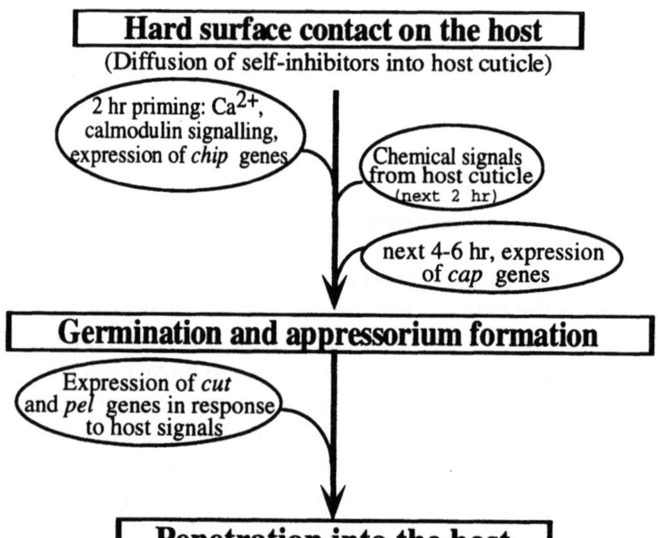

Fig. 11. Hypothetical scheme of the molecular events in the early stages of fungal interactions with plant cuticle. *chip* = *Colletotrichum* hard surface induced protein; *cap* = *Colletotrichum* appressorium genes

microbes that occupy the aerial plant surfaces, as illustrated by the example of *P. mendocina* that produces cutinase as indicated in Sect. 10.1. Fungal conidia have chemicals that prevent them from germinating and differentiating until they reach a favorable ecological niche such as a plant surface. These chemicals, called self-inhibitors, are most often lipophylic molecules [146]. Upon contact with the host surface, these inhibitors can diffuse into the lipophylic cuticle and thus relieve the self-inhibition [147] permitting transcription of genes – such as calmodulin gene – that play important roles in pathogenesis [148]. Contact of fungal conidia with the host surface itself can induce the expression of fungal genes involved in the early events that are crucial for the germination and differentiation of the infection structures required for successful penetration into the plant host. Differential display approaches are beginning to reveal the nature of some of these genes [149, 150]. As the conidia of pathogens carry small amounts of cutinase [151], small amounts of cutin monomers are likely to be generated during the early phases of fungal interaction with the plant surface. Such cutin monomers, as well as certain soluble cuticular components [152], were found to trigger the differentiation of the infection structure (appressorium) in some fungi [153, 154]. The next step is actual penetration into the host. Since cutin is the major structural component of the cuticle, it is a major physical barrier to penetration by the invading pathogens, particularly fungal pathogens.

How pathogens penetrate into the host through the polyester barrier has been debated for the better part of a century [119]. This process was thought to be mediated merely by the physical force of growth of the fungus, mainly by the osmotic force generated in the infection structure, the appressorium. This penetration has also been suggested to be assisted by enzymes secreted by the infection peg. Since cutin is the first barrier, one of the key enzymes involved in this process was postulated to be cutinase. However, direct examination of the role of such enzymes was not possible until cutinase was purified in the 1970s [13, 100]. With the availability of the purified enzyme, its cDNA [155], and gene [156], it became possible to investigate the role of cutinase in the penetration through the cuticular layer. The evidence that suggests that cutinase may be important in this process can be summarized as follows [27]: (a) Pathogenic fungi produce and secrete cutinase targeted at the penetration point and during actual infection of the host such an enzyme is produced as detected immunocytochemically. (b) Specific inhibition of cutinase by chemicals or antibodies including monoclonal antibodies prevents infection. (c) Cutinase-deficient mutants have significantly reduced virulence but infection can be restored with exogenous cutinase. (d) Pathogens that cannot infect a host without a breached cuticle (wound) can be genetically engineered to provide cutinase-producing capability and such engineered organisms can infect intact hosts, without requiring a breached cuticle. (e) Knocking out cutinase gene decreases virulence of organisms that have a single cutinase gene. However, gene knockout can lead to misleading conclusions when multiple cutinase genes are present [157, 158]. When a laboratory strain that contains only one cutinase gene was knocked out, drastic decrease in virulence was actually observed [159]. In many reports where cutinases have been knocked out, virulence decrease was not easily detected [109, 160–162], but the extent to which all cutinase activity was eliminated in the fungus invading the host has not been established. Most probably, the penetration of the host is not mediated exclusively by either physical force or by enzymatic degradation of the polymer. Depending on the host-pathogen system, the relative importance of the physical force and enzymatic degradation in the penetration process probably varies a great deal. For example, in the case of *Magnaportha griseae* the great turgor pressure produced in the appressorium may play a major role in gaining access into the host [163]. On the other hand, in fungi that do not form appressoria and exert less turgor pressure, the enzymatic degradation of the barrier may be of crucial importance. In such cases, cutinase-targeted approaches may be of practical value [164]. In fact, selective inhibition of cutinase by antibodies and chemical inhibitors, including suicide inhibitors, was found to protect many plant organs from infection by their pathogens [4]. Application of a cutinase inhibitor showed protection of the fruits from lesion formation in a papaya field [165, 166].

The cutin monomers produced by the fungus may act as early alarm signals of fungal attack and trigger defense reactions in the host. Cutin monomers could induce alkalanization of the medium when plant cell cultures were treated with cutin monomers, possibly indicative of the early phase of the defense reaction [167]. Hypocotyl segments produced oxidative burst (H_2O_2) when treated with cutin hydrolysate, indicative of the defense reaction triggered by cutin monomers [168]. Treatment of intact plants with cutin monomers, albeit at fairly high concentrations, was reported to protect the plants against fungal attack [169]. Thus, the polymer may act not only as a physical barrier against fungal infection but also serve as the sentry that sends early signals to alert the host.

12.1.3
Regulation of Cutinase Gene Transcription

How a fungus that is in contact with an insoluble polymer senses the nature of the polymer present in the environment to trigger the induction of the appropriate extracellular hydrolytic enzyme is an intriguing question. One possibility is that the microorganism might, upon starvation, secrete very small amounts of hydrolytic enzymes and the products generated from the extracellular material actually present in the immediate environment would then be transported into the microbe and cause the induction of the appropriate hydrolytic enzyme [170]. If such a hypothesis is valid, the fungus might perceive that it is resting on cutin by the mere presence of small amounts of cutin hydrolysate in the medium. In fact, glucose-grown *F. solani* f. *pisi* was found to secrete cutinase after depletion of glucose when cutin hydrolysate was added. Glucose repressed transcription of cutinase gene [170].

Conidia from pathogenic fungi that land on a plant surface carry low levels of cutinase to sense the contact with the host, and the cutin monomers generated upon contact of the conidia with the plant can be the inducers that allow the fungus to produce enough quantities of cutinase to gain access into the host (Fig. 12). This postulate was supported by many lines of evidence. Highly pa-

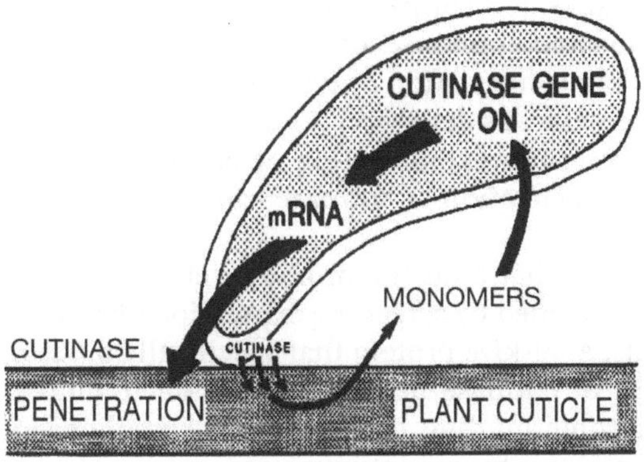

Fig. 12. Schematic representation of how the plant cuticle induces cutinase in a fungal spore

thogenic species and isolates of fungi were found to have cutinase on their conidia [151]. Induction of cutinase gene expression by contact with cutin and activation of transcription of cutinase gene by the unique monomers of cutin have been demonstrated [171]. Cloning of fungal cutinase gene led to identification of the promoter elements required for transcriptional activation by hydroxy fatty acids [172]. The promoter activity was tested by transforming *F. solani* f. *pisi* with the 5′-flanking region of cutinase gene fused to a reporter gene, chloramphenicol acetyl transferase (CAT). Deletion and mutation of the various segments of the cutinase gene and a measurement of the effects of such alterations on the promoter activity revealed several important regulatory elements in the cutinase promoter [173]. A G-rich region at –340 bp that is needed for inducible expression, a silencer element at – 287 bp that keeps cutinase gene repressed until induced, and a palindromic region at – 170 bp that is absolutely essential for transcriptional activation by the cutin monomers were identified. The palindromic region was found to be composed of two overlapping palindromes. Mutation analysis showed that it is the palindrome 2 that is absolutely essential for inducible expression of cutinase gene [172]. Binding of cutinase promoter to fungal nuclear proteins was demonstrated [172, 174]. Binding of different promoter fragments to nuclear proteins showed that two separate segments of the promoter, one in palindrome 2 and another closer to the transcription initiation site, bound nuclear proteins [172]. The binding to the palindrome was found only in the nuclear extracts from glucose-depleted cultures, whereas the other was found in nuclei from both glucose-containing and glucose-depleted cultures. The palindrome region was found to bind to a 50 kDa protein. Phosphorylation appears to be involved in the activation of cutinase gene expression by cutin monomers. For example, nuclear proteins bound to the cutinase promoter, but this binding was prevented by dephosphorylation of the nuclear extract [174]. Phosphorylation of a 50-kDa protein catalyzed by a nuclear preparation required the presence of cutin monomer. This cutin monomer-dependent phosphorylation was found to be necessary for activation of cutinase transcription in a nuclear preparation [173]. In this system protein kinase inhibitors prevented cutinase transcription activation by hydroxy fatty acids. Western blot and immunoblot analyses showed that cutin monomers were required for phosphorylation of tyrosine residues and a 50-kDa tyrosine-phosphorylated protein was detected [173]. This phosphorylation was found to be involved in transcriptional activation of cutinase as anti-phosphotyrosine antibodies inhibited activation of cutinase transcription induced by cutin monomers.

Transacting factors that bind the promoter elements found in the cutinase promoter have been cloned by screening expression libraries with concatenated promoter elements. A 50-kDa protein that specifically binds palindrome 1 (and not palindrome 2) of cutinase promoter was cloned and expressed in *E. coli* [175]. This palindrome binding protein (PBP) autophosphorylates [173]; the biological function of this kinase activity remains to be elucidated. The cutinase transcription factor (CTF) that binds palindrome 2 that is essential for transcriptional activation of cutinase gene by hydroxy fatty acids exists in two forms, both of which have been cloned – CTF1α and CTF1β. When expressed, CTF1α specifically binds palindrome 2 [176]. CTF1α is a 101-kDa protein which contains a Cys_6Zn_2 binuclear cluster motif sharing homology to the Cys_6Zn_2 binuclear cluster DNA-binding domains of transcription factors from yeast and filamentous fungi. When CTF1α is expressed in yeast it can transactivate the native cutinase promoter fused to CAT gene. Mutation of palindrome 2 but not palindrome 1 abolished this transactivation. Thus, CTF1α positively acts in vivo by binding selectively to palindrome 2 of cutinase gene promoter. When PBP and CTF1α are introduced together into yeast, PBP acts as a repressor by inhibiting the transactivation of the native cutinase promoter by CTF1α. (D. Li and P.E. Kolattukudy, manuscript in preparation) Deletion and mutational analysis in the yeast transactivation system showed that the C-terminal 50 amino acids, the first *N*-terminal nuclear localization signal, and the Cys_6Zn_2 binuclear cluster domain are essential for CTF1α to function as a positive acting factor. CTF1β is a 98-kDa protein. Like CTF1α, CTF1β also contains Cys_6Zn_2 binuclear cluster motif. CTF1β expressed in *E. coli* shows specific binding to palindrome 2. When fused to GAL4 DNA binding domain and introduced into yeast promoter strain, this protein is able to turn on the expression of β-galactosidase reporter gene. CTF1β expressed in yeast was not able to transactivate native cutinase gene promoter fused to CAT gene, unlike CTF1α. However, CTF1β was able to transactivate the cutinase gene promoter in which palindrome 1 is mutated, indicating that CTF1β may function with the promoter of another cutinase gene or when palindrome 1 is unoccupied. (D. Li, T. Sirakova and P.E. Kolattukudy, manuscript in preparation)

The induction of cutinase by cutin monomers produced by a constitutively expressed cutinase indicate that there are multiple cutinase genes. In fact, most pathogenic *F. solani* f. *pisi* strains showed multiple cutinase genes even by Southern hybridization of genomic DNA [157, 158]. Recently, three cutinase genes were cloned from *F. solani* f. *pisi* and sequenced (T. Sirakova and P.E. Kolattukudy, unpublished). *Cut*1 is the inducible gene that requires CTF1α. *Cut*2 and *cut*3 have two nucleotide substitutions in palindrome 1 whereas palindrome 2 is identical to that of *cut*1. The two nucleotide substitution pre-

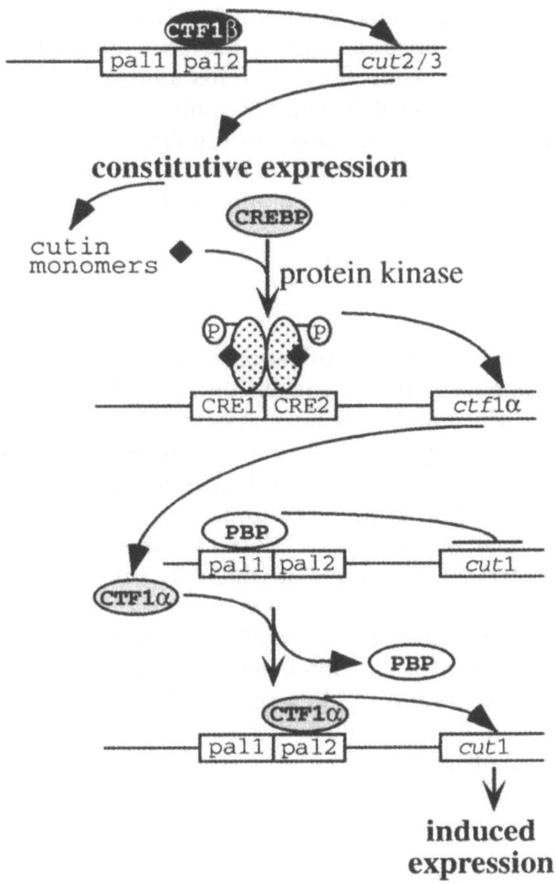

Fig. 13. Postulated mechanism by which cutin monomers produced by constitutive expression of *cut2/3* induce *cut1*

vents palindrome 1 of *cut2/cut3* from binding PBP as can be demonstrated by gel retardation analysis. Therefore we postulate that CTF1β activates cutinase *cut2/cut3* transcription in a constitutive fashion as palindrome 1 region is unoccupied (Fig. 13). On the other hand, *cut1* has PBP bound to palindrome 1 and therefore CTF1α binding is prevented, possibly by competition by and/or lack of CTF1α in the uninduced cells. Upon addition of hydroxy fatty acids, CTF1α may be induced, which then outcompetes or replaces PBP from *cut1* promoter and binds to palindrome 2, and induces transcription. According to this hypothesis CTF1α gene expression may be the key step that is regulated by hydroxy fatty acids. Therefore, CTF1α gene was cloned and sequenced (D. Li and P.E. Kolattukudy, unpublished). The promoter segment of *ctf1α* shows two potential *cis* elements, CRE1 and CRE2. CRE1, located at –203 to –221, is a palindrome sequence resembling steroid hormone response elements, specifically the GRE consensus element. CRE2, located at – 160 to – 181, is a direct repeat resembling the thyroid hormone response element. These elements were concatenated and were used to screen expression libraries. This approach yielded a clone that encodes a protein that binds CRE1. This CRE1 binding protein (CREBP) may be a steroid receptor-like protein and, if so, the regulation of cutinase gene expression by the hydrophobic cutin monomers may bear analogy to gene transcriptional activation by steroid hormones in mammals.

12.2 Function of Suberin

Chemical and ultrastructural identification of suberized walls found in a variety of plant organs leads to the conclusion that whenever a plant needs to erect a barrier to diffusion, during normal development or in response to environmental stress, either physical, chemical, or biological, the plant resorts to deposition of suberin on the cell walls of the barrier layer [80]. Suberin deposition during wound healing of underground storage organs has been demonstrated [2]. As summarized elsewhere [3], trees respond to wounding by depositing a series of barrier zones, some of which are suberized layers. Wounding of roots of *Abiesbalsamea* and *Tsugacanadensis* also causes the formation of periderm with heavily suberized cell walls. Chemical and ultrastructural examination of the barrier zone laid down in *Quercus suber* in response to wounding is found to be suberized. Wounding causes suberization, not only on tubers and other organs that are normally protected by suberin but also on fruit and leaves that are normally protected by cutin, leading to the conclusion that suberization is a universal response to wounding in all plant organs [80]. Attack by parasites and insects have been found to cause the formation of suberized barriers.

Suberization also results from stress. Salt stress due to mineral deficiencies can result in changes in suberization. For example, magnesium deficiency caused the hypodermis and endodermis of *Zea mays* roots to have more heavily suberized cell walls [81], whereas iron deficiency in *Phaseolus vulgaris* led to a drastic decrease in suberin deposition in the roots [84]. Physical impedence to root growth causes increased suberization in the walls of endodermal cells in *Hordeum vulgaris* roots [177]. Cold stress can also cause suberization to protect the plants from freezing damage [178, 179]. Suberin deposition in the bark of *Vitis vinifera* was shown to protect the plant from freezing damage. Growth at low temperature was also found to induce the deposition of aliphatic components in the epidermis and in the mestome sheet in *Secale cereale* [180].

Fungal attack can cause suberization that makes an impermeable chemical and physical barrier to further fungal ingress. Many reports indicate that the presence of heavily suberized periderm protects the older plant organs whereas very young organs that do not have such barriers are more susceptible to pathogen attack. Not only does pre-existing suberin act as a barrier, but many reports suggest that fungal or viral attack elicits the deposition of suberin in cell walls and thus limits infection. Such examples have been discussed elsewhere [3]. In many cases the barrier layers were subjected to microscopic or cytochemical examination. In the absence of chemical evidence, the barrier layers formed by fungal infection have been variously designated as lignin, suberin, lignin-suberin-like, etc. In one case firm identification of the barrier layer was made. Chemical analysis of the depolymerization products of the vascular coating induced by *Verticilium alboatrum* in tomatoes showed that the wall coating is suberin [83]. The amount of aliphatic suberin components deposited on the walls of resistant tomatoes was found to be many times higher than that in the susceptible near-isogenic line. Thus, the degree of suberization induced by the fungal attack may be related to the degree of resistance to fungal attack.

To erect diffusion barriers during development, suberin is laid down in various internal locations to seal off specific regions of the plant. There are many examples where only microscopic identification has been done [3]. The following are some cases where both ultrastructural and chemical studies have identified suberin. Chemical analysis of casparian band of the endodermis of *Sorghum bicolor* and other monocots showed that this layer is suberized [181, 183]. Suberized layers are also found in the mestome sheet and bundle sheath cell walls of grasses. These layers have been chemically characterized in *Z. mays* and *S. cereale* [180, 184]. Suberin is deposited in the chalazal region of the seed coat when the seed reaches maturity as a means of sealing off the seed where vascular tissue was attached to the developing seed as demonstrated with citrus seeds [20]. Chemical and ultrastructural characterization of crystal idioblast cell walls from *Agave americana* showed that these walls are suberized [185]; presumably the diffusion of calcium oxalate into the rest of the plant is prevented by such a barrier layer. In all of these cases, the suberized walls appear to be erected to protect the rest of the plant from potentially toxic components.

The cell walls of the periderm layer are made highly hydrophobic by the deposition of suberin so that these walls can be made into an effective diffusion barrier by the deposition of soluble waxes. There is evidence that the soluble waxes play the most dominant role as a good diffusion barrier while the polymer matrix probably constitutes a nearly impermeable barrier that prevents pathogen entry. Although removal of suberin-associated waxes to examine their role in diffusion is not physically possible, specific inhibition of wax biosynthesis with trichloroacetate could be achieved without affecting suberin polymer synthesis. This lack of wax prevented development of diffusion resistance in wound-healing potato slices, strongly suggesting that waxes constitute the major diffusion barrier in the suberized cell walls [186]. This conclusion was supported by evidence obtained from the bark of mature *Betula pendula* trees [187].

13
Potential Commercial Applications

The major site of occurrence of polyesters in plants is the cuticle where the insoluble polymer constitutes a major barrier. This material present in the skin of fruit constitutes a significant part of the waste produced in fruit processing. The very large quantities of the cutin-rich agricultural wastes that are produced may be a source of high value polyesters. Since the insoluble polymer is relatively refractory to most treatments, it would be possible to remove the cellular debris attached to the cuticle by physical, chemical, or biological processes; then the recovery of the insoluble polymer could be practical. This material may then be subjected to chemical or enzymatic hydrolysis to recover the polyfunctional cutin monomers. Such monomers can be used to make polymers of predetermined degrees of polymerization to yield liquid or soft material that can be used for high value lubricants, or it can be made into larger polymers for other uses in high value products. Since the hydroxy fatty acid polymer would be relatively inert from an immunological point of view, biomedical applications of such polymers should be examined. Since cutinase has many potential commercial applications, the cutin-rich agricultural waste may also be used as a fermentation medium for the production of cutinase. The potential of generating such high value products from agricultural waste has not been adequately explored.

Suberin is also a significant component of waste. Cork from the cork industry is already used for a variety of applications, particularly those which involve insulation. This industry produces 280,000 tons of cork annually, of which almost one-third is rejected as cork dust and low granulometric fraction [35]. Such materials are currently used for combustion for energy production. Potato processing and processing of other tubers and roots covered with suberized layers also generate huge amounts of suberized material as waste material. Disposal of the peels that are generated in such processing industry are considered a major problem. It has been used as landfills and sometimes as cattle feed. Even though cattle can utilize part of the energy contained in the suberin, the polyester domains from these materials may be used for more valuable purposes. The potential of recovering such hydroxy fatty acids and dicarboxylic acids for production of high value products has not been fully explored. Some efforts have been made to synthesize suberin-based polyurethanes [188].

The other potential use of plant polyesters is to use plants to generate large amounts of polyesters by genetic engineering. Low levels of polyester formation have been achieved in plants by the expression of bacterial poly-β-hydroxybutyrate-generating enzymes. When the molecular mechanisms used in the biosynthesis of plant polyester are fully elucidated and the genes involved are cloned, it might be possible to use such plant genes to produce polyesters internally in the storage organs of plants. If such approaches succeed, plants might be used as a renewable resource for polyester production for commercial purposes.

Acknowledgement. The work for the author's laboratory was supported in part by the United States National Science Foundation.

References

1. Kerstiens G (ed) (1996) Plant cuticles: an integrated functional approach. BIOS Scientific Publishers, Oxford UK
2. Kolattukudy PE (1987) Lipid derived defensive polymers and waxes and their role in plant-microbe interaction. In: Stumpf PK (ed) The biochemistry of plants: vol 9 – lipids: structure and function. Academic Press, New York, p 291

3. Kolattukudy PE, Espelie KE (1989) Chemistry, biochemistry and function of suberin and associated waxes. In: Rowe J (ed) Natural products of woody plants, chemicals extraneous to the lignocellulosic cell wall. Springer, Berlin Heidelberg New York, p 304
4. Kolattukudy PE, Kämper J, Kämper U, González-Candelas L, Guo W (1994) Fungus-induced degradation and reinforcement of defensive barriers of plants. In: Petrini O, Ouellette G (eds) Alteration of host walls by fungi. APS Press, St. Paul MN, p 67
5. Kolattukudy PE (1980) Cutin, suberin and waxes. In: Stumpf PK (ed) The biochemistry of plants: vol 4 – lipids: structure and function. Academic Press, London, p 571
6. Holloway PJ (1982) The chemical constitution of plant cutins. In: Cutler DF, Alvin KL, Price CE (eds) The plant cuticle. Academic Press, London, p 45
7. Kolattukudy PE (1981) Structure, biosynthesis and biodegradation of cutin and suberin. In: Briggs WR (ed) Annual reviews of plant physiol, vol. 32. Annual Reviews, Palo Alto CA, p 539
8. Ray AK, Chen Z-A, Stark RE (1998) Phytochemistry 49:65
9. Walton TJ, Kolattukudy PE (1972) Biochemistry 11:1885
10. Kolattukudy PE, Walton TJ (1973) Prog Chem Fats Other Lipids 8:121
11. Kolattukudy PE, Agrawal VP (1974) Lipids 9:682
12. Holloway PJ (1974) Phytochemistry 13:2201
13. Purdy RE, Kolattukudy PE (1975) Biochemistry 14:2832
14. Crisp CE (1965) The biopolymer cutin. PhD Thesis, University of California, Davis CA
15. Ray AK, Stark RE (1998) Phytochemistry 48:1313
16. Kolattukudy PE, Espelie KE (1985) Biosynthesis of cutin, suberin, and associated waxes. In: Higuchi T (ed) Biosynthesis and biodegradation of wood components. Academic Press, New York p 161
17. Bernards MA, Lewis NG (1998) Phytochemistry 47:915
18. Deas AHB, Baker EA, Holloway PJ (1974) Phytochemistry 13:1901
19. Kolattukudy PE (1974) Biochemistry 13:1354
20. Espelie KE, David RW, Kolattukudy PE (1980) Planta 149:498
21. Caldicott AB, Simoneit BRT, Eglinton G (1975) Phytochemistry 14:2223
22. Espelie KE, Köller W, Kolattukudy PE (1983) Lipids 32:13
23. Espelie KE, Dean BB, Kolattukudy PE (1979) Plant Physiol 64:1089
24. Deas AHB, Holloway PJ (1977) The intermolecular structure of some plant cutins. In: Tevini M, Lichtenthaler HK (eds) Lipids and lipid polymers in higher plants. Springer, Berlin Heidelberg New York, p 293
25. Kolattukudy PE (1977) Lipid polymers and associated phenols, their chemistry, biosynthesis, and role in pathogenesis. In: Loewus FA, Runeckles VC (eds) Recent advances in phytochemistry. Plenum Publishing, New York, p 185
26. Zlotnik-Mazori T, Stark RE (1988) Macromolecules 21:2412
27. Kolattukudy PE (1996) Biosynthetic pathways of cutin and waxes, and their sensitivity to environmental stress. In: Kerstiens G (ed) Plant cuticles: an integrated functional approach. BIOS Scientific Publishers, Oxford UK, chap 3
28. Tegelaar EW, Wattendorff J, deLeeuw JW (1993) Rev Palaeobot Palynol 77:149
29. Villena JF, Casado CG, Luque P, Heredia A (1996) J Exptl Bot 47 (suppl):56
30. Villena JF, Dominguez E, Stewart D, Heredia A (1999) Planta 208:181
31. Kolattukudy PE, Walton TJ, Kushwaha RPS (1973) Biochemistry 12:4488
32. Riley RG, Kolattukudy PE (1975) Plant Physiol 56:650
33. Hunt GM, Baker EA (1980) Phytochemistry 19:1415
34. Mohan R, Vijayan P, Kolattukudy PE (1993) Plant Mol Biol 22:475
35. Cordeiro N, Belgacem MN, Silvestre AJD, Neto CP, Gandini A (1998) Int J Biol Macromol 22:71
36. Brieskorn CH, Binnemann PH (1975) Phytochemistry 14:1363
37. Matzke K, Riederer M (1991) Planta 185:233
38. Bernards MA, Lopez ML, Zajicek J, Lewis NG (1995) J Biol Chem 270:7382
39. Negrel J, Pollet B, Lapierre C (1996) Phytochemistry 43:119
40. Schmutz A, Jenny T, Amrhein N, Ryser U (1993) Planta 189:453

41. Moire L, Schmutz A, Buchala A, Yan B, Stark RE, Ryser U (1999) Plant Physiol 119:1137
42. Garbow JR, Ferrantello LM, Stark RF (1989) Plant Physiol 90:783
43. Stark RE, Garbow JR (1992) Macromolecules 25:149
44. Neto CP, Rocha J, Gil A, Cordeiro N, Esculcas AP, Rocha S, Delgadillo I, Pedrosa de Jesus JD, Correia FA (1995) J Solid State NMR 4:143
45. Gil AM, Lopes M, Rocha J, Neto CP (1997) Int J Biol Macromol 20:293
46. Stark RE, Sohn W, Pacchiano RA Jr, Al-Bashir M, Garbow JR (1994) Plant Physiol 104:527
47. Cordeiro N, Aurenty P, Belgacem MN, Gandini A, Neto CP (1997) J Colloid Interface Sci 187:498
48. Tegelaar EW, Hollman P, van der Vegt ST, Leeuw JW, Holloway PJ (1995) Org Geochem 23:239
49. Heinen W, van den Brand I (1963) Z Naturforsch B 18:67
50. Kolattukudy PE (1970) Plant Physiol 46:759
51. Kolattukudy PE (1970) Biochem Biophys Res Commun 41:299
52. Kolattukudy PE, Walton TJ (1972) Biochemistry 11:1897
53. Soliday CL, Kolattukudy PE (1977) Plant Physiol 59:1116
54. Pinot F, Salaun J-P, Bosch H, Lesot A, Mioskowski C, Durst F (1992) Biochem Biophys Res Commun 184:183
55. Pinot F, Bosch H, Alayrac C, Mioskowski C, Vendais A, Durst F, Salaun JP (1993) Plant Physiol 102:1313
56. Tijet N, Helvig C, Pinot F, Le Bouquin R, Lesot A, Durst F, Salaun J-P, Benveniste I (1998) Biochem J 332:583
57. Walton TJ, Kolattukudy PE (1972) Biochem Biophys Res Commun 46:16
58. Soliday CL, Kolattukudy PE (1978) Arch Biochem Biophys 188:338
59. Espelie KE, Dean BB, Kolattukudy PE (1979) Plant Physiol 64:1089
60. Croteau R, Kolattukudy PE (1974) Arch Biochem Biophys 162:458
61. Croteau R, Kolattukudy PE (1974) Arch Biochem Biophys 162:471
62. Croteau R, Kolattukudy PE (1975) Arch Biochem Biophys 170:61
63. Blée E, Schuber F (1990) J Biol Chem 265:12,887
64. Blée E, Schuber F (1993) Plant J 4:113
65. Blée E, Stahl U, Schuber F, Stymne S (1993) Biochem Biophys Res Commun 197:778
66. Croteau R, Kolattukudy PE (1975) Arch Biochem Biophys 170:73
67. Kiyosue T, Beetham JK, Pinot F, Hammock BD, Yamaguchi-Shinozaki K, Shinozaki K (1994) Plant J 6:259
68. Stapleton A, Beetham JK, Pinot F, Garbarino JE, Rockhold DR, Friedman M, Hammock BD, Belknap WR (1994) Plant J 6:251
69. Blée E, Schuber F (1992) J Biol Chem 267:11,881
70. Blée E, Schuber F (1995) Eur J Biochem 230:229
71. Croteau R, Kolattukudy PE (1974) Biochemistry 13:3193
72. Kolattukudy PE (1973) Lipids 8:90
73. Dean BB, Kolattukudy PE (1977) Plant Physiol 59:48
74. Agrawal VP, Kolattukudy PE (1977) Plant Physiol 59:667
75. Agrawal VP, Kolattukudy PE (1978) Arch Biochem Biophys 191:452
76. Agrawal VP, Kolattukudy PE (1978) Arch Biochem Biophys 191:466
77. Schmidt A, Grimm R, Schmidt J, Scheel D, Strack D, Rosahl S (1999) J Biol Chem 274:4273
78. Roberts E, Kutchan T, Kolattukudy PE (1988) Plant Mol Biol 11:15
79. Espelie KE, Franceschi VR, Kolattukudy PE (1986) Plant Physiol 81:487
80. Dean BB, Kolattukudy PE (1976) Plant Physiol 58:411
81. Pozuelo JM, Espelie KE, Kolattukudy PE (1984) Plant Physiol 74:256
82. Soliday CL, Dean BB, Kolattukudy PE (1978) Plant Physiol 61:170
83. Robb J, Lee S-W, Mohan R, Kolattukudy PE (1991) Plant Physiol 97:528
84. Sijmons PC, Kolattukudy PE, Bienfait HF (1985) Plant Physiol 78:115
85. Mohan R, Kolattukudy PE (1990) Plant Physiol 921:276

86. Sherf B, Bajar AM, Kolattukudy PE (1993) Plant Physiol 101:201
87. Hankin L, Kolattukudy PE (1971) Plant Soil 34:525
88. Patty BR, Chandra AK (1981) Plant Soil 61:419
89. Sebastian J, Chandra AK, Kolattukudy PE (1987) J Bacteriol 169:131
90. Sebastian J, Kolattukudy PE (1988) Arch Biochem Biophys 263:77
91. Sebastian J (1987) Discovery and purification of cutinase and cloning of cutinase gene from a *Pseudomonas* sp. cohabitting with an apparently nitrogen-fixing *Corynebacterium* sp. in the phyllosphere. PhD thesis, Washington State University, Pullman WA
92. Kolattukudy P, Poulose AJ (1996) Cutinase cleaning compositions. US Pat 5,512,203
93. Kolattukudy P, Poulose AJ (1996) Enzymes as agricultural chemical adjuvants. US Pat 5,545, 547
94. Lin T-S, Kolattukudy PE (1980) Physiol Plant Pathol 17:1
95. Fett WF, Wijey C, Moreau RA, Osman SF (1999) J Applied Microbiol 86:561
96. Heinen W, De Vries H (1966) Arch Mikrobiol 54:331
97. Heinen W, De Vries H (1966) Arch Mikrobiol 54:339
98. Shishiyama J, Araki F, Akai S (1970) Plant Cell Physiol 11:937
99. van den Ende G, Linskens HF (1974) Ann Rev Phytopathol 12:247
100. Purdy RE, Kolattukudy PE (1975) Biochemistry 14:2824
101. Kolattukudy PE, Purdy RE, Maiti IB (1981) Cutinases from fungi and pollen. In: Lowenstein JM (ed) Methods in enzymology. Academic Press, New York, p 652
102. Kolattukudy PE (1984) Cutinases from fungi and pollen. In: Borgstrom B, Brockman H (eds) Lipases. Elsevier Science Publishers, Amsterdam, p 471
103. Lin T-S, Kolattukudy PE (1976) Biochem Biophys Res Commun 72:243
104. Lin T-S, Kolattukudy PE (1980) Eur J Biochem 106:341
105. Holt GD, Hart GW (1986) J Biol Chem 261:8049
106. Jackson SP, Tijan R (1998) Cell 55:125
107. Trail F, Köller W (1990) Physiol Mol Plant Pathol 36:495
108. Yao C, Köller W (1995) Molec Plant Microbe Interact 8:122
109. Köller W, Yao C, Trial F, Parker DM (1995) Can J Bot 73:S1109
110. Gindro K, Pezet R (1999) FEMS Microbiol Lett 171:239
111. Bonnen AM, Hammerschmidt R (1989) Physiol Mol Plant Pathol 35:463
112. Salinas J (1992) Function of cutinolytic enzymes in the infection of Gerbera flowers by *Botrytis cinerea*. PhD Thesis, University of Utrecht, Utrecht
113. Murphy CA, Cameron JA, Huang SJ, Vinopal RT (1996) Appl Environ Microbiol 62: 456
114. Jendrossek D, Schirma A, Schelegel HG (1996) Appl Microbiol Biotechnol 46:451–463
115. Jendrossek D (2000) (Chapter in this volume)
116. Purdy RE, Kolattukudy PE (1973) Arch Biochem Biophys 159:61
117. Köller W, Kolattukudy PE (1982) Biochemistry 21:3083
118. Köller W, Allan CR, Kolattukudy PE (1982) Pesticide Biochem Physiol 18:15
119. Kolattukudy PE (1985) Ann Rev Phytopathol 23:223
120. Prompers JJ, Hilbers CW, Pepermans HAM (1999) FEBS Lett 456:409–416
121. Martinez C, De Geus P, Lauwereys M, Matthyssens G, Cambillau C (1992) Nature 356:615
122. Martinez C, Nicolas A, van Tilbeurgh H, Egloff M-P, Cudrey C, Verger R, Cambillau C (1994) Biochemistry 33:83
123. Prompers JJ, Groenewegen A, Hilbers CW, Pepermans HAM (1999) Biochemistry 38:5315
124. Prompers JJ, van Noorloos B, Mannesse MLM, Groenewegen A, Egmond MR, Verheij HM, Hilbers CW, Pepermans HAM (1999) Biochemistry 38:5982
125. Brown AJ, Kolattukudy PE (1978) J Agric Food Chem 26:1263
126. Brown AJ, Kolattukudy PE (1978) Arch Biochem Biophys 190:17
127. Knox RB, Clarke A, Harrison S, Smith P, Marchalonis JJ (1976) Proc Natl Acad Sci USA 73:2788
128. Heslop-Harrison Y (1977) Ann Bot 41:913

129. Linskens HF (1975) The physiological basis of incompatibility in angiosperms. In: Duckett JG, Racey PA (eds) The biology of the male gamete. Academic Press (for the Linnean Society of London), London, p 143
130. Heinen W, Linskens HF (1961) Nature 191:1416
131. Linskens HF, Heinen W (1962) Z Bot 50:338
132. Shaykh M, Kolattukudy PE, Davis R (1977) Plant Pysiol 60:907
133. Maiti IB, Kolattukudy PE, Shaykh M (1979) Arch Biochem Biophys 196:412
134. Hiscock SJ, Dewey FM, Coleman JOD, Dickinson HG (1994) Planta 193:377
135. Parameswaren N, Wilhelm GE (1979) Eur J For Pathol 9:103
136. Ofong AU, Pearce RB (1994) Eur J For Pathol 24:316
137. Zimmerman W, Seemuller E (1984) Phytopathol Z 110:192
138. Schultz E, Chamuris GP, Dallabrida S (1996) Mycologia 88:831
139. Fernando G, Zimmermann W, Kolattukudy PE (1984) Physiol Plant Pathol 24:143
140. Faulds CB, Williamson G (1993) Biotechnol Appl Biochem 17:349
141. Edwards D, Abbott GD, Raven JA (1996) Cuticles of early land plants: a palaeoecophysiological evaluation. In: Kerstiens G (ed) Plant cuticles: an integrated functional approach. BIOS Scientific Publishers, Oxford UK, chap 1
142. Schreiber L, Kirsch T, Riederer M (1996) Diffusion through cuticles: principles and models. In: Kerstiens G (ed) Plant cuticles: an integrated functional approach. BIOS Scientific Publishers, Oxford UK, chap 4
143. Rieder M, Schreiber L (1995) Waxes: the transport barriers of plant cuticles. In: Hamilton RJ (ed) Waxes: chemistry, molecular biology and functions. The Oily Press, Dundee, p 131
144. Schönherr J, Baur P (1996) Effects of temperature, surfactants and other adjuvants on rates of uptake of organic compounds. In: Kerstiens G (ed) Plant cuticles: an integrated functional approach. BIOS Scientific Publishers, Oxford UK, chap 6
145. Barnes JD, Cardoso-Vilhena J (1996) Interaction between electromagnetic radiation and the plant cuticle. In: Kerstiens G (ed) Plant cuticles: an integrated functional approach. BIOS Scientific Publishers, Oxford UK, chap 7
146. Macko V (1981) Inhibitors and stimulants of spore germination and infection structure formation in fungi. In: Turian G, Holh HR (eds) The fungal spore morphogenetic controls. Academic Press, New York, p 565
147. Hegde Y, Kolattukudy PE (1997) Physiol Molec Plant Pathol 51:75
148. Liu Z-M, Kolattukudy PE (1999) J Bacteriol 181:3571
149. Liu Z-M, Kolattukudy PE (1998) J Bacteriol 180:3592
150. Kim Y-K, Li D, Kolattukudy PE (1998) J Bacteriol 180:5144
151. Köller W, Allan CR, Kolattukudy PE (1982) Physiol Plant Pathol 20:47
152. Podila GK, Rogers LM, Kolattukudy PE (1993) Plant Physiol 103:267
153. Francis SA, Dewey FM, Gurr SJ (1996) Physiol Mol Plant Pathol 49:201
154. Gilbert RD, Johnson AM, Dean RA (1996) Physiol Molec Plant Pathol 48:335
155. Soliday CL, Flurkey WH, Okita TW, Kolattukudy PE (1984) Proc Natl Acad Sci USA 81:3939
156. Soliday CL, Dickman MB, Kolattukudy PE (1989) J Bacteriol 171:1942
157. Kolattukudy PE, Crawford MS (1987) The role of polymer degrading enzymes in fungal pathogenesis. In: Nishumura S, Vance CP, Doke N (eds) Molecular determinants of plant diseases. Springer, Berlin Heidelberg New York, p 75
158. Woloshuk CP (1986) Cutinase of *Fusarium solani* f. sp. *pisi*: mechanism of induction and relatedness to other *Fusarium* species. PhD thesis, Washington State University, Pullman WA
159. Rogers LM, Flaishman MA, Kolattukudy PE (1994) Plant Cell 6:935
160. Sweigard JA, Chumley FG, Valent B (1992) Mol Gen Genet 232:183
161. van Kan JA, van't Klooster JW, Wagemakers CA, Dees DC, van der Vlugt-Bergmans CJ (1997) Mol Plant Microbe Interact 10:30
162. Crowhurst RN, Binnie SJ, Bowen JK, Hawthorne BT, Plummer KM, Rees-George J, Rikkerink EH, Templeton MD (1997) Mol Plant Microbe Interact 10:355

163. Howard RG, Valent B (1996) Annu Rev Microbiol 50:491
164. Maiti IB, Kolattukudy PE (1979) Science 205:507
165. Kolattukudy PE, Ettinger WF, Sebastian J (1987) Cuticular lipids in plant-microbe interactions. In: Stumpf PK, Mudd BD, Ness WD (eds) The metabolism, structure, and function of plant lipids. Plenum Publishing, New York, p 473
166. Kolattukudy PE, Kim Y, Li D, Liu ZM, Rogers L (2000) Early molecular communication between *Colletotrichum gloeosporioides* and its host. In: Prusky D, Freemann S, Dickman MB (eds) Collelotrichum, Host specificity, pathology and host-pathogen interaction. American Phytopathological Society Press, St. Paul MN, p 78
167. Schweizer P, Felix G, Buchala A, Müller C, Métraux J-P (1996) Plant J 10:331
168. Fauth M, Schweizer P, Buchala A, Markstadter C, Riederer M, Kato T, Kauss H (1998) Plant Physiol 117:1373
169. Schweizer P, Jeanguenat A, Whitacre D, Métraux JP, Mössinger E (1996) Physiol Mol Plant Pathol 49:103
170. Lin T-S, Kolattukudy PE (1978) J Bacteriol 133:942
171. Woloshuk CP, Kolattukudy PE (1986) Proc Natl Acad Sci USA 83:1704
172. Kämper J, Kämper U, Rogers LM, Kolattukudy PE (1994) J Biol Chem 269:9195
173. Kolattukudy PE, Rogers LM, Li D, Hwang C-S, Flaishman MA (1995) Proc Natl Acad Sci USA 92:4080
174. Bajar A, Podila GK, Kolattukudy PE (1991) Proc Natl Acad Sci USA 88:8208
175. Li D, Kolattukudy PE (1995) J Biol Chem 270:11,753
176. Li D, Kolattukudy PE (1997) J Biol Chem 272:12,462
177. Wilson AJ, Robards AW (1978) Protoplasma 95:225
178. Meiering AG, Paroschy JH, Peterson RL, Hostetter G, Neff A (1980) Am J Enol Vitic 31:81
179. Paroshy JH, Meiering AG, Peterson RL, Hostetter G, Neff A (1980) Am J Enol Vitic 31:227
180. Griffith M, Huner NPA, Espelie KE, Kolattukudy PE (1985) Protoplasma 125:53
181. Espelie KE, Kolattukudy PE (1979) Plant Physiol 63:433
182. Zeier J, Schreiber L (1997) Plant Physiol 113:1223
183. Zeier J, Schreiber L (1998) Planta 206:349
184. Espelie KE, Kolattukudy PE (1979) Plant Sci Lett 15:225
185. Espelie KE, Wattendorff J, Kolattukudy PE (1982) Planta 155:166
186. Soliday CL, Kolattukudy PE, Davis RW (1979) Planta 146:607
187. Vogt E, Schönherr J, Schmidt HW (1983) Planta 158:294
188. Cordeiro N, Belgacem MN, Gandini A, Neto CP (1997) Ind Crops Prod 6:165

Received: December 1999

[illegible]

Polyesters from Microorganisms

Young Baek Kim [1], Robert W. Lenz [2]

[1] Polymer Science and Engineering Department, PaiChai University, 439–6 Doma-2-dong, Seoku, Daejon, 302–735, Korea
E-mail: ybkim@mail.paichai.ac.kr

[2] Polymer Science and Engineering Department, Silvio O. Conte National Center for Polymer Research, University of Massachusetts at Amherst, Box 34530, Amherst, MA 01003–1530, USA

Bacterial polyesters have been found to have useful properties for applications as thermoplastics, elastomers, and adhesives and are biodegradable and biocompatible. Poly(3-hydroxyalkanoates) (PHAs) and poly(β-malate) are the most representative polyesters synthesized by microorganisms. PHAs containing a wide variety of repeating units can be produced by bacteria, including those containing many types of pendant functional groups which can be synthesized by microorganisms that are grown on unnatural organic substrates. Poly(β-malate) is of interest primarily for medical applications, especially for drug delivery systems. In this chapter, the bacterial production and properties of poly(3-hydroxyalkanoates) and poly(β-malate) are described with emphasis on the former.

Keywords. Bacterial polyester, Medium-chain-length poly(3-hydroxyalkanoates), *Pseudomonas oleovorans*, *Pseudomonas putida*, Functional poly(3-hydroxyalkanoates), Short-chain-length poly(3-hydroxyalkanoates)

Advances in Biochemical Engineering/
Biotechnology, Vol. 71
Managing Editor: Th. Scheper

1 Occurrence and Function of Polyhydroxyalkanoates (PHAs)

The polyesters synthesized by microorganisms were first investigated by Lemoigne, who isolated the reserve polymers in *Bacillus megaterium* in 1925 and identified the polymer to be poly(3-hydroxybutyrate), poly(3HB) [1]. The chemical structure of poly(3HB) is shown in Fig. 1.

Poly(3HB) was long thought to be the only type of polyester produced by microorganisms until polyesters containing 3-hydroxyalkanoate units longer than 3-hydroxybutyrate (3HB) were isolated from microorganisms in sewage sludge in 1974 [2]. That polymer contained both 3HB and 3-hydroxyvalerate (3HV) units as the major repeating units, with 3-hydroxyhexanoate (3HHx) and possibly 3-hydroxyheptanoate (3HHp) units as minor components. Another PHA extracted from marine sediments was found to contain at least 11 3-hydroxyalkanoate (3HA) units, the principal ones being 3HB and 3HV [3]. Later, PHAs containing 3HB, 3HHx, and 3-hydroxyoctanoate (3HO) units were discovered in sewage sludge [4]. The compositions of various PHAs found in specific microbes and in environmental samples are shown in Table 1.

A wide variety of microorganisms can synthesize PHAs, and many of the types of microorganisms capable of accumulating PHAs are listed in Table 2 along with the carbon sources that were found to support PHA production in each microorganism. More than 90 repeating units have been found incorporated into PHAs [5]. Most of the microorganisms are capable of accumulating PHAs in large amounts, varying from 30% to 80% of their cellular dry weight. All of these PHAs are formed within the cell's cytoplasm as granule inclusions, which can be observed under the light microscope as refractile bodies [6–8]. The shape of the PHA granules is usually spherical, and the size varies according to organism.

Investigations on the changes of number and size of granules in *Ralstonia eutropha* (formerly *Alcaligenes eutrophus*) grown under nitrogen-limited conditions at different growth times showed that the number of granules per cell was fixed at the earliest stages of polymer accumulation [9], and polymer production ceased when a poly(3HB) content of approximately 80 wt% was attain-

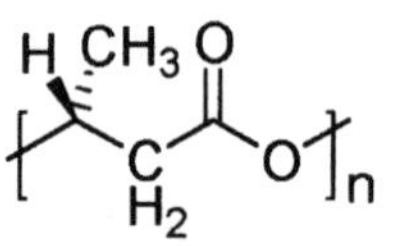

Fig. 1. The structure of poly(3-hydroxybutyrate)

Table 1. Compositions of various PHAs found in specific microbes and in environmental samples

Source	3HA Units[a]								
	3HB	3HV	3HHx	3HHp	3HO	3HN	3HD	3HUD	3HDOD
R. rubrum	×	×	×	×					
Rb. sphaeroidis	×	×							
P. oleovorans	×	×	×	×	×	×	×	×	×
P. cepacia	×	×							
R. eutropha	×	×							
B. megaterium	×	×							
Aphanotheco	×	×							
Microvoleus	×	×							
Marine sediments	×	×	×	×	×				
Sewage	×		×		×				
Sewage sludge	×	×	×	×					

[a] See text for abbreviations.

Table 2. The accumulation of PHAs in various microorganisms

Genus	Group[a]	PHA wt%	Substrate from which PHA is produced
Acinetobacter	10	<1	Glucose
Aphanothece	CB	<1	NS
Aquaspirillum	6	ND	NS
Azospirillum	6	57	3-hydroxybutyrate
Axobacter	7	73	Glucose
Bacillus	15	25	Glucose
Beggiatoa	2	57	Acetate
Beijerinckia	7	38	Glucose
Caulobacter	4	36	Glucose/glutamate/yeast extract
Chloroflexus	1	<1	Yeast extract/glycylglycine
Chlorogloea	CB	10	Acetate, CO_2
Chromatium	1	20	Acetate
Chromobacterium	8	37	Glucose/peptone
Clostridium	15	13	Trypton/peptone/glucose
Derxia	7	26	Glucose
Ectothiorhodospira	1	ND	NS
Escherichia[b]	8	ND	Prypton/glucose/yeast extract
Gampnosphaeria	CB	ND	ND
Haemophilus[b]	8	ND	Brain-heart-infusion
Halobacterium	13	38	Glucose
Hyphomicrobium	4	ND	Methanol, glucose
Lamprocystis	1	ND	NS
Lampropedia	10	ND	NS
Leptothrix	3	67	Pyruvate
Methylobacterium	7	47	Methanol
Methylocystis	ND	70	Methane
Methylosinus	7	25	Methane
Micrococcus	14	28	Pentone/trypton

Table 2 (continued)

Genus	Group[a]	PHA wt%	Substrate from which PHA is produced
Micrococieus	CB	<1	NS
Microcystis	CB	ND	ND
Moraxella	10	ND	NS
Mycoplana	17	ND	Methanol
Nitrobacta	12	ND	NS
Nitrococcus	12	ND	NS
Nocardia	17	14	Butane
Oceanospirillum	6	ND	NS
Paracoccus	10	ND	NS
Photobacterium	8	ND	NS
Pseudomonas	7	67	Methanol
Ralstroni3	7	96	Glucose
Rhizobium	7	57	Methanol l
Rhodobacter	1	60	Acetate
Rhodospirillum	1	47	Acetate
Sphaerotilus	3	45	Glucose/peptone
Spirillum	6	40	Lactate
Spirulina	CB	6	CO_2
Streptomyces	17	4	Glucose
Syntrophomonas	9	30	Crotonate
Thiobacillus	12	ND	Glucose
Thiocapse	1	ND	NS
Thiocystis	1	ND	NS
Thiodictyon	1	ND	NS
Thiopedia	1	ND	NS
Thiosphaera	1	ND	Acetone, CO_2
Vibrio	8	ND	NS
Xanthobacter	7	ND	NS
Zoogloea	7	ND	Yeast extract/casamino acids

[a] After Bergey's manual.
[b] PHB found in cell membranes.
ND(S): not detected (selected).

ed even though the poly(3HB) synthase activity remained high. This result suggested that physical constraints operated and the cell was unable to accommodate more polymer, despite the availability of substrate and active synthase.

The average molecular weight of the poly(3HB) in the granules is generally between 10^5 and 10^6 (number average MW) depending on the microorganism with the MWs differing for different microorganisms [10, 11]. The X-ray diffraction studies showed that PHA granules were amorphous in vivo [12–14].

PHAs are synthesized as intracellular energy and carbon storage materials. Therefore, bacterial cells with a high content of poly(3HB) are better able to survive than cells with a low poly(3HB) content [15]. Poly(3HB) also serves as an endogenous carbon and energy source for sporulation in *Bacillus* species and cyst formation in *Azobacter* species [16–18] and the majority of bacteria investigated accumulate PHAs in response to a nutrient limitation. The reported nutrient deficiencies in the growth medium, which induced polymer production for various microorganisms, are listed in Table 3.

PHAs can be classified into several types by the structure of the repeating units in the polymers. In this chapter, PHAs will be classified as illustrated in Fig. 2, and each type of PHA will be discussed in the corresponding section.

In Fig. 2, PHAs are largely classified as short-chain-length PHAs [poly (HASCL)] and medium-chain-length PHAs [poly(HALMCL)] by the sizes of the repeating units in the polymers. Poly(HASCL)s contain only 3HB and 3HV units while poly(HAMCL)s contain only 3-hydroxyalkanoate units longer than 3HV. Poly(HASCL)s and poly(HAMCL)s have very different properties as will be discussed in the corresponding sections below, but some bacteria (very few) act as both poly(HASCL)-producing and poly(HAMCL)-producing microorganisms. PHAs containing both types of repeating units are classified as hybrids of poly(HASCL) and poly(HAMCL) in Fig. 2 [19–22]. Some of these microorganisms were shown to synthesize both poly(HASCL)s and poly(HAMCL)s as separate polymers [19, 23, 24]. Furthermore, both 4-hydroxybutyrate (4HB) [25] and 5-hydroxyvalerate (5HV) [26] units can be incorporated into poly(HASCL)s by feeding the microorganisms appropriate organic substrates.

Table 3. List of limiting compounds leading to PHA formation

Limiting compound	Organism
Ammonium	*Alcaligenes latus* *Pseudomonas oleovorans* *Pseudomonas cepacia* *Ralstonia eutrophus* *Rhodobacter sphaeroides* *Speudomonas sp. K.* *Methylocystus oarvus* *Thiosphaera pantotropha* *Rhizobium ORS 571*
Carbon	*Spirillum sp.* *Hyphomicrobium sp.* *Azospirillum brasiliense*
Iron	*Pseudomonas sp. K.*
Magnesium	*Pseudomonas sp. K.*
Oxygen	*Azotobacter vinelandii* *Azobacter beijerinckii* *Rhizobium ORS571*
Phosphate	*Rhodospirillum rubrum* *Rhodobacter sphaeroidis* *Caulobacter crescentus* *Pseudomonas oleovorans*
Potassium sulfate	*Bacillus thuringiensis* *Pseudomonas sp. K.* *Pseudomonas oleovorans* *Rhodospirillum rubrum* *Rhodobacter sphaeroids*

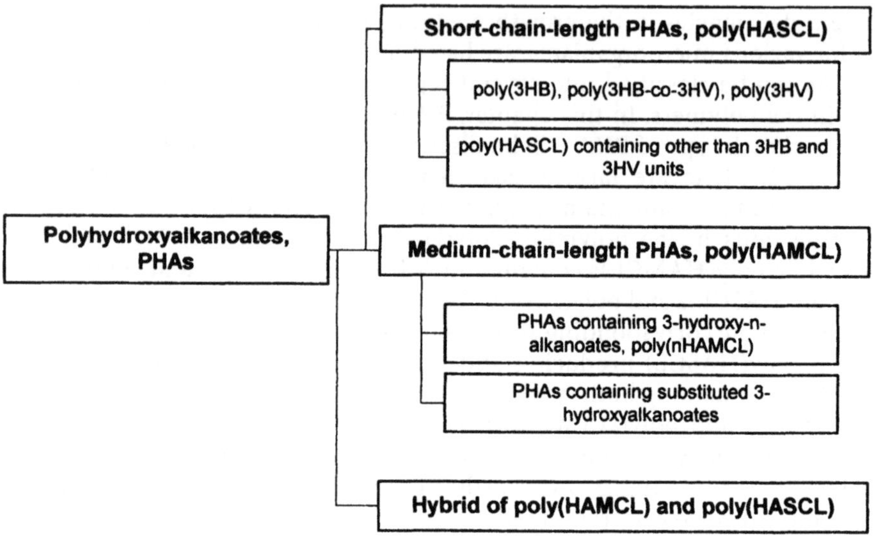

Fig. 2. Classification of polyhydroxyalkanoates

Mcl-PHAs are also classified according to the type of substituent in the alkyl in the side chains so that poly(HAMCL)s containing unsubstituted *n*-alkyl groups are identified as medium-chain-length poly(3-hydroxy-*n*-alkanoate) [poly (nHAMCL)]. Both poly(nHAMCL) and PHAs containing unusual groups are described in Sect. 3.

2 Short-Chain-Length Poly(3-hydroxyalkanoates) [poly(HASCL)s]

2.1 Poly(3-hydroxybutyrate), Poly(3-hydroxybutyrate-co-3-hydroxyvalerate), and Poly(3-hydroxyvalerate)

Most of the naturally occurring reserve polymer granules contain only poly(3HB), which is a thermoplastic with good mechanical properties similar to those of polypropylene. However, as a plastic, poly(3HB) is brittle due to its high degree of crystallinity and its highly spherulitic morphology, and its melting temperature (180 °C) is close to its thermal degradation temperature (200 °C). These poor properties for processing and applications were found to be offset by the incorporation of 3HV units into the polyester to lower the melting temperature and modify its morphology. As a result, the resulting copolymers, poly(3-hydroxybutyrate-*co*-3-hydroxyvalerates) [poly(3HB-*co*-3HV)] have significantly lower processing temperatures and much higher flexibility than poly(3HB), and these polyesters have been marketed under the trade name of "Biopol".

Of the many types of bacteria that produce poly(3HB) [27], *R. eutropha* is the most intensively investigated. *R. eutropha* produces poly(3HB) when it is grown with various organic substrates such as ethanol and glucose, and poly(3HB-*co*-3HV)s are produced by *R. eutropha* grown with glucose in the presence of either propionic acid or valeric acid as a co-substrate; the properties of poly(3HB-*co*-3HV)s containing various amounts of 3HV units have been described [28]. A copolymer containing as much as 90 mol% 3HV units can be obtained from *R. eutropha* grown with only valeric acid [29] but a PHA containing 100 mol% of 3HV unit has been synthesized by *Chromobacterium violaceum* [30]. Poly(3HB-*co*-3HV) containing 95 mol% of 3HV was synthesized by *Pseudomonas* sp. HJ-2 grown solely with valeric acid [24], and a microorganism that synthesizes poly(3HB-*co*-3HV) copolymers when grown with only glucose has been separated [31].

The pathway and regulation of poly(3HB) synthesis in *R. eutropha*, *Zoogloea ramigera*, and *Azotobacter beijerinckii* have been studied intensively [32–34]. In most microorganisms poly(3HB) is synthesized from acetyl coenzyme A (acetyl-CoA) by a sequence of three enzymatic reactions. The 3HV units are produced by an enzyme catalyzed coupling of acetyl-CoA and propionyl-CoA, which can be produced from either propionic acid or valeric acid through β-oxidation and deacetylation. However, propionic acid lowers the PHA yield when this carboxylic acid is used in high concentration because of its toxicity. Consequently, only poly(3HB-*co*-3HV) containing 3HV units less than 20 mol% has been produced on the large scale. The PHAs synthesized by *R. eutropha* grown with glucose in the presence of high concentrations of propionic acid contained mixtures of poly(3HB) and poly(3HB-*co*-3HV) [24].

Rhodococcus ruber [35], *Nocardia corallina* [36] and *R. eutropha* SH69 accumulate PHAs containing 3HV in the absence of typical 3HV precursors such as propionate or valerate. A mutant strain of *N. corallina*, in which the gene encoding the large subunit of methylmalonyl-CoA mutase was disrupted, produced poly(3HB-*co*-3HV) with only very low 3HV contents from glucose, indicating that the propionyl-CoA necessary for the production of 3HV was most likely produced by the methylmalonyl-CoA pathway [37]. However, the mutant produced poly(3HB) and poly(3HB-*co*-3HV) when it was grown with glucose and valeric acid, respectively. These results indicated that this microorganism synthesized repeating units by both the fatty acid degradation pathway and the traditional poly(3HB) synthetic pathway.

Poly(3HB-*co*-3HV)s have been investigated intensively, but most of the available data are for poly(3HB-*co*-3HV)s containing less than 25 mol% of 3HV units. There are only a limited number of reports on the properties of poly(3HB-*co*-3HV) containing more than 50 mol% of 3HV units, and reports on the poly(3HB-*co*-3HV) containing 3HV units more than 70 mol% are very rare [38]. Nevertheless, it is known that poly(3HB-*co*-3HV)s containing such high fractions of 3HV units form plastics that are as brittle as those from poly(3HB) homopolymer. However, these reports lack the information on the microstructures of the polymers, and recent results showed that poly(3HB-*co*-3HV) isolated from *Pseudomonas* sp. HJ-2 that contained 3HV (70–95 mol% of 3HV units) can form plastics that are flexible with maximum elongation ratios higher than 300% [23]. Poly(3HB-*co*-3HV) containing more than 20 mol% of 3HV units can be fabricated into films and fibers with different elasticities by controlling the conditions, and the maximum elongation ratio attained was between 300% and 1200% with ultimate strengths of 8–12 MPa.

2.2
Short-Chain-Length-PHAs Containing Repeating Units other than 3HB and 3HV

Terpolymers of 3HB, 3HV, and 5-hydroxyvalerate (5HV) were obtained from *R. eutropha* grown with mixtures of 5-chloropentanoic acid and pentanoic acid [25], and PHAs containing 3HB units and 4-hydroxybutyrate (4HB) units have been produced by *R. eutropha* grown with mixtures of glucose and γ-butyrolactone [26]. The mole fraction of 4HB unit in these PHAs varied as a function of the concentration of carbon sources and pH of the culture. PHAs containing 5HV or 4HB were less crystalline and were biodegraded more readily than poly(3HB) or poly(3HB-*co*-3HV).

A new type of PHA containing 3-hydroxypivalic acid has been synthesized by *Rhodococcus ruber* NCIMB 40126, *R. opacus* MR 22, and *N. corallina* N°724 grown with 3-hydroxypivalic acid [39]. The PHAs produced by these microorganisms grown with various mixtures of organic substrates containing 3-hydroxypivalic acid were either poly(3-hydroxypivalate), or copolymers of 3HB and 3-hydroxypivalate (3HPv), or terpolymers of 3HB, 3HV, and 3HPv.

Polymers containing 3-hydroxy-4-pentenoate, 3HB, and 3HV were synthesized by *Rhodospirillum rubrum* and a strain of *Burkholderia* [40, 41]. The repeating units found in this PHA included 3HB, 3HV, and 3-hydroxypentenoate (3HP), and the amount of 3HP units was as high as 19%, but no other poly(HASCL)s containing functional group have been reported to date.

3
Medium-Chain-Length PHAs [poly(HAMCL)s]

PHAs containing 3-hydroxy-*n*-alkanoates (3HA) longer than 3HV form elastomers with relatively low melting temperatures, low glass transition temperatures and low degrees of crystallinity. With only a few exceptions, poly-(HAMCL)s are obtained as copolymers containing from two to six different types of 3HA units. *P. oleovorans* and *P. putida* have been the most intensively investigated microorganisms of those synthesizing poly(HAMCL)s, and both can produce a wide variety of poly(HAMCL)s from various organic substrates.

The pathway of poly(HAMCL) production in *P. oleovorans* and *P. putida* is very different from that for poly(3HB). These microorganisms produce PHAs when they are grown with relatively long carbon sources including alkanes, alkenes, and carboxylic acids as carbon sources for PHA production. A generalized pathway for poly(HAMCL) production in *P. oleovorans* grown with long organic substrates is illustrated in Fig. 3.

In the pathway, the 3HA-CoA is produced from the carbon source by the fatty acid β-oxidation route so that the 3HA-CoA so formed can be utilized either for the production of a PHA directly or for the production of acetyl-CoA, which results in the formation of 3HA-CoA containing two carbons less than the original 3HA. 3HA-CoA thus formed can similarly be utilized for the production of

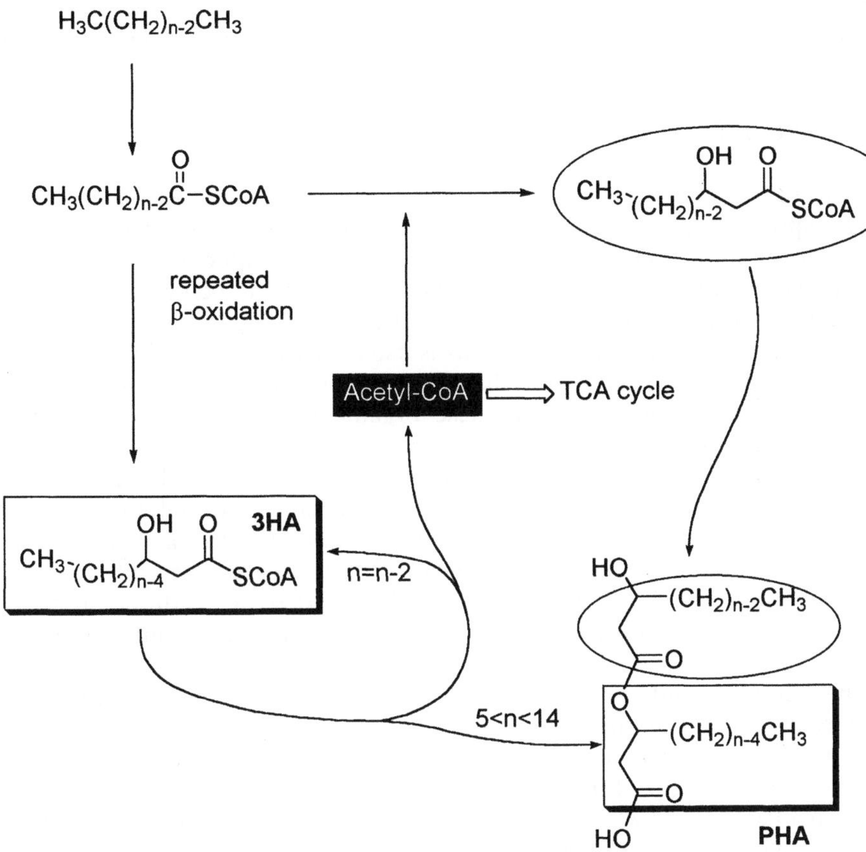

Fig. 3. General pathway for PHA synthesis in *Pseudomonas oleovorans*

either a PHA or acetate. If the microorganism can polymerize more than one kind of 3HA, the resulting PHA is a copolymer.

There are also microorganisms that can produce poly(HAMCL)s when grown with substrates that are much different from those discussed above, such as glucose [42–44]. The 3HA monomers produced by these microorganisms were most likely obtained from intermediates of the de novo fatty acid synthesis route [45]. PHAs synthesized from unrelated organic substrates are described in Sect. 3.1.2.

In this section, poly(HAMCL)s are largely subdivided into two types based on the structures of side groups; these are PHAs containing only 3-hydroxy-*n*-alkanoates and PHAs containing substituted 3-hydroxy-*n*-alkanoates.

3.1 Medium-Chain-Length-PHAs Containing only 3-Hydroxy-*n*-alkanoates [poly(nHAMCL)s]

3.1.1 *Poly(nHAMCL)s From Higher Alkanes and Alkanoic Acids*

Substrates that contain alkane moieties higher than pentane and *n*-pentanoic acid support mcl-PH*n*A production by *P. oleovorans* and *P. putida* while organic substrates lower than hexane and *n*-hexanoic acid support cell growth but they do not support PHA production. The growth with *n*-alkanes is carried out in heterogeneous systems consisting of aqueous and organic phases [46–49] while the growth with *n*-alkanoic acids is usually carried out in homogeneous systems. *n*-Alkanoic acids longer than hexadecanoic acid do not support PHA production, presumably because these organic substrates are insoluble in aqueous medium.

The fact that short alkanes or alkanoic acids do not support PHA production most likely indicates that *P. oleovorans* and *P. putida* cannot polymerize 3HAs shorter than 3HHx [50–52]. The repeating unit compositions of PHAs synthesized by *P. oleovorans* and *P. putida* grown with various alkanoic acids are listed in Table 4 [51, 52]. Table 4 shows that the PHAs synthesized by different microorganisms grown with the same organic substrate have very similar compositions.

The repeating unit compositions in Table 4 show that the 3HO unit is the main repeating unit in the PHAs prepared from an alkanoic acid which has an even number of carbon atoms and is higher than heptanoic acid, while the 3-hydroxynonanoate (3HN) unit is the main repeating unit in the PHAs prepared from an alkanoic acid which has an odd number of carbons and is higher than octanoic acid. Results in various reports on the production of poly(nHAMCL) indicate that *P. oleovorans* and *P. putida* prefer 3HO and 3HN monomers for polymer production and 3HA monomers longer than 3-hydroxydodecanoate (3HDD), and 3HA monomers shorter than 3HHx, are poorly polymerized by these microorganisms. The preference for 3HO and 3HN units of *P. oleovorans* and *P. putida* is very strong considering that the 3HO unit is formed from hexadecanoic acid and tetradecanoic acid only after several deacetylation reactions.

As shown in Table 4, 3HA units containing two carbons more than the carbon source can also be found in the PHAs. For example, PHAs synthesized by *P. putida* and *P. oleovorans* grown with *n*-hexanoic acid contain significant amounts of 3HO units. This chain extension presumably takes place by the reaction between acetyl-CoA and the alkanoic acid through the fatty acid synthesis process. The high preference for 3HO and 3HN units of these microorganisms can also be seen from the fact that the amounts of 3HO and 3HN units are significantly high in the PHAs synthesized from hexanoic acid and heptanoic acid, from which 3HO and 3HN monomers can be formed only after one chain extension reaction. A study carried out using alkanoic acids labeled with ^{13}C showed that PHA precursors could be generated by elongation of fatty acids with an acetyl-CoA molecule [53]. It can also be seen in Table 4 that small

Table 4. Repeating unit compositions of PHAs prepared from *n*-alkanoic acids by *P. putida* and *P. oleovorans*

Carbon Source	3HB	3HV	3HHx	3HHp	3HO	3HN	3HD	3HUD	3HDOD	3HTRD	3HTED
Hexanoic acid[a]			86.1		10.0		3.9				
Hexanoic acid[b]	3	< 1	72		22						
Heptanoic acid[a]		3.8		92		4.2					
Heptanoic acid[b]		7	< 1	86	< 1	7					
Octanoic acid[a]	1.6		13.2		85.2						
Octanoic acid[b]			10.7		83.3		6				
Nonanoic acid[a]		6.2		35.6		58.2					
Nonanoic acid[b]		1.7		25.3		70.1		2.9			
Decanoic acid[a]			14.6		54.4		30.9				
Decanoic acid[b]											
Undecanoic acid[b]		1.9		21.8	0.4	50.2		25.7			
Dodecanoic acid[b]			6.6	0.9	40.0	3.2	32.8		16.5		
Tridecanoic acid[b]			1.0	22.9	8.6	45.6	5.7	10.0	1.4	5.0	
Tetradecanoic acid[b]		NR	7.3		55.2		27.8		7.8		1.9
Hexadecanoic acid[b]		0.2	7.2	0.4	51.2	0.8	28.8	0.3	8.6		

[a] *P. putida.*
[b] *P. oleovorans.*

amounts of 3HA units having one more or one less carbon than the major 3HA units can be incorporated in the PHAs, but it is not understood how these repeating units are produced. As discussed above, with a few exceptions PHAs synthesized by microorganisms, especially by *P. oleovorans* and *P. putida*, are copolymers.

Polymers containing long side chains have properties with common features which originate from ordering of side chains. These polymers crystallize in sheet-like layers, as shown in Fig. 4, with the side chains extending from the polymer backbone. Because of this arrangement, these polymers are called "comb-like" polymers. The fiber repeat length, d_3 in Fig. 4, depends on the interaction between the side chains and the conformation of the main chain while the interplanar spacing, d_4 in Fig. 4, is constant regardless of the type of main chain and the degree of interaction between side chains. The melting transition and glass transition temperatures of comb-like polymers decrease as the length of the side chain increases, above a critical length, these transition temperatures can increase again.

Wide angle X-ray diffraction, WAXD, studies on the PHAs from octanoic acid (PHA-OCT), nonanoic acid (PHA-NON), and decanoic acid (PHA-DEC) showed that these polymers had the same features as synthetic comb-like polymers [54]. PHA-OCT is the most crystalline ($\Delta Hm = 8.3$ cal/g) and PHA-HEX (the PHA from hexanoic acid) is amorphous, while PHA-HEP (the PHA from heptanoic acid) has a very low degree of crystallinity ($\Delta Hm = 1.3$ cal/g). The degrees of crystallinity of PHA-OCT and PHA-NON were determined to be 20 and 30%,

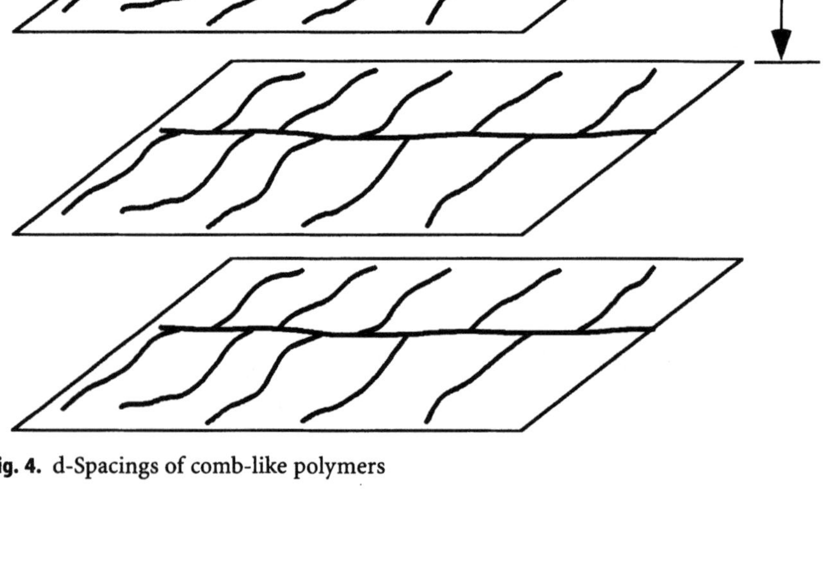

Fig. 4. d-Spacings of comb-like polymers

respectively, by a WAXD study [54]. The crystallization and mechanical properties of PHA-OCT copolymers have been investigated in detail, and the equilibrium melting temperature of PHA-OCT homopolymer was estimated to be 68 °C [55].

The melting transition temperature and the heat of fusion of poly(nHAMCL) prepared from various alkanoic acids are collected in Table 5. The glass transition temperatures of poly(nHAMCL)s are usually between –30 °C and –20 °C.

Two parameters, the length of the side chains and the inhomogeneity of repeating units, are responsible for the lower transition temperatures of the PHA containing side chains shorter than the critical length, but it is difficult to separate the effects of these two parameters on the melt transition temperatures. PHA-OCT and PHA-NON have higher melting temperatures than PHA-HEX and PHA-HEP. In addition, PHA-OCT has a higher melting temperature than PHA-NON, which might be explained by the higher homogeneity of repeating unit composition of this PHA than that of PHA-NON (Table 4). However, the monomer compositions of PHA-HEX and PHA-HEP in Table 4 show that the higher T_m of PHA-NON than those of PHA-HEX and PHA-HEP is not caused by the higher homogeneity of repeating unit composition. Therefore, the higher T_m of PHA-NON than that of either PHA-HEX or PHA-HEP is most likely due to the effect of the length of the side chains which indicates that the critical length for the side chains of PHAs for increasing T_m is that of pentyl (C5), but the PHAs produced from longer *n*-alkanoic acids have lower transition temperatures because they are copolymers with more types of repeating units in higher amounts than in PHA-NON or PHA-OCT as seen in Table 4.

3.1.2
Poly(nHAMCL)s From Non-Alkyl Organic Substrates

P. putida and some microorganisms [42–44] are capable of synthesizing poly(nHAMCL)s from non-alkyl based organic substrates, especially from glucose. *P. putida* grown with glucose produced PHAs containing both saturated and unsaturated 3HA units, and the seven types of 3HA units found in the PHA are sequential intermediates in the fatty acid synthetic pathway of bacteria. Therefore, the 3HA units in these PHAs are most likely produced by de novo

Table 5. Melting temperature and heats of fusion of poly(nHAMCL)

Carbon source	T_m	ΔH_m, cal/g
Hexanoic acid	–	–
Heptanoic acid	45	1.3
Octanoic acid	56	5.34
Nonanoic acid	50.0	4.37
Decanoic acid	54	4.9
Undecanoic acid	45.0	3.92
Dodecanoic acid	46.2	4.46
Tridecanoic acid	46.3	1.91
Tetradecanoic acid	43.7	3.19

fatty acid synthesis. The intermediates in the fatty acid synthesis are (*R*)-3-hydroxyalkanoate-acyl carrier proteins that can be converted to substrates for the PHA production.

P. putida grown with hexanoic acid contained approximately 75, 11, and 10 mol% of 3HHx, 3HO, and 3HD units and also small amounts of four unsaturated repeating units. The mechanism for the formation of 3HO unit was investigated by ^{13}C NMR study, which showed that the most of 3HO units found in this PHA were formed by the reaction of hexanoic acid with acetyl-CoA [53]. These results confirmed that *P. putida* produces 3HA units by fatty acid synthesis pathway, through a β-oxidation and chain elongation process.

Analysis of a *P. putida* mutant and complementation of this mutant with the cloned *pha* locus revealed that the PHA polymerase genes necessary for PHA synthesis from octanoic acid were also responsible for PHA formation from glucose. Furthermore, the poly(nHAMCL)s produced from glucose contained 3HD unit as the main repeating unit while poly(nHAMCL)s synthesized from long organic substrates contained 3HO unit as the main repeating unit, and the fractions of 3HD unit in PHAs synthesized from non-alkyl organic substrates are usually as high as 70%. These PHAs also contain small amounts of 3HHx, 3HO, and 3HDOD units, but the repeating unit compositions of PHAs synthesized from non-alkyl organic substrates can be altered by the growth conditions. For example, the PHA synthesized by *P. putida* grown with glucose contained 3HA units containing carbon-carbon double bond, and the fraction of 3HA units containing carbon-carbon double bond increased from 10% to 20%, while the fraction of monomers longer than 3HD increased from 18% to 28% when the growth temperature was decreased from 30 °C to 15 °C.

Pseudomonas sp. HJ-2 synthesized PHAs containing 3HD unit as the main repeating unit when it was grown with dicarboxylic acids containing an even number of carbon atoms; the composition of the PHA was also affected by the growth conditions. These results suggest that 3HA units found in these PHAs are most likely formed by the fatty acid synthesis pathway. Interestingly, this microorganism synthesized only poly(3HB) when it was grown with dicarboxylic acid containing odd number of carbons. The DSC analysis of a PHA containing 66% of 3HD, 23% of 3HO, and 10% of 3HDOD, which was produced by *Pseudomonas* sp. HJ-2 when grown with fructose, showed thermal transition temperatures which were almost identical to those of PHO and PHN.

3.2
Medium-Chain-Length-PHAs Containing Substituted 3-Hydroxyalkanoates

Synthesis of PHAs by microorganisms grown with non-natural carbon sources have been investigated intensively. The major interest has been to understand the relationship between the composition of the carbon source and the composition of the repeating units in the polymer so produced.

Various organic substrates containing substituents have been fed to microorganisms to obtain poly(HAMCL)s containing the corresponding substituents, but as mentioned above there has been only one example of poly(HASCL) containing functional group synthesized by microorganism (see Sect. 2.2). In this section, poly(HAMCL)s are grouped by the nature of substituent in the polymers.

3.2.1
PHAs Containing Unsaturated Units

Witholt et al. were the first to report that *P. oleovorans* produced PHAs containing carbon-carbon double bonds when this microorganism was grown with *n*-alkenes such as 7-octene, 8-nonene, and 10-decene [56] and the content of unsaturated 3HA units was controlled by controlling the composition of organic substrate mixtures. When *P. oleovorans* was grown solely with 7-octene, 8-nonene, or 9-decene, the resulting polymer contained 55%, 39%, and 46% of unsaturated 3HA units, respectively, with the remaining units containing alkyl substituents. They suggested that the saturated 3HA units in the PHAs were produced by the β-oxidation process from the unsaturated ends of the organic substrates.

A PHA containing virtually 100% of unsaturated 3HA units was synthesized by *P. oleovorans* grown with 10-undecenoic acid (10UND$^=$) [57]. PHA-10UND$^=$ contained 3-hydroxy-10-undecenoate (3HUD$^=$), 3-hydroxy-8-nonenoate (3HN$^=$), and 3-hydroxy-6-heptenoate (3HHp$^=$) units. The 3HN$^=$ unit was the main repeating unit, indicating that 3HA with 9 carbon atoms was still preferred when the 3HA contained unsaturated group.

The synthesis of PHAs by *P. oleovorans* grown with various mixtures of NA and 10UND$^=$ was investigated thoroughly [51, 57]. It was found that the fraction of unsaturated 3HA units in PHAs were commonly higher than the fraction of 10UND$^=$ in the organic substrate mixtures of NA and 10UND$^=$. *P. oleovorans* consumed 10UND$^=$ and NA in the organic substrate at almost the same rates even though this microorganism grew much more slowly, and with a much lower PHA yield, when it was grown solely with 10UND$^=$ than when it was grown solely with NA. The composition of PHAs synthesized from *P. oleovorans* grown with various mixtures of NA and 10UND$^=$ did not change according to the growth time and the resulting polymers proved to be random copolymers of saturated and unsaturated 3HA units.

The fact that 10UND$^=$ was consumed by *P. oleovorans* at the same rate that NA was consumed, and the total fraction of unsaturated 3HA units produced from 10UND$^=$ in the polymer was higher than that of saturated 3HA units produced from NA, suggested that incorporation of 3HA units formed from the organic substrate yielding lower PHA yield and poorer growth (10UND$^=$ in this case) could be enhanced when a carbon source yielding higher PHA yield and better growth (NA in this case) is present. These observations led to the production of PHAs containing a variety of functional repeating units as will be discussed in Sect. 3.2.3.

An X-ray diffraction study of PHA-10UND$^=$ showed that this PHA was amorphous; however, the DSC thermogram of the polymer showed a small but clear endotherm at approximately 38 °C. The glass transition temperature of PHA-10UND$^=$ is approximately –46 °C which is significantly lower than those of the poly(nHAMCL)s. Both the melting temperature and the glass transition temperature of PHAs from mixtures of NA and 10UND$^{=}$ decreased according to the content of unsaturated repeating units. PHA-10UND$^{=}$ is soft, sticky, and flows slowly at room temperature. The carbon-carbon double bonds in the PHAs synthesized from organic substrate mixtures containing 10UND$^{=}$ can be further modified in various chemical reactions [51, 58], as described elsewhere here. PHAs containing carbon-carbon double bonds were also synthesized from *P. oleovorans* grown with 3-hydroxy-7-octenoate and 3-hydroxy-6-octenoate [59], and PHAs synthesized from various plant oils [60, 61], tallow [62], and unrelated organic substrates such as glucose [43] contained unsaturated repeating units. The amounts of 3HA units containing carbon-carbon double bonds were relatively low (<10%) and these repeating units were usually longer than 3HUD. Recently, PHAs containing unsaturated repeating units were synthesized by *P. oleovorans* grown on mixtures of NA with either 6-allyloxyhexanoic acid (6AHA) or 11-allyloxyundecanoic acid (11AHUA). The PHA obtained from mixtures containing 6AHA included 4-allyloxy-3-hydroxybutyrate as the main repeating unit while the PHA obtained from mixtures containing 11AHUA included 5-allyloxy-3-hydroxyvalerate unit as the main repeating unit [63].

PHAs containing carbon-carbon triple bonds synthesized by *P. oleovorans* and *P. putida* grown with 10-undecynoic acid (10-UND$^{\equiv}$) have been reported [64]. The amount of carbon-carbon triple bond could be controlled between 0% and 100%, but the yield of the PHA containing 100% carbon-carbon triple bond was very low. The repeating units formed from 10-UND$^{\equiv}$ were 3-hydroxy-8-nonynoate (3HN$^{\equiv}$) and 3-hydroxy-6-heptynoate (3HHp$^{\equiv}$) units in the amounts of 26 mol% and 74 mol%, respectively. The glass transition temperatures of PHAs synthesized from mixtures of NA and UND$^{\equiv}$ increased from –30 °C to –20 °C as the content of carbon-carbon triple bond increased from 0% to 100%. These polymers were crosslinked when cations such as Co^{2+} and Pt^{2+} were added.

The general structures and repeating unit compositions of PHAs synthesized from various organic substrates containing carbon-carbon unsaturated repeating units are shown in Fig. 5 and Table 6.

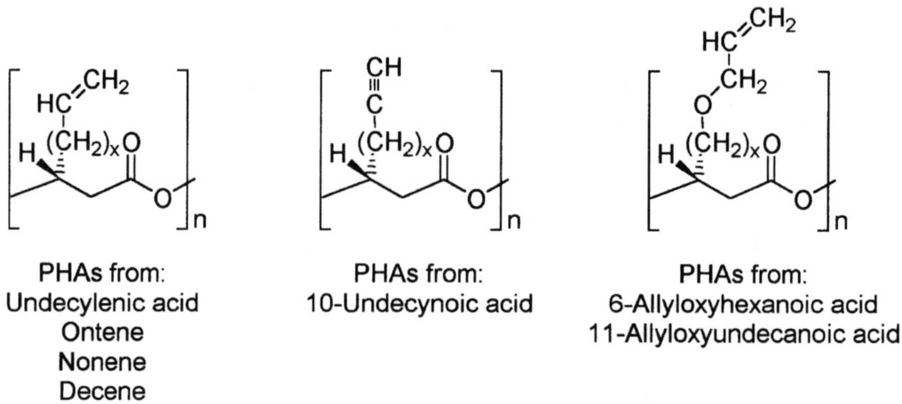

Fig. 5. Structures of unsaturated PHAs

Table 6. Repeating unit composition of PHAs biosynthesized by microorganisms grown with various unsaturated carbon sources

Carbon substrate	Repeating Unit	Relative Amount (%)
Undecylenic acid	3-Hydroxy-10-undecenoate	30
	3-Hydroxy-8-nonenoate	62
	3-Hydroxy-6-heptenoate	8
10-Undecynoic acid	3-Hydroxy-10-undecynoate	26
	3-Hydroxy-8-nonynoate	74
3-Hydroxy-7-octenoic acid[a]	3-Hydroxy-7-octenoate	62
	3-Hydroxy-5-hexenoate	2
3-Hydroxy-6-octenoic acid[b]	3-Hydroxy-6-octenoate	76
	3-Hydroxy-5-hexenoate	5
Octene[b]	3-Hydroxy-7-octenoate	49
	3-Hydroxy-5-hexenoate	6
Nonene[b]	3-Hydroxy-8-nonenoate	39
Decene[b]	3-Hydroxy-9-decenoate	12
	3-Hydroxy-7-octenoate	34
11-Allyloxyundecanoic acid[c]	5-Allyloxy-3-hydroxyvalerate	40
	7-Allyloxy-3-hydroxyheptanoate	12
6-Allyloxyhexanoic acid[d]	4-Allyloxy-3-hydroxybutyrate	50
	6-Allyloxy-3-hydroxyhexanoate	13

[a] Equimolar mixture with NA.
[b] Saturated units were also produced.
[c] Mixture containing 25 mol% of NA.
[d] Equimolar mixture with NA.

PHAs containing both carbon-carbon double and carbon-carbon triple bonds were synthesized by *P. oleovorans* and *P. putida* grown with various mixtures of 10UND$^{=}$ and 10UND$^{\equiv}$ [24]. PHAs obtained from mixtures containing these two organic substrates contained 3HA units containing carbon-carbon triple bonds in higher fractions than the fractions of 10-UND$^{\equiv}$ in the organic substrate mixtures. These results are similar to those obtained from cells grown with mixtures of NA and 10-UND$^{=}$. The compositions of PHAs synthesized from these two microorganisms were similar but the content of unsaturated repeating units were always higher in PHAs synthesized by *P. oleovorans* than in those synthesized by *P. putida*.

3.2.2
PHAs Containing Units with Aromatic Groups

PHAs containing an aromatic group were first produced with *P. oleovorans* by growth on 5-phenylvaleric acid (5PVA) [65]. The PHA synthesized from 5PVA was found to be a homopolymer of 3-hydroxy-5-phenylvalerate (3HPV) units, poly(3-hydroxy-5-phenylvalerate) [poly(3HPV)], with the structure shown in Fig. 6.

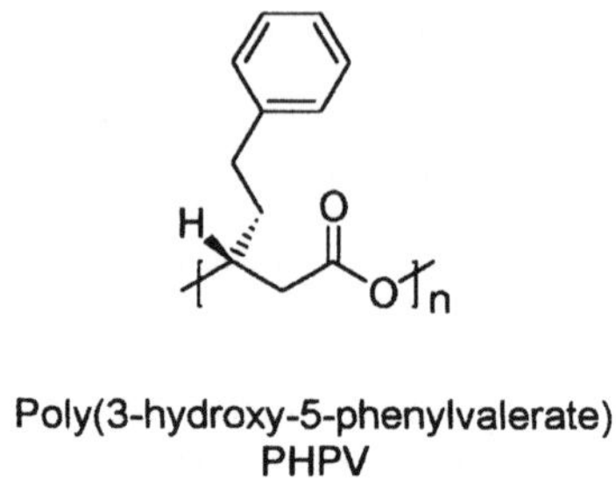

Fig. 6. Poly(3-hydroxy-5-phenylvalerate)

The glass transition temperature of poly(3HPV) is approximately 15°C, and no noticeable melting endotherm was detected by DSC analysis. X-ray diffraction analysis of poly(3HPV) also showed that poly(3HPV) was amorphous, although films of this polymer is opaque, and poly(3HPV) shows no birefringence when observed with a polarizing microscope. However, a recent X-ray study of poly(3HPV) showed that sharp diffraction peaks appear at low angles when poly(3HPV) was oriented with d-spacing values as large as 68 Å [66].

The production of a homopolymer from 5PVA showed that the 3HPV units were produced directly from 5PVA by β-oxidation. It was expected that 3-hydroxy-3-phenylpropionate (3HPP) could be produced from 3HPV by deacetylation, but the production of poly(3HPV) indicates that, if formed, 3HPP could not be polymerized. *P. oleovorans* pre-grown with glucose produced poly(3HPV) from 5PVA while *P. putida* pre-grown with glucose produced only negligible amount of poly(3HPV) [24]. However, when both of these microorganisms were pre-grown with nonanoic acid (NA) they produced poly(3HPV) from 5PVA. These results indicated that *P. putida* needed to be induced to be able to synthesize a PHA from 5PVA. PHAs containing *p*-methylphenyl group were synthesized by *P. oleovorans* grown with 5-(*p*-methylphenyl)valeric acid and 8-(*p*-methylphenyl)octanoic acid. Unlike poly(3HPV), these PHAs containing *p*-methylphenyl groups were crystalline with melting and glass transition temperatures of 94°C and 17°C, respectively [67, 68].

PHAs containing phenoxy groups were synthesized by *P. oleovorans* and *P. putida* [69, 70]. Repeating unit compositions of PHAs synthesized by *P. oleovorans* grown with alkanoic acids containing phenoxy groups are listed in Table 7. It is remarkable that 3-hydroxy-4-phenoxybutyrate (3HPB) unit was the main repeating unit in PHAs from 6-phenoxyhexanoic acid and 8-phenoxyoctanoic acid. This result, that 3HPB was the main repeating unit, was unexpected considering the structure of 3HPB. In 3HPB, the polar oxygen atom is located close to the acid group, and it is believed that the presence of the polar group near the carboxylic acid can inhibit polymerizability. The structure of 3HPB can be considered to be similar to that of 4-(1(or 2)-propenoxy)-3-hydroxybutyrate, which has eight atoms, and the formation of 3-hydroxy-5-phenoxypentanoate (3HPPN) unit as the main repeating unit in the PHA from 11-phenoxyundecanoic acid may occur because 3HPPN is similar in structure to 5-(1(or 2)-propenoxy)-3-hydroxypentanoate, which has nine atoms. The fact that

Table 7. Repeating unit composition of PHAs biosynthesized by *P. oleovorans* grown with various phenoxyalkanoic acids

Carbon substrate	Repeating Unit	Relative Amount (%)
6-Phenoxyhexanoic acid	3-Hydroxy-4-phenoxybutyrate	76
	3-Hydroxy-6-phenoxyhexanoate	24
8-Phenoxyoctanoic acid	3-Hydroxy-4-phenoxybutyrate	65
	3-Hydroxy-6-phenoxyhexanoate	30
	3-Hydorxy-8-phenoxyoctanoate	5
11-Phenoxyundecanoic acid	3-Hydroxy-5-phenoxypentanoate	91
	3-Hydroxy-7-phenoxyheptanoate	9
	3-Hydroxy-9-phenoxynonanoate	Trace
6-(*p*-Methylphenoxy) hexanoic acid	3-Hydroxy-4-(*p*-methylphenoxybutyrate)	27
	3-Hydroxy-6-(*p*-methylphenoxyhexanoate)	73
8-(*p*-Methylphenoxy) octanoic acid	3-Hydroxy-4-(*p*-methylphenoxybutyrate)	30
	3-Hydroxy-6-(*p*-methylphenoxyhexanoate)	70
8-(*m*-Methylphenoxy) octanoic acid	3-Hydroxy-4-(*m*-methylphenoxybutyrate)	30
	3-Hydroxy-6-(*m*-methylphenoxyhexanoate)	70

3HPPN unit, which is produced after three deacetylation steps, is the main repeating unit in the PHA synthesized from 11-phenoxyundecanoic acid again shows that *P. oleovorans* prefers repeating units containing eight or nine atoms.

A PHA production study on *P. oleovorans* and *P. putida* grown with alkanoic acids containing *p*-, *m*-, and *o*-methylphenoxy group showed that the alkanoic acids containing *p*-methylphenoxy group supported PHA production best. Almost no PHA was obtained from *o*-methylphenoxyalkanoic acids, while a small amount of PHA was obtained from (*m*-methylphenoxy)alkanoic acids. The PHA containing *p*-methylphenoxy group was crystalline with melting and glass transition temperatures of 14 °C and 97 °C, but the PHA containing *m*-methylphenoxy groups was amorphous and had a glass transition temperature of 6 °C.

Studies on the synthesis of PHAs by *P. putida* and *P. oleovorans* grown with carbon sources containing *p*-nitrophenoxy and *p*-cyanophenoxy group showed that *P. putida* synthesized PHAs containing a significant amount (up to 24%) of *p*-cyanophenoxy group while *P. oleovorans* synthesized PHA containing less than 3% *p*-cyanophenoxy group [71], and the contents of nitrophenoxy groups in the PHAs were lower than 6%. The PHA formed from *p*-cyanophenoxyhexanoic acid contained only 3-hydroxy-6-(*p*-cyanophenoxy)hexanoate units.

3.2.3
PHAs Containing Units with Halogen Groups

PHAs containing Br [72], F [73, 74], and Cl [75] have been synthesized by *P. oleovorans* grown with organic substrates containing the corresponding halogen. The PHA production by *P. oleovorans* grown with organic substrates containing halogen groups have a common unusual feature, which is that *P. oleovorans* does not produce PHAs when a halogenated organic substrate is fed as a sole carbon source. Instead, this microorganism can only produce PHAs containing significant amounts of halogenated 3HA units when it is grown with a co-substrate that can support cell growth and PHA production in good yields. For example, 6-bromohexanoic acid (6BrHA), 8-bromooctanoic acid (8BrOA), and 11-bromoundecanoic acid (11BrUA) support only cell growth by *P. oleovorans* without PHA production when they are used as sole carbon source, but this bacterium can produce PHAs containing significant amounts of brominated 3HA units when it is grown with mixtures of NA and one of the brominated carboxylic acid mentioned above. The growth rate and the dry cell yield increased in the order 6BrHA, 8BrOA, and 11BrUA. Similarly, *P. oleovorans* grown with only fluorinated or chlorinated alkanes did not produce PHA, but this microorganism produced PHAs containing significant amounts of units with halogen groups, when it was fed with a co-substrate such as octane or nonane. This type of polymer production is termed cometabolism. The detailed mechanism for the production of PHAs by cometabolism is not understood, but various of organic substrate mixtures are known to support such production of PHAs, and many of the PHAs containing unusual functional groups discussed in next section were synthesized by cometabolism.

The organic substrates and repeating units found in PHAs synthesized by *P. oleovorans* grown with mono-halogenated organic substrate are collected in Table 8. The synthesis of PHAs containing multi-fluorinated 3HA units was investigated by growing *P. oleovorans* with hexanoic, nonanoic, and undecanoic acids in which all the protons attached to C6 and higher Cs were replaced by F [74]. Tetradecafluorinated undecanoic acid did not support PHA production, but approximately 17% of fluorinated repeating units could be incorporated into a PHA when *P. putida* was grown with mixtures of NA and nonafluorinated NA. The heat of fusion of the PHAs so produced increased significantly (3.2 cal/g to 8 cal/g) while the melting temperature increased slightly (48 °C to

Table 8. Carbon substrates and amount of halogenated repeating units in PHAs biosynthesized by *P. oleovorans* grown with mixtures containing halogenated carbon substrate

Carbon substrate	Halogenated 3HA unit	Reported amount (mol %)
Nonane + 1-Fluorononane	3-Hydroxy-9-fluorononanoate	0–24
Octane + 8-Chlorooctane	3-Hydroxy-6-chlorohexanoate	0–10
	3-Hydroxy-8-chlorooctanoate	0–60
Nonanoic acid + 6-Bromohexanoic acid	3-Hydroxy-4-bromobutyrate 3-Hydroxy-6-bromohexanoate	0–25
Nonanoic acid + 8-Bromooctanoic acid	3-Hydroxy-6-bromohexanoate 3-Hydroxy-8-bromooctanoate	0–25
Nonanoic acid + 11-Bromoundecanoic acid	3-Hydroxy-7-bromoheptanoate 3-Hydroxy-9-bromononanoate	0–38

51 °C) when approximately 12.4% nonafluoro-3-hydroxynonanoate was present in the polymer. The repeating units found in these PHAs included trifluoro-3-hydroxybutyrate, heptafluoro-3-hydroxyoctanoate, and nonafluoro-3-hydroxynonanoate. The thermal properties and the PHA composition changed with growth time, which indicated that the multi-fluorinated 3HA units were not distributed in the PHA randomly but chains, or chain segments, existed with relatively high multi-fluorinate 3HA units.

PHAs containing bromine can be prepared by chemical modification of PHAs containing unsaturated repeating units. For example, bromination of PHA-10UND$^{=}$ or PHA-10UND$^{\equiv}$ proceeded to completion rapidly to yield polymers with increased glass transition temperatures [76].

3.2.4
Other Unusual PHAs

Investigations on PHA production by, and growth of, *P. oleovorans* fed with various organic substrates showed that organic substrates can be classified into three types. Organic substrates in Group A support both cell growth and PHA production, organic substrates in Group B support cell growth without PHA production, and organic substrates in Group C do not support either cell growth or PHA production. However, many of organic substrates in Group B are incorporated into PHAs when they are co-fed with a substrate in Group A through cometabolism, as briefly described in Sect. 3.2.3 for the production of PHAs containing brominated 3HA units. Examples of substrates belonging to Group B that are incorporated into PHAs through co-metabolism by *P. oleovorans* are 5-methyloctanoic acid, 6-methyloctanoic acid, 4-cyclohexylbutyric acid, 5-cyclohexylpentanoic acid, 6-bromohexanoic acid, 8-bromooctanoic acid, 11-bromoundecanoic acid, 11-cyanoundecanoic acid, 6-methoxyhexanoic acid, 8-methoxyoctanoic acid, 6-ethoxyhexanoic acid, 8-ethoxyoctanoic acid, and 12-hydroxydodecanoic acid [51, 63, 77, 78].

Some of the organic substrate mixtures that support PHA production by *P. oleovorans* through cometabolism do not support PHA production by *P. putida*. *P. oleovorans* can produce PHAs containing a fairly wide variety of functional groups through cometabolism, but *P. putida* is much less adaptable. For example, *P. putida* did not produce brominated PHAs when it was grown with mixtures of brominated alkanoic acids and NA [79], so PHA production by *P. oleovorans* and *P. putida* was compared by feeding these two microorganisms with the same organic substrates. In general, *P. putida* was much less able than *P. oleovorans* to produce PHAs containing functional groups. For example, *P. putida* was able to produce PHAs containing 3HA units from 10UND$^{=}$ or 10UND$^{\equiv}$ only when NA is present as a co-substrate, while *P. oleovorans* did not require NA for that purpose. The only example in which *P. putida* produced PHAs containing unusual functional group in higher yields than *P. oleovorans* was with some organic substrates containing aromatic substituents [79].

As noted above, no PHA was formed when *P. oleovorans* was grown solely with 5- and 6-methyloctanoate, but PHAs containing 5-methyl-3-hydroxyoctanoate and 6-methyl-3-hdyroxyoctanoate units were obtained by cometabolism when this microorganism was grown in the presence of OA [77, 80]. In that case, the amounts of methyl branched 3HA units were as high as 47 and 30 mol% when *P. oleovorans* was grown with 5- and 6-methyloctanoic acid, respectively, and OA. An interesting observation was that stereoselective polymerization might have taken place with 5-methyloctanoic acid. In contrast to these results, PHAs containing methyl branched side chains could be produced by *P. oleovorans* when it was grown solely with either 6-methylnonanoic acid, 7-methylnonaoic acid, 8-methylnonanoic acid, 7-methyldecanoic acid, 9-methyldecanoic acid or 2,6-dimethylhexanoic acid [80]. PHAs containing 65% of 3-hydroxy-7-methyl-6-octenoate and 35% of 3-hydroxy-5-methylhexanoate units were produced by *P. citronellolis* grown with D,L-citronellol [81].

11-Cyanoundecanoic acid supported PHA production by *P. oleovorans* through cometabolism when it was co-fed with NA, and PHAs containing cyano group are of interest in that the cyano group can induce polar interaction with other polar groups [51]. The content of cyano group was approximately 30% when an equimolar mixture of NA and 11-cyanoundecanoic acid was used and the repeating units produced from this substrate were most likely 7-cyano-3-hydroxyheptanoate and 9-cyano-3-hydroxynonanoate units.

PHAs containing methoxy, ethoxy and methoxyethoxy groups have been synthesized by *P. oleovorans* grown with organic substrates containing methyl, ethyl, and methoxyethyl ethers of ω-hydroxy-*n*-hexanoic, ω-hydroxy-*n*-octanoic, and ω-hydroxy-*n*-undecanoic acids [63]. Methoxy and ethoxy ethers of 11-hydroxy-*n*-undecanoic acid supported PHA production when they were used as sole organic substrates. The PHA synthesized from 11-methoxyundecanoic acid contained 88% of 7-methoxy-3-hydroxyheptanoate (7M3HHp) and 12% of 9-methoxy-3-hydroxynonanoate (7M3HN), while the PHA synthesized from 11-ethoxyundecanoic acid contained 56% of 5-ethoxy-3-hydroxypentanoate (5E3HP) and 44% of 7-ethoxy-3-hydroxyheptanoate (5E3HHp). These results show that *P. oleovorans* prefers to produce PHA containing repeating units having nine or eight atoms as is 7M3HHp and 5E3HP. Methyl and ethyl ethers of 6-hydroxy-*n*-hexanoic acid and 8-hydroxy-*n*-octanoic acids supported PHA production though cometabolism when NA was used as a co-substrate. The major repeating units from each of these organic substrates were ω-alkoxy 3HA units that had eight or nine atoms. PHAs containing alkoxy groups did not precipitate from methanol but formed milky suspension indicating that these polymers are more polar. Contact angle measurement using water droplets also showed that the hydrophilicity of the PHAs containing alkoxy groups were higher than that of PHA-NON.

PHA production by *P. oleovorans* fed with organic substrates containing cyclohexyl group was investigated using 4-cyclohexylbutyric acid and 5-cyclohexylvaleric acid [51, 78]. Both organic substrates support cell growth with production of a very small amount of PHA. When these organic substrates were co-fed with NA, *P. oleovorans* produced PHAs containing 4-cyclohexyl-3-hydroxybutyrate and 5-cyclohexyl-3-hydroxyvalerate units, respectively. The content of 3HA units containing cyclohexyl group was approximately 10% when an equimolar mixture of 4-cyclohexylbutyric acid and NA was used as the carbon source. The content of cyclohexyl group could be controlled by altering the composition of the organic substrates.

PHAs containing ester groups in the side chain were synthesized by *P. oleovorans* grown with methyl and ethyl esters of octanoic acid, nonanoic acid, and decanoic acid [51, 82]. The 3HA units present in PHAs obtained from methyl octanoate included methyl esters of 3-hydroxysuccinic acid, 3-hydroxyadipic acid, methyl 3-hydroxysuberic acid, 3-hydroxysebacic acid, 3HHx, and 3HO units. The PHAs obtained from methyl decanoate contained the same repeating units except for the methyl ester of 3-hydroxysebacic acid. 3HO units were the main repeating units in PHAs produced from methyl octanoate and methyl decanoate. Methyl and propyl esters of heptanoic acid also supported PHA production by *P. oleovorans*. The PHA produced from methyl heptanoate contained 3HV, 3HHp, 3HO, 3HN, and approximately 3 mol % of a 3HA unit containing a methyl ester. The PHA from ethyl heptanoate contained no 3HA units with ester groups. The PHA from propyl heptanoate did not contain the 3HV, 3HHp, 3HO, and 3HN units, but instead this polymer contained only repeating units having propyl esters.

P. oleovorans grown with an equimolar mixture of the monobenzyl ester of sebacic acid and NA produced a PHA containing 4% benzyl ester groups [51]. The same microorganism grown with 1:1 (v/v) mixture of nonane and benzyl-10-undecenoate produced a PHA containing approximately 5 mol% of benzyl ester and 17 mol% of unsaturated units. *P. oleovorans* grown with 1:1 (v/v) mixture of nonane and benzyl-10-undecenoate produced a PHA containing approximately 7 mol% of methyl ester and 60 mol% of unsaturated units.

Pseudomonas aeruginosa 44T1 grown with euphorbia or castor oil produced PHAs containing 7–15% of 3HA units containing epoxy and hydroxy groups [83]. PHAs containing epoxy groups were also prepared by epoxidation of PHAs containing carbon-carbon double bonds [58].

Recombinant strains of *Pseudomonas putida* GPp104 and *R. eutropha* which harbored plasmid pHP1014::E156 with the PHA-biosynthesis genes of *Thiocapsa pfennigii* produced PHAs containing 3HB, 3HHx, and 4-hydroxyhexanoate (4HHx) when they were grown solely with 4-hydroxyhexanoic acid [21]. PHAs containing 4-hydroxyheptanoate (4HHp), 4-hydroxyoctanoate, and 5-hydroxyhexanoate (5HHx) were also synthesized by the same recombinant strain of *Pseudomonas putada* GPp104 grown with 4-hydroxyheptanoic acid, 4-hydroxyoctanoic acid, and 5-hydroxyhexanoic acid, respectively [84].

4 PHAs Containing Repeating Units of both Scl-PHA and Mcl-PHA

R. rubrum [85], *A. caviae* 440 [86], *Pseudomonas* sp. 61–3 [19], *Pseudomonas* sp. A33, and other isolates of aerobic bacteria [87] can produce PHAs containing both 3HB and 3HV and some 3HA units longer than 3HV. *A. caviae* produced only a poly(3HB-*co*-3HV) copolymer, which contained approximately 95% of 3HV units when grown with alkanoic acids containing an odd number of carbon atoms. The same microorganism produced PHAs containing 3HB and 3HHx units when grown with alkanoic acids containing odd number of carbons. The melting temperature of the PHAs containing 3HB and 3HHx units decreased from 151 °C to 52 °C as the content of 3HHx increased from 5% to

25%. *Pseudomonas* sp. 61-3 grown with glucose produced mixtures of poly(3HB) and PHAs containing 3HB, 3HHx, 3HO, 3HD, 3HDOD, 3-hydroxypentadecanoic acid (3H5DD), and 3-hydroxyheptadecanoic acid (3H7TD). No melting temperature was detected from this PHA. *Pseudomonas* sp. A33 produced PHAs containing 3HB (13–15%), 3HHx (2.6–3.5%), 3HO (13.8–15.7%), and 3HD (65.7–70.5%) when grown with either 1,3-butanediol or 3-hydroxybutyric acid. It was suggested in this study that *A. caviae* produces 3HB and 3HHx units from enoayl-CoA derivatives produced from the long chain substrates by fatty acid oxidation [88, 89]. That is, crotonyl-CoA, hexenoyl-CoA and octenoyl-CoA were produced by deacetylation and β-oxidation from the carbon source. Crotonyl-CoA and hexenoyl-CoA were converted to 3-hydroxybutyryl-CoA and 3-hydroxyhexanoyl-CoA while octenoyl-CoA was not converted to 3-hydroxyoctanoyl-CoA. The production of PHAs with both repeating units of poly(HASCL) and poly(HAMCL) is not well understood, and some microorganisms can produce mixtures of poly(HASCL)s and poly(HAMCL)s instead of copolymers. In these cases there may be more than one type of polymerase in the cell.

5
Microstructures of PHAs Biosynthesized from Various Mixtures of Organic Substrates

As discussed in the examples given in the previous sections, the PHAs produced by bacteria can be either homopolymers, random copolymers, or mixtures either of homopolymers or copolymers. The types of PHAs present can often be identified by fractionation methods based on the different solubilities in various solvents. For example, poly(nHAMCL) and copolymers containing significant amount of 3nHA are soluble in alkanes such as *n*-heptane and *n*-hexane, while poly(HASCL)s and poly(HAMCL)s containing various functional groups such as olefin, aromatic, and carbon-carbon triple bonds are not. Poly(3HB-*co*-3HV)s containing significant amount of 3HV are soluble in acetone while poly(3HB-*co*-3HV) containing small amount of 3HV are not soluble in acetone, and the temperature can be adjusted for selective dissolution of component polymers [90].

The distribution of repeating units in poly(HASCL)s has been determined by the use of ^{13}C NMR spectroscopy. However, a fractionation study on a series of poly(3HB-*co*-3HV) samples showed that the apparent distribution of repeating units in poly(3HB-*co*-3HV) determined by ^{13}C NMR spectra might be misleading [91]. Furthermore, ^{13}C NMR spectroscopy is not useful to elucidate the distribution of repeating units in poly(HAMCL)s because the long side chains in the units do not affect the chemical shift values of the carbonyl carbon in the main chain. A statistical method was developed that correlates the molecular weights of partially hydrolyzed PHAs to the distribution of the repeating units [92–96].

There have been only a few examples of homopolymers synthesized by microorganisms. They are poly(3HB), poly(3HV), poly(3HPV), poly(3-hydroxyheptanoate) (PHHp), and poly(4HB) [97, 98]. Many microorganisms are known that produce poly(3HB), but only one microorganism was reported to produce poly(3HV). The production of poly(3HPV) was described above. The production of PHHp was first reported to be synthesized by *P. oleovorans* [99] grown with heptanoic acid, but copolymers were obtained in other studies [24, 100, 101] suggesting that the compositions of PHAs synthesized by *P. oleovorans* might be affected by the growth conditions. *Pseudomonas* sp. HJ-2 produced PHHp when grown with heptanoic acid under controlled conditions.

All poly(3HB-*co*-3HV)s synthesized by *R. europha* that were characterized for sequence distribution were found to be random copolymers as indicated by ^{13}C NMR spectrometry. However, as noted above, the fractionation of some poly(3HB-*co*-3HV) copolymers revealed that the sequence distributions determined by ^{13}C NMR might not be reliable. In that study, poly(3HB) that had a diad sequence indicative of a random poly(3HB-*co*-3HV), as determined by ^{13}C NMR, was fractionated into various copolymers that had significantly different compositions.

The nature of poly(HAMCL)s synthesized from mixtures of organic substrates seems to be determined by the nature of organic substrates. To date, all of the PHAs synthesized by *P. oleovorans* and *P. putida* grown with various organic substrate mixtures have been found to be random copolymers. The microstructure of PHAs, which were synthesized by mixtures of NA and either a brominated carboxylic acid or undecylenic acid, were investigated by FAB-mass spectrometry of the partially hydrolyzed samples. These PHAs proved to be random copolymers, and thermal analysis of PHAs synthesized from mixtures of NA and either undecylenic acid or 10-undecynoic acid also showed that these polymers were random copolymers. A selective crosslinking of the PHA through olefin groups by diborane followed by extraction also indicated that the PHAs from mixtures of 10UND$^{=}$ and NA were most likely random copolymers.

The first example of the production of the physical mixtures of PHAs from mixtures of organic substrate was in the PHA synthesized by mixtures of 5PVA and NA. Mixtures of these organic substrates yielded PHAs containing PHAs that had compositions sufficiently different to be fractionated by the different solubility in *n*-hexane, and the fractionation studies of the PHAs showed that two types of PHAs were present. One was the PHA containing more than 70% of unsubstituted 3HA units formed from NA and the other was a PHA containing more than 70% of 3HPV units. A recent study showed that each fraction was a random copolymer of 3HPV and unsubstituted 3HA units. Surprisingly, no poly(3HPV) was produced when mixtures of NA and 3HPV were fed to *P. putida*.

The PHA synthesized by mixtures of 5-(*p*-methylphenyl)valeric acid and 8-(*p*-methylphenyl)octanoic acid were also found to be mixtures of PHAs, and all of the PHAs produced to date from mixtures of aliphatic carboxylic acid and a organic substrate containing aromatic groups were mixtures of PHAs containing significantly different compositions. PHAs of different compositions can be fractionated and each fraction seems to be random copolymers containing 3HA units formed from both organic substrates. It was concluded in these studies that the microorganism consumes the different substrates at different rates and this selectivity is responsible for the production of PHAs of different compositions. As a result, the composition of the PHA produced at any given time is probably determined by the amount of substrates present in the microorganism.

6
Poly(*β*-malate)

Poly-*β*-malate is a PHA produced by a variety of microorganisms the structure of which is shown in shown in Fig. 7 [102–105].

Unlike the PHAs described above, this polyester is not produced as an intracellular storate granule, but it is found both within the cell and extracellularly, depending on the microorganism. A black yeast, *Aureobasidium sp.*, produces poly-*β*-malate as an extracellular, water-soluble biopolyester from glucose [103], but the fungus *Penicillium cyclopium*, and the slime mold *Physarum polycephalam*, produce it in small amounts as an intracellular polymer, which serves as an inhibitor for various enzymes [105]. In all cases, however, the polyester has only units of L-malic acid, which are esterified only at the *β*-carboxyl position [105, 106]. That is, the units of the polyester are formed in both a stereospecific (isotactic) and regiospecific manner, but the polymer has also been found with some branching on the pendent carboxyl group [107]. At neutral pH, which is the common condition for most organisms within which poly-*β*-malate is formed, the pendent carboxyl groups are completely ionized and the polymer is highly water-soluble.

Poly-*β*-malate is readily degraded completely to L-malic acid under both acid and base conditions [108], and it can also be hydrolyzed by enzymes within the cell [105, 106]. Recently, several bacteria were isolated which were able to utilize poly-*β*-malate as sole carbon source for growth [109]. Because the polymer is biodegradable and bioadsorbable, it is of considerable interest for pharmaceutical applications, especially in controlled-release drug delivery systems [97, 98]. Chemical routes to poly-*β*-malate are expected to provide materials with various properties [110].

The unbranched polymer produced by *P. polycephalum* and related *Physarum* strains has a weight average molecular weight between 40,000 and 60,000 Daltons and a polydispersity of 1.5–3.0 depending on the culture conditions and the age of the samples [111]. The acid form of poly-*β*-malate does not show either a T_g or a T_m in the solid state, by DSC analysis, below its thermal decomposition temperature of 185 °C.

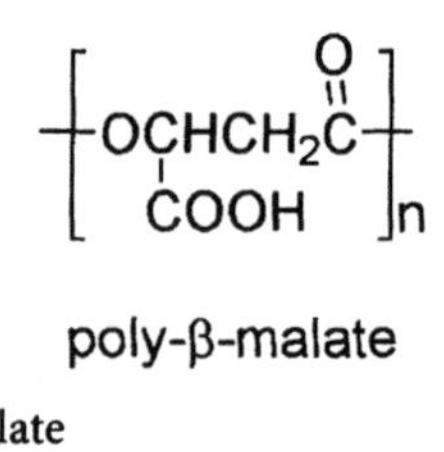

Fig. 7. The structure of poly-*β*-malate

References

1. Lemoigne M (1925) Ann Inst Pasteur 39:144
2. Wallen LL, Rohwedder WK (1974) Environ Sci Technol 8:576
3. Findlay RH, White DC (1983) Appl Environ Microbiol 45:71
4. Odham G, Runlid A, Westerdahl G, Marden P (1986) Appl Environ Microbiol 52:905
5. Steinbüchel A, Valentin HH (1995) FEMS Microbiol Lett 128:219
6. Shively JM (1974) Ann Rev Microbiol 28:167
7. Dawes EA, Senior P (1973) Adv Microb Physiol 10:135
8. Clayton RK, Sistrom WR (1978) In: Merrick JM (ed) The photosynthetic bacteria metabolism of reserve materials. Plenum Press, New York
9. Ballard DG, Holmes PA, Senior PJ (1987) Recent advances in mechanistic and synthetic aspects of polymerization. Reidel Publishing
10. Ellar D, Lundgren DG, Okamura K, Marchessault RH (1968) J Mol Biol 35:489
11. Anderson AJ, Dawes EA (1990) Microbiol Rev 54:450
12. Bernard GN, Sanders JKM (1988) FEBS Lett 231:16
13. Bernard GN, Sanders JKM (1989) J Biol Chem 264:3286
14. Kawaguchi Y, Doi Y (1990) FEMS Microbiol Lett 79:151
15. Tal S, Okon Y (1985) Can J Microbiol 31:608
16. Slepecky RA, Law JH (1961) J Bacteriol 82:37
17. Stevenson LH, Socolofsky MD (1966) J Bacteriol 91:304
18. Sillman CE, Casida LE (1986) Soil Boil Biochem 18:23
19. Kato M, Bao HJ, Kang CK, Fukui T, Doi Y(1996) Appl Microbiol Biotechnol 45:363
20. Fukui T, Kichise T, Yoshida Y, Doi Y (1997) Biotechnol Lett 19:1093
21. Lee EY, Jendrossek DJ, Schirmer A, Choi CY, Steinbüchel A (1995) Appl Microbiol Biotechnol 42:901
22. Steinbüchel A, Wiese S (1992) Appl Microbiol Biotechnol 37:691
23. Ramsay BA, Saracovan I, Ramsay JA, Marchessault RH (1992) Appl Environ Microbiol 58:744
24. Rhee YH, Chung JW, Kim YB (1999) Chungnam National University (unpublished data)
25. Doi Y, Tamaki A, Kunioka M, Soga KJ (1987) J Chem Soc Chem Commun 8:1635
26. Doi Y, Tamaki A, Kunioka A, Soga K (1988) Appl Microbiol Biotechnol 28:330
27. Brandl H, Gross RA, Lenz RW, Fuller RC (1990) Adv Biochem Eng/Biotechnol 41:77
28. Doi Y, Kuniock M, Nakamure Y, Soga K (1986) Macromolecules 2860
29. Doi Y, Tamaki A, Kunioka A, Soga K (1987) Makromol Chem Rapid Comm 8:631
30. Steinbüchel A, Debzi E-M, Marchesault RH, Timm A (1993) Appl Microbiol Biotechnol 39:443
31. Yoon JS, Park SK, Kim YB, Maeng HY, Rhee YH (1996) J Microbiology 34:279
32. Doi Y, Kawaguchi Y, Nakamura Y, Kunioka M (1989) Appl Environ Microbiol 55:2932
33. Ploux O, Masamune S, Salsh CT (1988) Eur J Biochem 174:177
34. Senior PJ, Dawes EA (1973) Biochem J 134:225
35. Anderson AJ, Haywood GW, Williams DR, Dawes EA (1990) In: Dawes EA (ed) Novel biodegradable microbial polymers. Kluwer Academic Publishers, Dordrecht, p 119
36. Valentin HE, Dennis D (1996) Appl Environ Microbiol 62:372
37. Williams DR, Anderson AJ, Dawes EA, Ewing DF (1994) Appl Microbiol Biotechnol 40:717
38. Doi Y (1990) Microbial Polyesters. VCH, New York, p 117
39. Füchtenbusch B, Fabritius D, Wältermann M, Steinbüchel A (1998) 159:85
40. Rodriguez MFA, da Silva LF, Gomez JGC, Valentin HE, Steinbüchel A (1995) Appl Microbiol Biotechnol 43:880
41. Ballistreri A, Montaudo G, Impallomeni G, Lenz RW, Ulmer HW, Fuller RC (1995) Macromolceules 28:3664
42. Haywood GW, Anderson AJ, Ewing DF, Dawes EA (1990) Appl Environ Microbiol 56:3354

43. Huijberts GNM, Eggink G, de Waard P, Huisman GW, Witholt B (1992) Appl Environ Microbiol 58:536
44. Timm A, Steinbüchel A (1990) Appl Environ Microbiol 56:3360
45. Schweizer E (1989) Biosynthesis of fatty acids and related compounds. In: Ratledge C, Wilkinson SG (eds), Microbial lipids, vol. 2. Academic Press, London, p 3–50
46. Kellerhals MB, Hazenberg W, Witholt B (1999) Enzyme Microb Tech 24:111
47. Laane C, Boeren S, Vos K, Veeger C (1987) Biotech Bioeng 30:81
48. Schmid A, Sonnleitener B, Witholt B (1998) Biotech Bioeng 60:10
49. Rajagopal AN (1996) Enzyme Microb Technol 19:606
50. Huisman GW, Wonink E, Meima R, Kazemier B, Terpstra P, Witholt B (1991) J Biol Chem 266:2191
51. Kim YB (1991) PhD. thesis. University of Massachusetts at Amherst
52. Yoon SC, Song JJ, Kim TU (1994) In: Doi Y, Fukuda K (eds) Biodegradable plastics and polymers. Elsevier, Amsterdam London New York Tokyo, p 400
53. Huijberts GNM, De Rijk TC, De Waard P, Eggink G (1994) J Bacterol 176:1661
54. Marchessault RH, Monasterios CJ, Morin FG, Sundararajan PR (1990) Int J Biol Macromol 12:158
55. Gagnon KD, Lenz RW, Farris RJ, Fuller RC (1992) Macromolecules 25:3723
56. Lagaveen RG, Huisman GW, Preusting H, Ketelaar P, Eggink G, Witholt B (1988) Appl Environ Microbiol 54:2924
57. Kim YB, Lenz RW, Fuller RC (1995) J Pol Sci 33:1367
58. Park WH, Lenz RW, Goodwin S (1998) Macromolecules 21:1480
59. Fritzsche K, Lenz RW, Fuller RC (1990) Int J Biol Macromol 12:85
60. Tan IKP, Kumar KS, Theanmalar M, Gan SN, Gordon B III (1997) Appl Microbiol Biotechnol 47:207
61. Eggink G, van der Wal Hetty, Huijberts GNM, de Waard P (1993) Industrial Crops and Products 1:157
62. Cromwick AM, Foglia T, Lenz RW (1996) Appl Microbiol Biotechnol 46:464
63. Theodore K, Kim YB, Chang HN (1999) PaiChai University Manuscript (in preparation)
64. Kim YB, Rhee YH, Kim DY (1998) Macromolecules 31:4760
65. Fritzsche K, Lenz RW, Fuller RC (1990) Makromol Chem 191:1957
66. Kim YB, Kim DY, Rhee YH (1999) Macromolecules 32:6058
67. Curley JM, Hazer B, Lenz RW, Fuller RC (1996) Macromolecules 29:1762
68. Hazer B, Lenz RW, Fuller RC (1996) Polymer 37:5951
69. Kim YB, Heo GS, Han SH, Rhee YH, Kim JS (1996) Macromolecules 29:3432
70. Song JJ, Yoon SC (1996) Appl Environ Microbiol 62:536
71. Kim O, Gross RA, Rutherford DR (1995) Can J Microbiol 41:32
72. Kim YB, Lenz RW, Fuller RC (1992) Macromolecules 25:1852
73. Abe C, Tama Y, Nakamura Y, Doi Y (1990) Polym Commun 31:404
74. Kim O, Gross RA, Hammer WJ, Newmark RA (1996) Macromolecules 29:4752
75. Doi Y, Abe C (1990) Macromolecules 23:3705
76. Kim YB, Jung TK (1999) PaiChai University Manuscript (in preparation)
77. Fritzsche K, Lenz RW, Fuller RC (1990) Int J Biol Macromol 12:92
78. Andújar M, Aponte MA, Diaz E, Schroder E (1997) Macromolecules 30:1611
79. Kim DY, Rhee YH, Kim YB (2000) PaiChai University Manuscript (in preparation)
80. Hazer B, Lenz RW, Fuller RC (1994) Macromolecules 27:45
81. Choi MH, Yoon SC (1994) Appl Environ Microbiol 60:3245
82. Scholz C, Fuller RC, Lenz RW (1994) Macromol Chem Phys 195:1405
83. Eggink G, de Waard P, Huijberts (1995) Can J Microbiol 41:14
84. Valentin HE, Schönebaun A, Steinbüchel A (1996) Appl Microbiol Biotechnol 46:261
85. Brandl H, Knee EJ, Fuller RC, Gross RC, Lenz RW (1989) Int J Miol Macromol 11:49
86. Doi Y, Kitamura S, Abe H (1995) Macromolecules 28:4822
87. Lee EY, Jendrossek D, Schirmer A, Choi CY, Steinbüchel A (1995) Appl Microbiol Biotechnol 42:901
88. Fukui T, Doi Y (1997) J Bacteriol 179:4821

89. Fukui T, Shiomi N, Doi Y (1998) J Bacteriol 180:667
90. Mitomo H, Morishita N, Doi Y (1995) Polymer 36:2573
91. Yoshie N, Menju H, Sato H, Inoue Y (1995) Macromolecules 28:6516
92. Ulmer HW, Gross RA, Posada M, Weishach P, Fuller RC, Lenz RW (1994) Macromolecules 27:1675
93. Ballistreri A, Garozzo D, Giuffrida M, Impalloment G, Montaudo G (1989) Macromolecules 22:2107
94. Ballistreri A, Montaudo G, Garozzo D, Giuffrida M, Montaudo MS (1991) Macromolecules 24:1231
95. Ballistreri A, Montaudo G, Impallomei G, Lenz RW, Kim YB, Fuller RC (1990) Macromolecules 23:5059
96. Ballistreri A, Montaudo G, Giuffrida M, Lenz RW, Kim YB, Fuller RC (1992) Macromolecules 25:1845
97. Kimura H, Yoshida Y, Doi Y (1992) Biotechnol Lett 14:445
98. Saito Y, Doi Y (1994) J Biol Macromol 16:99
99. Huisman GW, DeleeuWO, Eggink G, Witholt B (1989) Appl Environ Microbiol 55:1949
100. Gross RH, DeMello C, Lenz RW, Brandl H, Fuller RC (1989) Macromolecules 22:1106
101. Scholtz C, Fuller RC, Lenz RW (1994) Macromolecules 27:2886
102. Cammas S, Guerin P, Girault JP, Holler E, Gache Y, Vert M (1993) Macromolecules 26:4681
103. Nagata N, Nakahara T, Tabuchi T (1993) Biosci Biotech Biochem 54:638
104. Liu S, Steinbüchel A (1996) Appl Microbiol Biotechnol 46:273
105. Korherr C, Roth M, Holler E (1995) Can J Microbiol 42 (Suppl 1):192
106. Nagata N, Nakahara T, Tabuchi T, Morita R, Brewer JR, Fujishige S (1993) Polymer J 25:585
107. Fischer H, Erdmann S, Holler E (1989) Biochemistry 28:5219
108. Windisch C, Miller S, Reisner H, Angerer B, Achammer G, Holler E (1992) Cell Biol Int Rep 16:1211
109. Gödde C, Liegergesell M, Steinbüchel A (1999) FEMS Microbiol Lett 173:365
110. Vert M (1998) Polymer Degrad Stabil 59:169
111. Willibald B, Holler E (1996) Paper 3/01. In: Witholt B (Ed) Preprints International Symposium of Bacterial Polyhydroxyalkanoate '96, Davos, Switzerland

Received: January 2000

Biochemical and Molecular Basis of Microbial Synthesis of Polyhydroxyalkanoates in Microorganisms

Alexander Steinbüchel, Silke Hein

Institut für Mikrobiologie, Westfälische Wilhelms-Universität Münster, Correnssstraße 3, 48149 Münster, Germany
E-mail: steinbu@uni-muenster.de

Intensive research on the physiology, biochemistry, and molecular genetics of the metabolism of polyhydroxyalkanoates (PHA) during the last 15 years has revealed a dramatic increase of our knowledge on the biosynthesis of these polyesters in bacteria. This mainly very basic research has revealed several new, hitherto not described enzymes and pathways. In addition, many genes encoding the enzymes of these pathways and in particular the key enzyme of PHA biosynthesis, PHA synthase, were cloned and characterized at a molecular level. This knowledge was utilized to establish PHA biosynthesis in many prokaryotic and eukaryotic organisms, which were unable to synthesize PHAs, and to apply the methodology of metabolic engineering, thus opening new perspectives for the production of various PHAs by fermentation biotechnology or agriculture in economically feasible processes. This contribution summarizes the properties of PHA synthases and gives an overview on the genes for these enzymes and other enzymes of PHA biosynthesis that have been cloned and are available. It also summarizes our current knowledge on the regulation at the enzyme and gene level of PHA biosynthesis in bacteria.

Keywords. Polyhydroxyalkanoic acids, Microbial polyesters, PHA, PHA synthase, Metabolic engineering, PHA granules, *Ralstonia eutropha*, *Pseudomonas aeruginosa*

Advances in Biochemical Engineering/
Biotechnology, Vol. 71
Managing Editor: Th. Scheper

List of Symbols and Abbreviations

3HA	3-Hydroxyalkanoic acid
3HB	3-Hydroxybutyric acid
4HB	4-Hydroxybutyric acid
3HD	3-Hydroxydecanoic acid
3HDD	3-Hydroxydodecanoic acid
3HHp	3-Hydroxyheptanoic acid
3HHx	3-Hydroxyhexanoic acid
3HN	3-Hydroxynonanoic acid
3HO	3-Hydroxyoctanoic acid
3HP	3-Hydroxypropionic acid
3HPE	3-Hydroxy-4-pentenoic acid
3HPV	3-Hydroxyphenylvaleric acid
3HV	3-Hydroxyvaleric acid
4HV	4-Hydroxyvaleric acid
5HV	5-Hydroxyvaleric acid
MCL	medium-carbon chain length
PHA	Polyhydroxyalkanoic acids
PHB	Poly(3-hydroxybutyric acid)
SCL	short-carbon chain length

1 Introduction

Many bacteria are able to synthesize polyesters of hydroxyalkanoic acids and to accumulate these water insoluble polyhydroxyalkanoic acids (PHA) in the cytoplasm as inclusions and as storage compounds for energy and carbon; these inclusions are referred to as PHA granules [1]. At present approximately 130 different hydroxyalkanoic acids are known as constituents of these bacterial storage polyesters [2, 3]. Besides these polyesters, the only other polyesters existing in living organisms are polymalic acid, which is a water soluble polyester occurring only in lower eukaryotic organisms [4, 5], and cutin and suberin which are water insoluble polyesters occurring in plants [6].

Accumulation of PHAs in the bacterial cell usually occurs if a carbon source is provided in excess, and if at least one other nutrient, which is essential for growth, has been depleted [7], i.e., if growth is imbalanced (see also [8]). PHAs are synthesized by diverting either central intermediates of the carbon metabolism or derivatives from precursor substrates, which are provided as carbon source for the growth of the bacteria, to hydroxyacyl-CoA thioesters. The latter are then polymerized by PHA synthases that become bound to the surface of PHA granules together with other proteins. Since the cloning of the PHA operon of *Ralstonia eutropha* approximately 12 years ago, more than 50 PHA synthase structural genes and also many other genes related to PHA biosynthesis from various other bacteria have been cloned, and many of them were analyzed at a molecular level (Table 1). These genes comprise the structural genes for (i) PHA synthases, (ii) granule-associated proteins, and (iii) enzymes, which catalyze the formation of hydroxyacyl-CoA thioesters, as well as probably also genes for (iv) proteins that have a regulatory function. The availability of these genes and detailed knowledge of the biochemistry of the enzymes as well as of the structural and regulatory proteins will be essential to establish functional active PHA biosynthesis pathways in organisms that are themselves not able to accumulate PHAs but are considered as more suitable for commercial production of PHAs either by fermentative processes or by cultivation of transgenic crops. To achieve this, knowledge about putative regulatory events affecting the activity of the enzymes or the expression of these enzymes or other proteins might be important.

PHAs are biodegradable, being hydrolyzed by extracellular PHA depolymerases and subsequently utilized as sources of carbon and energy by many bacteria and fungi [9, 10]. In addition, they share some interesting features. From some of them, particularly the homopolyester poly(3-hydroxybutyrate), poly(3HB), a copolyester of 3-hydroxybutyrate and 3-hydroxyvalerate, poly (3HB-*co*-3HV) and PHAs consisting of 3-hydroxyoctanoate, 3-hydroxydecanoate and a few other medium chain-length 3-hydroxyalkanoates, poly($3HA_{MCL}$), various materials have been manufactured, and applications in various areas have been revealed [11]. Commercial production of PHAs is so far only possible by fermentative biotechnological processes. Two chapters of this issue describe the current possibilities and achievements to produce the various PHAs by fermentation of wild type and recombinant bacteria [12, 13]. In addition, the generation of transgenic PHA accumulating plants has made significant progress [14], and also various strategies have been successfully used for the *in vitro* synthesis of PHAs [15]; however, production processes based on the latter are not yet economically feasible.

This review will provide an update of an overview which was published recently [53] and comprised the knowledge of PHA synthase structural genes and

Table 1. Alphabetic list of bacteria strains from which PHA synthase genes were cloned and characterized (updated from [53])

Bacterium	Accession no.	Number of PHA synthases	Ref.
Acinetobacter sp	L37761	1	[16]
Aeromonas caviae	D88825	1	[17]
Alcaligenes latus	AF078795	1	[18]
Alcaligenes sp. SH-69	U78047	1	[19]
Azorhizobium caulinodans	AJ006237	1	[20]
Bacillus megaterium	AF109909	1	[21]
Burkholderia cepacia	AF153086	1	[22]
Chromatium vinosum D	L01112	1	[23]
Chromobacterium violaceum	AF061446	1	[24]
Comamonas acidovorans	AB009273	1	[25]
Ectothiorhodospira shaposhnikovii	–	1	[26]
Lamprocystis roseopersicina 3112	–	1	[26]
Methylobacterium extorquens IBT6	L07893	1	[27]
Nocardia corallina	AF019964	1	[28]
Paracoccus denitrificans	D43764	1	[29]
Pseudomonas acidophila	–	1	[30]
Pseudomonas aeruginosa	X66592	2	[31]
Pseudomonas citronellolis	–	1	[32]
Pseudomonas fluorescens	–	1	[33]
Pseudomonas mendocina	–	2	[34]
Pseudomonas putida BM01	AF042276	2	[35]
Pseudomonas putida KT2442	–	2	[36]
Pseudomonas putida U	AF150670	2	[37]
Pseudomonas sp. DSMZ1650	–	1	[32]
Pseudomonas sp. GP4BH1	–	3	[32]
Pseudomonas sp. 61–3	ABO14757/ABO14758	3	[38]
Pseudomonas sp.	Z80158	1	[39]
Pseudomonas oleovorans	M58445	2	[36]
Ralstonia eutropha H16	M64341	1	[41]
Rhizobium etli	U30612	1	[42]
Rhodobacter capsulatus	–	1	[43]
Rhodobacter sphaeroides	L17049	1	[44]
Rhodococcus ruber PP2	X66407	1	[45]
Rhodospirillum rubrum Ha	AJ245888	1	[46]
Rhodospirillum rubrum ATCC25903	AF178117	1	[46]
Rickettsia prowazekii	AJ235273.1	2	[47]
Sinorhizobium meliloti 41	X93358	1	[48]
Synechocystis sp. PCC6803	Slr1830/29	1	[49]
Syntrophomanas wolfei	–	1	[50]
Thiocystis violacea 2311	L01113	1	[51]
Thiocapsa pfennigii 9111	X93599	1	[26]
Zoogloea ramigera	U66242	1	[52]
		$\Sigma = 53$	

the organization of these genes and of other genes related to PHA metabolism in the bacterial genomes. Emphasis will be on the enzymology of PHA synthases and on our current knowledge of the regulation of PHA synthesis and accumulation at the enzyme and gene level, which will be summarized. Regulation of PHA synthesis and accumulation at the physiological level will be considered in a different chapter of this book [8].

2 Enzymology of PHA Synthases

PHA synthases are the key enzymes of PHA biosynthesis. These enzymes catalyze the covalent linkage between the hydroxyl group of one and the carboxyl group of another hydroxyalkanoic acid. The substrates of PHA synthases are the coenzyme A thioesters of hydroxyalkanoic acids; there is no evidence that PHA synthases can utilize either free hydroxyalkanoic acids or other derivatives of hydroxyalkanoic acids. With respect to size, structure, and substrate specificity, three different types of PHA synthases (I, II, and III) can be distinguished (see below).

2.1 Sizes and Structures of PHA Synthases

With respect to size and structure, type I- and type II-PHA synthases are distinguished from type III-PHA synthases. Whereas type I- and type II-PHA synthases consist of only one type of subunit, type III-PHA synthases consist of two different types of subunits. The differences between type I- and type II-PHA synthases regarding the substrate specificity will be described in the next section.

Type I-PHA synthases are represented by the enzyme of *R. eutropha*, and the subunit exhibits a molecular weight of 64.317 Da and is encoded by $phaC_{Re}$. This type also occurs in the phototrophic nonsulfur purple bacteria and in most heterotrophic bacteria except the pseudomonads belonging to the rRNA homology group I. Type II-PHA synthase is represented by the PHA synthase 1 of *P. oleovorans* encoded by $phaC1_{Po}$ with the subunit exhibiting a molecular weight of 62.334 Da and seems to occur in all pseudomonads belonging to the rRNA homology group I. All PHA synthases belonging to these two types have a molecular weight in this range. The quaternary structures and the molecular weights of the native enzymes have so far not been revealed. Electron microscopic studies and biochemical studies do, however, indicate that the active form of the PHA synthases consists of aggregates of more than one subunit.

Type III-PHA synthase is represented by the enzyme of *Chromatium vinosum* and is encoded by $phaEC_{Cv}$. It consists of the two different subunits $PhaC_{Cv}$ and $PhaE_{Cv}$ exhibiting molecular weights of 39,730 and 40,525 Da, respectively [23]. The native PHA synthase, as isolated from recombinant strains of *E. coli*, exhibited a molecular weight of approximately 390 and 400 + 20 kDa as revealed in our laboratory [54] or 360 + 50 kDa but also 520 + 50 kDa as revealed in another laboratory [55]. The lower molecular weights for the holoenzyme are consistent with the occurrence of particles of 11.2–12.8 nm in diameter as revealed by electron microscopic studies of homogenous preparations of this enzyme [54] and also with the occurrence of protein complexes of approximately 10–11 nm diameter at the surface of PHA granules isolated from *C. vinosum* [56]. Altogether, these data indicate that the active PHA synthase of *C. vinosum* is probably a decamer of $PhaC_{Cv}$ and $PhaE_{Cv}$. The exact composition of the native enzyme could so far not been determined. One laboratory obtained data from which it could be concluded that $PhaC_{Cv}$ represents the minor component, with the holoenzyme likely to consist of four $PhaC_{Cv}$ and six $PhaE_{Cv}$ subunits [54]; in another laboratory data were obtained indicating that both subunits occur at an equimolar ratio [55]. $PhaC_{Cv}$ from *C. vinosum*, although much smaller than PhaC of type I- and type-II PHA synthases from other PHA accumulating bacteria, exhibited low but significant amino acid sequence homologies with the other PHA synthases and shared with them the highly conserved amino acid positions (see below). In contrast, $PhaE_{Cv}$ did not exhibit any significant homologies to other proteins. In crude extracts obtained from cells of recombinant strains of *E. coli*, poly(3HB) synthase activity was only measurable if both, $PhaC_{Cv}$ plus $PhaE_{Cv}$, were expressed together; expression of only $PhaC_{Cv}$ or only $PhaE_{Cv}$ did not result in the measurement of detectable enzyme activity and also did not confer the capability to synthesize PHA to *E. coli* [23]. When each subunit was cloned, expressed, and purified as an $(His)_6$-tagged construct, no detectable poly(3HB) synthase activity was measured with the $PhaE_{Cv}$-$(His)_6$-protein, and with the $PhaC_{Cv}$-$(His)_6$-protein the specific activity was only approximately 0.6% of that of the PhaEC holoenzyme. When both almost inactive protein preparations were combined, active poly(3HB) synthase was regained, and the specific activity was almost as high as of the wild type holoenzyme [55]. These experiments clearly show that type III-PHA synthases require two different proteins and that the enzyme is in practice completely or mostly inactive if $PhaE_{Cv}$ or $PhaC_{Cv}$, respectively, is lacking. The residual very low activity of $PhaC_{Cv}$ is only measurable if relatively high concentrations of the proteins are investigated. These experiments also suggest, that $PhaC_{Cv}$ is the catalytically active subunit; however, the function of $PhaE_{Cv}$ remains to be elucidated.

Aside from *C. vinosum*, type III-PHA synthases have so far been detected exclusively in the phototrophic purple sulfur bacteria such as *Thiocystis violacea* [51] and *Thiocapsa pfennigii* [26, 57] and in cyanobacteria such as *Synechocystis* sp. PCC6803 [49] or *Synechococcus* sp. MA19 [58]. In contrast, the photosynthetic nonsulfur purple bacteria possess type I-PHA synthases.

2.2 Substrate Specificities of PHA Synthases

PHA synthases exhibit a remarkably low substrate specificity. In general, the specificity of PHA synthases regarding the length of the hydroxyalkyl moiety of the coenzyme A thioester as well as of the position of the hydroxyl group, the presence of other substituents, and the position of double bonds, is low. In contrast, the stereospecificity of PHA synthases is strict: the hydroxyl carbon atom must have the R-configuration.

Nevertheless most PHA synthases incorporate either short carbon chain length hydroxyalkanoic acids (HA_{SCL}) comprising 3–5 carbon atoms or medium-carbon-chain length hydroxyalkanoic acids (HA_{MCL}) comprising more than 5 carbon atoms (up to 16) into PHAs. Type I- and type III-PHA synthases synthesize preferentially polyesters consisting of $3HA_{SCL}$; in addition to the coenzyme A thioesters of $3HA_{SCL}$, the respective thioesters of $4HA_{SCL}$ and $5HA_{SCL}$ are also frequently used as substrates by these PHA synthases. In contrast, type II-PHA synthases synthesize preferentially polyesters consisting of $3HA_{MCL}$. Whereas these enzymes are rather flexible regarding the occurrence of substituents at the hydroxyalkyl moiety as long as these substituents are not charged, they seem to be rather strict regarding the position of the hydroxyl group which must be at the carbon atom C3.

There are only very few PHA synthases which can incorporate $3HA_{SCL}$ as well as $3HA_{MCL}$ into the accumulated PHAs. Examples are the PHA synthases of *T. pfennigii* and *Aeromonas caviae*, which synthesize, for example, copolyesters of 3HB, 3HHx plus 3HO [26], or of 3HB plus 3HHx [59], from fatty acids, respectively. Recently it was found that the type I-PHA_{SCL} synthase of *R. eutropha* also conferred the incorporation of 3HHx [60] or of 3HO plus 3HD [61] into PHAs in addition to 3HB to certain recombinant strains of enterobacteria.

2.3
Primary Structures of PHA Synthases

An alignment of the amino acid sequences of 36 PHA synthases (*phaC* gene products) revealed significant homologies between all PHA synthases, and three clusters are evident from the phylogenetic tree shown in Fig. 1. First, the PHA_{MCL} synthases from the pseudomonads represent one cluster, which is in accordance with the substrate specificities of these enzymes. Furthermore, the PhaC1 proteins from these pseudomonads are clearly phylogenetically distinguished from the PhaC2 proteins. Interestingly, and as shown in the upper part of Fig. 1, the PHA_{SCL} synthases from the Gram-positive *N. corallina* and *R. ruber* seem to be more related to the PHA_{MCL} synthase from the pseudomonads than to those from other bacteria. Second, the PHA synthases from the phototrophic sulfur bacteria and the cyanobacterium *Synechocystis* sp. PCC6803 represent a second cluster as shown in the bottom part of Fig. 1. That type III-PHA synthases occur in these two groups of phototrophic bacteria is remarkable because a phylogenetic tree derived by maximum likelihood analysis of small-subunit rRNA sequences revealed no close phylogenetic relationship of bacteria belonging to the γ-subdivision of the purple bacteria and the cyanobacteria [49, 62]. Also interesting, PhaC1 from *R. prowazekii* and the PHA synthase from *B. megaterium* are more closely related to the PHA synthases of these phototrophic bacteria than to PHA synthases from other bacteria. Third, the phylogenetic relationships of all other PHA synthases, which are mostly type I-PHA_{SCL} synthases, are shown in the central part of Fig. 1.

The amino acid sequence similarities of all 36 PHA synthases were pairwise revealed, and the results of this comparison are compiled in Table 2. The data correspond well with phylogenetic tree shown in Fig. 1, and the similarities va-

Table 2. Percent identities of primary structures from 36 PHA synthases based on pairwise alignments

	1	2	3	4	5	6	7	8	9	10	11	12	13	14	15	16	17	18
1	100																	
2	83.7	100																
3	79.6	78.7	100															
4	80.7	80.7	81.6	100														
5	54.4	53.9	56.0	56.9	100													
6	54.2	53.3	54.7	55.5	76.4	100												
7	56.2	55.1	58.5	58.7	74.1	73.8	100											
8	53.7	52.4	55.5	54.4	71.1	74.8	69.3	100										
9	42.2	41.4	43.4	43.2	42.2	40.9	43.2	40.5	100									
10	37.1	36.5	38.2	37.1	36.7	36.0	37.4	38.0	36.6	100								
11	31.7	32.2	31.3	30.6	29.4	30.9	30.3	31.2	29.0	32.9	100							
12	31.1	30.9	30.2	31.1	31.8	32.5	32.7	31.6	30.4	33.0	73.8	100						
13	33.7	33.2	33.7	32.4	33.8	34.7	36.0	35.4	33.8	36.7	56.3	57.0	100					
14	32.0	32.7	33.4	33.9	33.7	35.0	33.9	35.0	32.7	36.6	55.1	57.9	56.6	100				
15	30.4	30.2	30.6	31.1	31.8	32.9	31.9	33.6	34.1	35.1	43.5	46.4	48.3	45.6	100			
16	33.8	35.0	35.1	34.4	34.1	35.3	35.3	37.3	33.1	37.9	41.7	41.7	47.1	45.1	51.9	100		
17	30.9	30.9	31.4	31.2	31.2	33.2	31.0	32.4	31.7	32.3	36.6	38.6	38.2	38.5	37.4	38.3	100	
18	31.8	30.7	31.2	30.5	33.2	33.5	33.2	33.9	33.7	33.7	37.2	38.4	41.6	40.9	43.0	42.9	45.3	100
19	31.3	32.6	34.5	33.8	31.1	31.6	32.0	32.1	31.8	34.1	35.6	36.5	38.3	37.9	37.4	41.3	36.8	37.3
20	29.1	28.9	31.5	29.7	29.1	30.0	30.9	31.1	28.3	32.7	36.5	36.0	37.9	36.6	37.7	39.3	34.1	36.4
21	31.0	30.2	31.9	30.4	29.8	29.3	29.6	29.8	29.6	30.7	32.0	33.5	35.2	34.1	34.4	36.4	32.8	35.0
22	34.0	33.4	34.2	34.5	33.9	34.5	33.9	34.1	35.9	35.7	35.7	38.2	42.3	37.6	38.6	41.1	36.3	38.4
23	33.8	34.7	35.2	35.0	32.1	33.2	33.9	34.2	34.4	34.9	34.2	35.5	38.8	36.0	35.4	40.2	34.7	35.7
24	35.3	35.7	37.6	37.6	35.1	36.0	35.7	36.2	35.9	39.0	38.3	39.7	44.2	40.8	42.8	44.5	38.0	40.8
25	35.2	34.3	36.5	34.5	35.0	34.8	36.1	35.7	36.2	36.0	33.5	35.4	39.4	36.1	37.9	42.4	35.6	38.6
26	35.4	34.5	34.9	34.7	33.6	32.1	34.5	34.1	35.9	35.6	32.9	34.9	38.8	35.8	36.9	38.9	34.5	38.1
27	32.8	32.4	33.7	33.2	33.8	31.8	33.8	33.5	34.9	35.1	34.7	35.7	40.5	34.2	36.1	40.0	35.6	37.9
28	31.6	33.1	33.1	33.3	33.4	33.4	34.3	33.8	34.5	33.4	34.4	35.0	37.5	33.6	32.9	38.3	37.1	36.8
29	27.9	29.3	29.7	29.1	30.0	31.6	32.2	32.4	27.4	31.8	33.9	34.7	38.0	37.3	38.8	38.4	34.0	35.1
30	21.8	20.6	20.9	21.8	23.3	21.8	23.0	22.7	25.0	22.8	21.7	22.0	22.8	24.6	22.5	25.1	21.4	24.3
31	21.7	20.3	20.0	22.0	22.3	21.2	22.0	22.0	22.9	21.3	22.8	21.9	23.3	24.8	23.1	25.1	22.5	25.7
32	22.8	21.9	21.6	23.3	23.9	23.1	24.2	23.1	25.1	22.9	22.6	21.8	23.2	24.1	23.8	26.4	22.9	25.5
33	19.6	19.6	19.6	20.1	21.2	19.8	19.6	18.5	21.6	20.6	19.7	20.5	21.0	22.4	19.4	23.8	21.6	23.8
34	21.3	20.4	20.7	20.7	20.4	21.0	20.1	21.5	23.8	21.7	21.8	21.8	24.6	24.0	23.7	24.0	22.8	25.1
35	15.4	14.8	16.6	15.7	16.6	16.0	16.0	16.6	15.4	16.5	16.5	15.6	16.8	16.8	19.0	18.0	17.1	21.0
36	32.2	34.1	34.6	34.5	35.0	34.6	33.8	39.9	33.8	35.3	33.5	32.9	38.9	35.4	39.0	39.1	36.2	35.4

The pairwise alignments of the primary structures of 36 PHA synthases and the calculation of identical amino acid residues were performed by using the program TREE [63]. 1, *Pseudomonas oleovorans* (PhaCl) [36]; 2, *Pseudomonas* sp. 61-3 (PhaCl) [37]; 3, *Pseudomonas aeruginosa* (PhaCl) [31]; 4, *Pseudomonas mendocina* (PhaCl) [34]; 5, *Pseudomonas mendocina* (PhaC2) [34]; 6, *Pseudomonas* sp. 61-3 (PhaC2) [37]; 7, *Pseudomonas aeruginosa* (PhaC2) [31]; 8, *Pseudomonas oleovorans* (PhaC2) [36]; 9, *Nocardia corallina* [28]; 10, *Rhodococcus ruber* PP2 [45]; 11, *Rhizobium etli* [42]; 12, *Sinorhizobium meliloti* 41 [48, 64]; 13, *Azorhizobium caulinodans* [20]; 14, *Methylobacterium extorquenz* IBT6 [27]; 15, *Rhodospirillum rubrum* Ha [46]; 16, *Rhodospirillum rubrum* ATCC25903 [46]; 17, *Acinetobacter* sp.

Table 2 (continued)

19	20	21	22	23	24	25	26	27	28	29	30	31	32	33	34	35	36
100																	
60.3	100																
55.7	59.8	100															
38.2	36.1	36.0	100														
34.3	34.4	32.9	71.0	100													
38.3	38.2	35.4	67.0	63.5	100												
35.4	33.6	34.2	60.6	56.0	59.4	100											
33.7	33.9	32.9	59.8	57.0	60.8	60.5	100										
34.5	34.7	32.6	55.8	52.2	56.9	57.9	55.8	100									
34.2	33.5	30.1	47.6	48.3	49.3	49.9	48.8	47.1	100								
34.2	34.6	30.3	33.3	31.9	37.8	31.8	33.9	34.5	33.8	100							
23.3	20.8	21.1	22.8	22.5	25.4	24.0	24.3	23.8	23.2	20.9	100						
23.6	21.8	21.0	23.6	22.8	24.5	22.5	24.2	24.0	23.7	22.0	87.3	100					
22.9	20.6	20.6	24.4	24.1	26.1	24.4	24.9	24.1	23.0	20.8	85.6	82.5	100				
22.3	18.0	19.4	22.7	21.4	23.3	21.9	24.7	23.4	22.2	21.8	55.8	56.5	54.5	100			
25.6	23.1	21.8	24.8	22.8	24.5	25.4	23.9	24.9	25.4	24.4	34.5	36.1	34.8	35.6	100		
16.5	14.3	15.9	20.5	19.6	20.2	19.0	20.2	21.2	19.6	20.3	23.1	23.8	22.6	23.1	20.2	100	
35.3	33.4	33.3	43.8	44.4	45.2	45.1	43.5	43.5	40.2	32.6	23.3	22.7	22.9	20.1	22.2	18.3	100

RA3849 [16]; 18, *Aeromonas caviae* FA440 [17]; 19, *Rhodobacter sphaeroides* [44]; 20, *Rhodobacter capsulatus* [43]; 21, *Paracoccus denitrificans* [29]; 22, *Alcaligenes* sp. SH-69 [52]; 23, *Comamonas acidovorans* [25]; 24, *Alcaligenes latus* [18]; 25, *Ralstonia eutropha* [44]; 26, *Burkholderia cepacia* [22]; 27, *Zoogloea ramigera* [19]; 28, *Pseudomonas* sp. 61-3 [37]; 29, *Rickettsia prowazekii* (phaC2) [47]; 30, *Chromatium vinosum* D [23]; 31, *Thiocystis violaceae* 2311 [51]; 32, *Thiocapsa pfennigii* 9111 [26]; 33, *Synechocystis* sp. PCC6803 [65]; 34, *Bacillus megaterium* [21]; 35, *Rickettsia prowazekii* (PhaC1) [47]; 36, *Chromobacterium violaceum* [24]

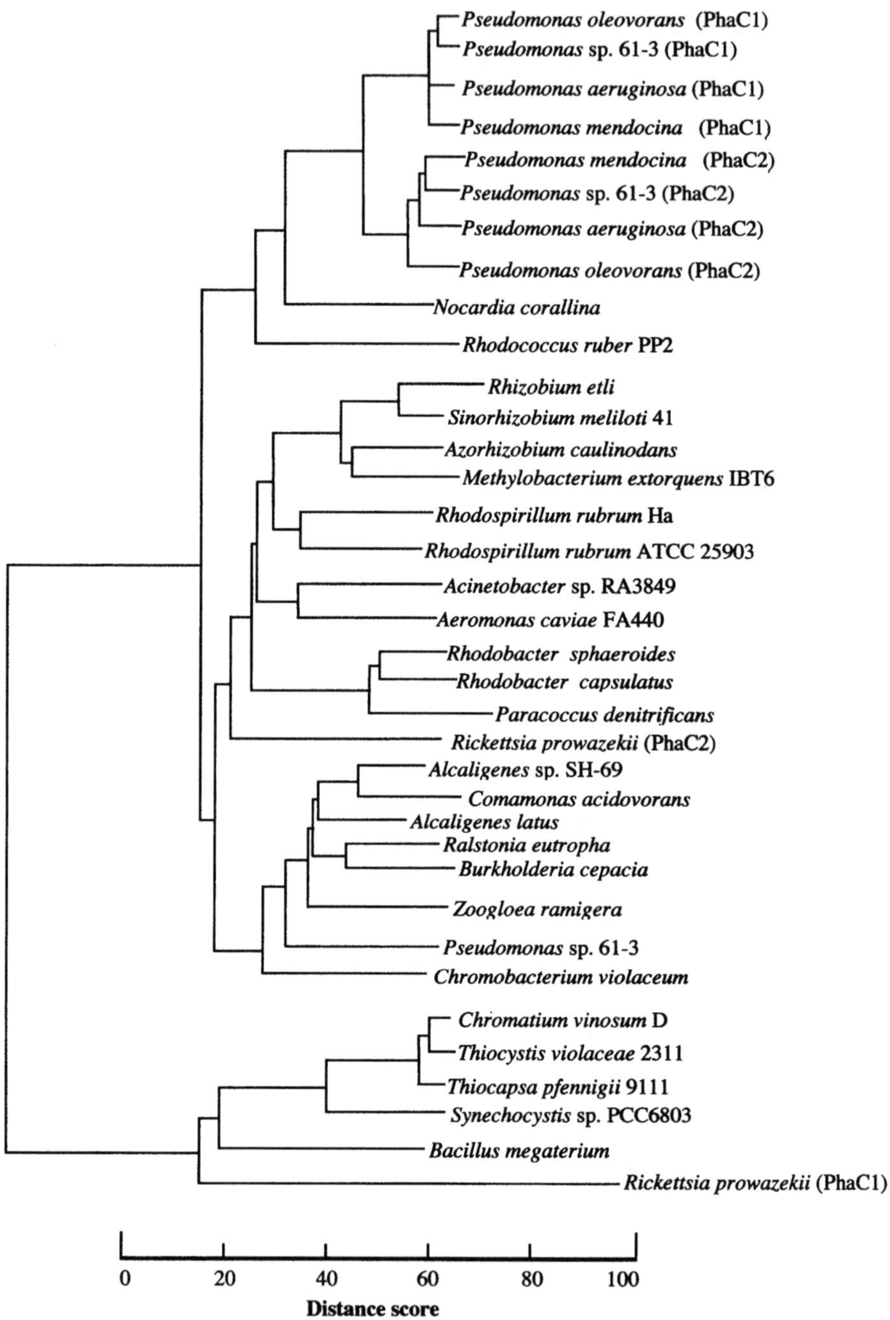

Fig. 1. Phylogenetic tree of 36 PHA synthases. The branching order and distance score were calculated by the program TREE as described by Feng and Doolittle [63]

ried between 14.3% and 87.3%. These data might, for example, be helpful in designing probes for hybridization experiments.

Whereas the alignment of 30 PHA synthases shown previously [53] revealed in total 15 highly conserved amino acid residues that were identical in any of these PHA synthases, only 8 highly conserved amino acid residues remained when the 36 PHA synthases were compared in the study (Fig. 2). Therefore, the number of amino acid residues probably exerting an important role in respect of the enzyme function was drastically reduced. With reference to the *R. eutropha* PHA synthase regarding the position of the amino acids, these 8 amino acid residues were Ser-260, Cys-319, Gly-322, Asp-351, Trp-425, Asp-480, Gly-507, and His-508. The amino acids Cys-319 and Gly-322 belong to the "G-x-C-x-G-G" motif that has an important function for the catalytic cycle of the enzyme (see below), and only in one PHA synthase (*R. ruber* PP2) is the first glycine and only in three PHA synthases (*R. sphaeroides*, *R. capsulatus*, *P. denitrificans*) is the second glycine of this motif replaced by serine or alanine, respectively (Fig. 2).

2.4 Catalytic Cycle of PHA Synthases

The current working hypothesis for the reaction mechanism of PHA synthases is based on a model noted by Merrick and coworkers [66] with two thiolates participating in the covalent catalysis of polyester synthesis. Synthesis of PHAs occurs most probably at thiolate groups [S_1H and S_2H] provided by the PHA synthases [E], with one thiolate functioning as a loading site and the other as the elongation site. According to the current model, the following reaction mechanism is proposed [55, 67, 68]: one thiol group [S_1H] receives from a coenzyme A thioester a hydroxyalkanoic acid with the latter becoming covalently bound to this thiol group [E-S_1-CO-Alkyl-OH] and the coenzyme A being released, whereas the growing polyester chain is bound to the second thiol group [E-S_2-poly$(HA)_n$-OH]. The latter is subsequently transferred to the free hydroxyl group upon a nucleophilic attack of the hydroxyl oxygen atom on the carbonyl carbon atom resulting in [E-S_1-poly$(HA)_{n+1}$-OH]. A subsequent transesterification of the elongated polyester chain from S_1 to S_2 results in [E-S_2-poly$(HA)_{n+1}$-OH] plus [E-S_1H], and the latter can now accept the next hydroxyalkanoic acid from a coenzyme A thioester. This catalytic cycle must be repeated several thousand times, otherwise the very high molecular weights of the formed polyesters cannot be explained. The catalytic cycle may be disrupted if the growing polyester chain is released from [E-S_2-poly(HA)-OH] by the nucleophilic attack of a hydroxyl group of a molecule that is not bound to [E] such as water. There is, for example, evidence that exogenous hydroxy-compounds like glycerol, polyethylene glycol or 1,3-propanediol also provide such a chain terminating hydroxyl group resulting in a PHA molecule that contains the respective hydroxy-compound covalently linked to the PHA molecules [69, 70].

Consistent with this catalytic cycle is the existence of one highly conserved cysteine residue, which occurs in any PHA synthase and which is Cys-319 in the PHA synthase of *R. eutropha* [71] and Cys-149 in the PHA synthase of *C. vino-*

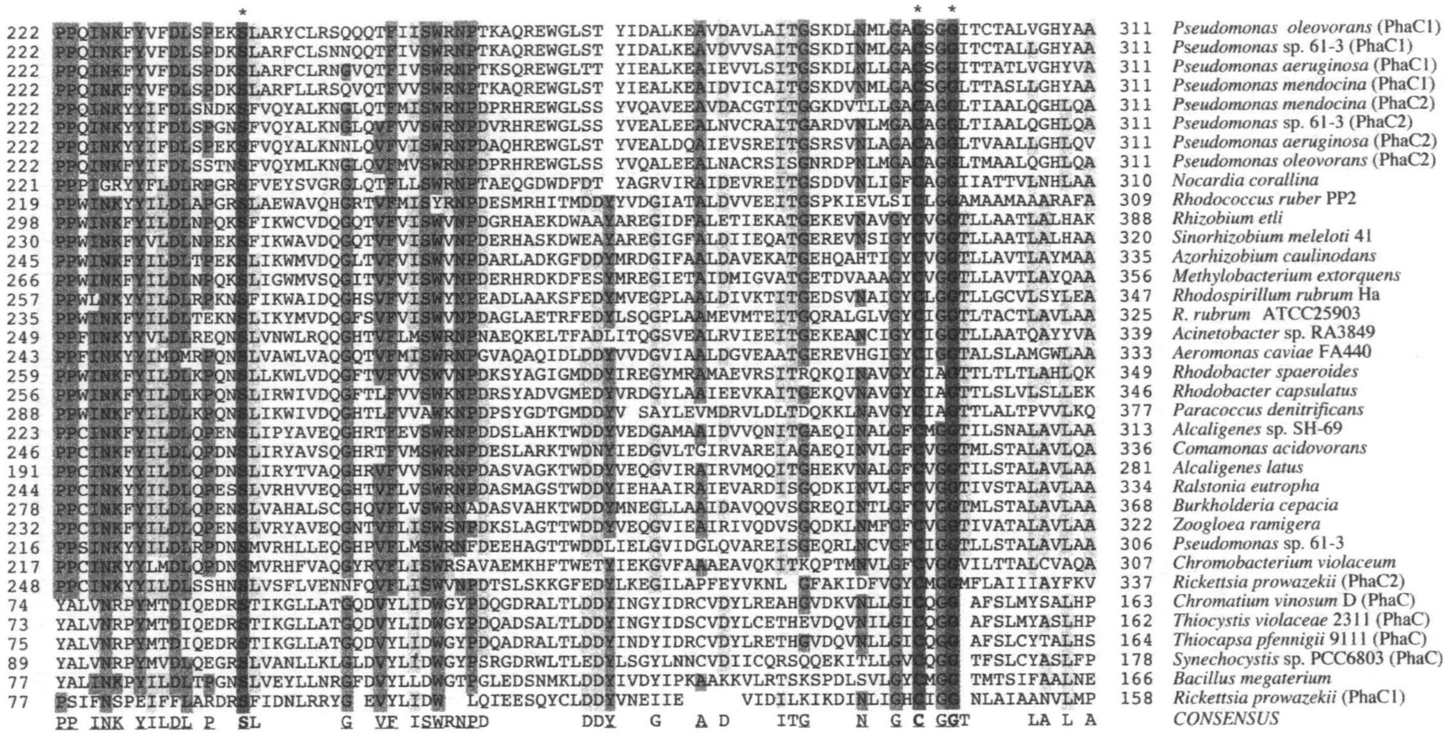

Fig. 2a–d. Multiple alignment of primary structures from 36 PHA synthases. A comparison of amino acid sequences derived from PHA synthase genes is shown. Amino acids are specified by the standard one-letter abbreviations. The consensus sequence represents amino acid residues (*shaded*) which are present in at least 50% of the PHA synthases. Highly conserved amino acids, which are present in at least 70% of the PHA synthases, are additionally *underlined* in the consensus sequence and the eight amino acid residues, which are present in all PHA synthases, are indicated as *bold letters*. See Table 1 for references

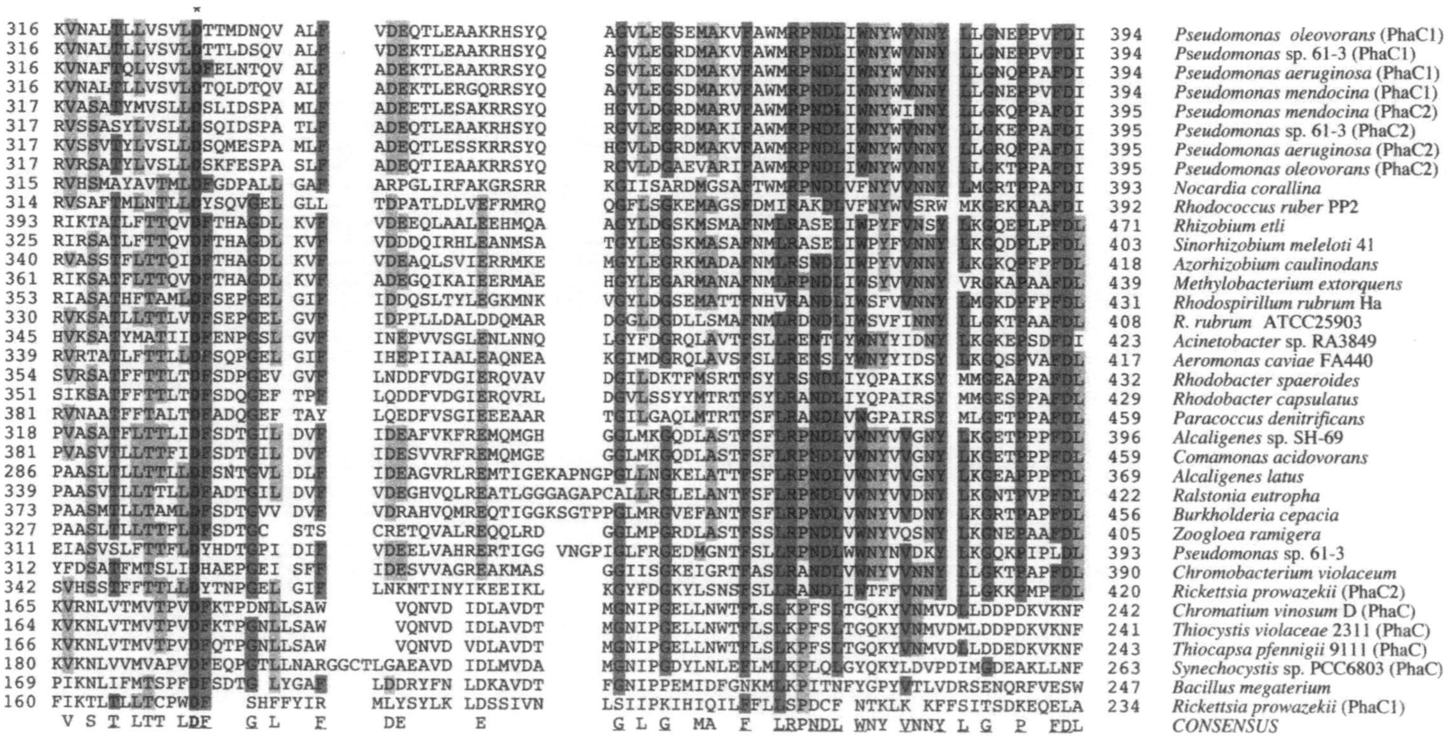

Fig. 2b (continued)

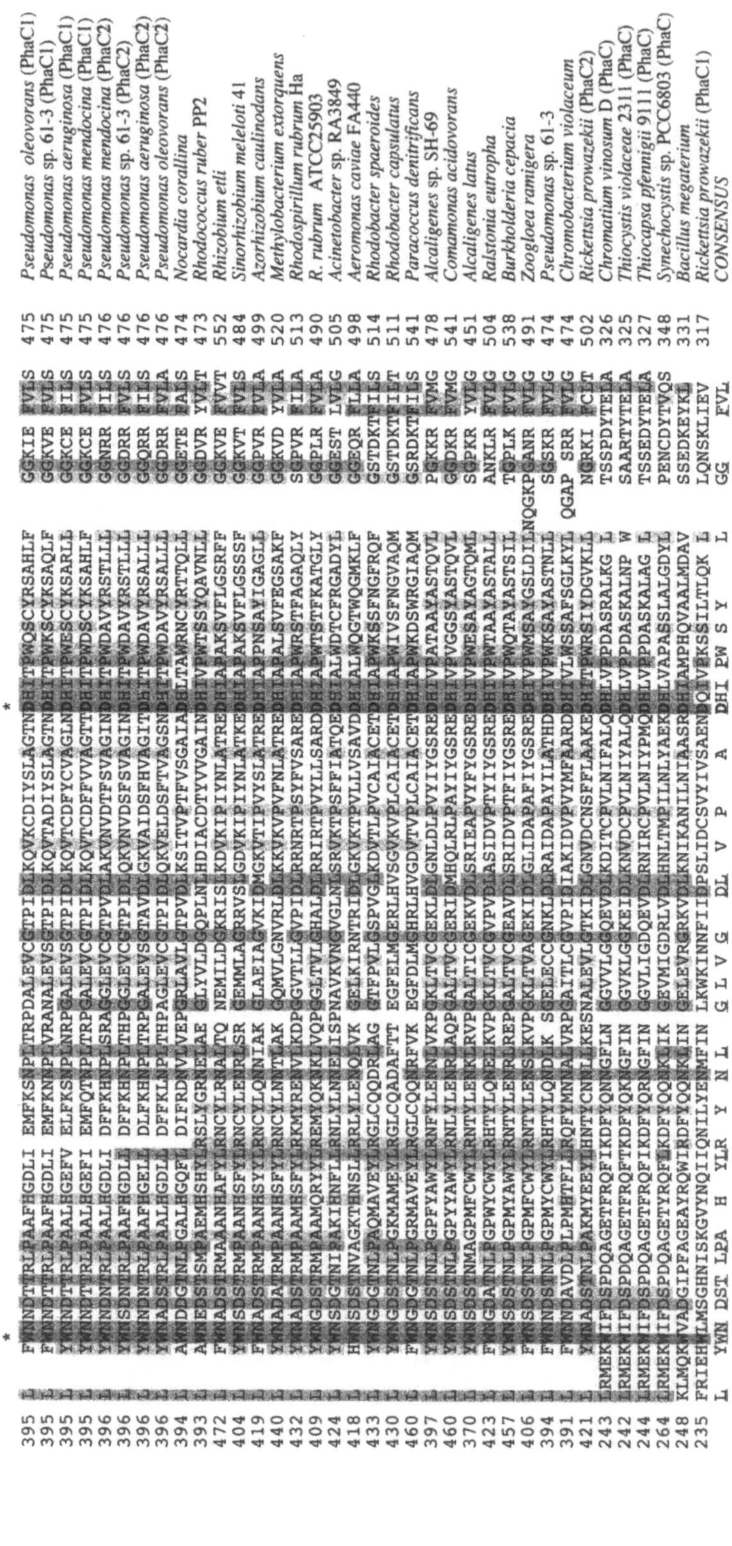

Fig. 2c (continued)

```
      **
476  NSGHIQSILNPPGNP        KARFM   TGADRPGDPVAWQENATKHADSWWLHWQSWLGERAGELEKAPTRLGN  RAYAAGEASPGTY 555  Pseudomonas oleovorans (PhaC1)
476  SSGHIQSILNPPGNP        KSRYM   TSTDMPATANEWQENSTKHTDSWWLHWQAWQAERSGKLKKSPTSLGN  KAYPSGEAAPGTY 555  Pseudomonas sp. 61-3 (PhaC1)
476  NSGHIQSILNPPGNP        KARFM   TNPELPAEPKAWLEQAGKHADSWWLHWQQWLAERSGKTRKAPASLGN  KTYPAGEAAPGTY 555  Pseudomonas aeruginosa (PhaC1)
476  NSGHIQSILNPPGNP        KARYM   TNSEMPLDPKAWQESSTKHADSWWLHWQSWLAERSGETKNAPRALGN  KKFPAGEAAPGTY 555  Pseudomonas mendocina (PhaC1)
477  NSGHIQSILNPPGNP        KANYY   ENTKLTSDPRAWYHDAQHQQGSWWPQWLEWIQARSGEQRETVMALGN  QNHPAMEAAPGTY 556  Pseudomonas mendocina (PhaC2)
477  NSGHIQSILNPPSNP        KSNYI   ENPKLSGDPRAWYYDGTHVEGSWWPRWLSWIQERSGTQRETLMALGN  QNYPPMEAAPGTY 556  Pseudomonas sp. 61-3 (PhaC2)
477  NSGHIQSILNPPGNP        KACYF   ENDKLSSDPRAWYYDAKREEGSWWPVWLGWLQERSGELGNPDFNLGS  AAHPPLEAAPGTY 556  Pseudomonas aeruginosa (PhaC2)
477  NSGHIQSIINPPGNP        KAYYL   ANPKLSSDPRAWLHDAKRSEGSWWPLWLEWITARSGPLKAPRSELGN  ATYPPLGPAPGTY 556  Pseudomonas oleovorans (PhaC2)
475  FSGHIASLVNPPGNP        KAHYW   TGGTPGPDPDAWLENAERQQGSWWQAWADWVLARGGEETAAPDAPGS  AHDRALDAAPGRY 554  Nocardia corallina
474  NGGHVAGAVNPPGKRVWFKAVGAPDAE    SGTPLPADPQVWDEAATRYEHSWWEDWTAWSNKRAGELVAPP AMGS  TAHPPLEDAPGTY 559  Rhodococcus ruber PP2
553  GSGHIAGVVNPP DK      RKYQFWT GGPAKG EYETWLEQASETPGSWWPHWQAWIETHDGRRVAARKPGG    DALNAIEEAPGSY 631  Rhizobium etli
485  GSGHIAGVVNPP AR      SKYQYWT GGAPKG DIETWMGKAKETAGSWWPHWQGWVERLDKRRVPARKAGG     PLNSIEEAPGSY 562  Sinorhizobium meleloti 41
500  GSGHIAGVVNPP VK      HKYQYWT GGPTGG DYDVWLKGAQEHKGSWWPDWAQWFSALHPDEVPAREPGG    SAFNPIEDAPGRY 578  Azorhizobium caulinodans
521  GSGHIAGVVAPPGPK      AKYGFRT GGPARG RFEDWVAAATEHPGSWWPYWYKWLEEQAPERVPARIPGT    GALPSLAPAPGTY 600  Methylobacterium extorquens
514  ASGHIAGVVNPP SA      NKYCFWT NTKKAK DAESWLEKASQTDGSWWTDWEAWVEKFTEGKVPARVPAM    PACRCL      AM 587  Rhodospirillum rubrum Ha
491  GSGHIAGVINPP AK      ARYGYWT NADTSL EAESWLEGATPHGGSWWPDWAAWAAGYAGPKVPARDPTK    APRPPLEDAPGSY 569  R. rubrum ATCC25903
506  ESGHVAGIVNPP SR      NKYGCYT NAAKFE NTKQWLDGAEYHPESWWLRWQAWVTPYTGEQVPARNLGN    AQYPSIEAAPGRY 584  Acinetobacter sp. RA3849
499  ESGHIAGIINPP AA      NKYGFWH NGAEAE SPESWLAGATHQGGSWWPEMMGFIQNRDEGSEPVPARVP    EE  GLAPAPGHY 575  Aeromonas caviae FA440
515  QSGHVAGIVNPP SR      NKYGHYT NEGPAG TPESFREGAEFHAGSWWPRWGAWLAERSGKQVPARQPGD    SKHPELAPAPGSY 593  Rhodobacter spaeroides
512  ESGHIAGIVNPP SK      DKYGHYT SAAPIA DHQVWKAQATYTKGSWWPRWGEWLAGHSGKKIPARAPGD    ATHPPLAPAPGTY 590  Rhodobacter capsulatus
542  ESGHIAGIVNPP SK      KKYGHYT SDAGFGQGEQHWLDKASHHEGSWWGRWGEWLARRAGGMVDARDPGE    G    FGPAPGLY 617  Paracoccus denitrificans
479  ASGHIAGVINPP AK      GKRSHWTRADGKFPGTLDQWLEGATEHPGSWWTDWSGWLKSHAGKQIAAPKAYGKGTKFKAIEPAPGRY 561  Alcaligenes sp. SH-69
542  ASGHIAGVINPP AK      KKRSYWLREDGQLPATLKEWQAGADEYPGSWWADWSPWLAEHGGKLVAAPKQYGKGREYTAIEPAPGRY 624  Comamonas acidovorans
452  ASGHIAGVINPP QK      KKRSYW  TNEQLDGDFNQWLEGSTEHPGSWWTDWSDWLKQHAGKEIAAPKTPGN  KTHKPIEPAPGRY 531  Alcaligenes latus
505  ASGHIAGVINPP AK      NKRSHWT ND ALPESPQQWLAGAIEHHGSWWPDWTAWLAGQAGAKRAAPANYGN  ARYRAIEPAPGRY 585  Ralstonia eutropha
539  ASGHIAGVINPP AK      KKRSFWV NDNDLPDAADDWFAGAAEQPGSWWPTWTEWLGQYGGRKVAPPAQAGS  AQFPVIEPAPGRY 619  Burkholderia cepacia
492  ASGHIAGVINSV AK      NKRTYWI  NDGGAADAQAWFDGAQEVPGSWWPQWAGFLTQHGGKKVKPKAKPGN  ARYTAIEAAPGRY 571  Zoogloea ramigera
475  ASGHIAGVINPP AK      QKRHYW  TNNRVTKNPETWFKNAEQHPGSWWNDWFTWLAGHSGERQPAVAHTGN  NKYPPLEPAPGRY 554  Pseudomonas sp. 61-3
475  ASGHIAGSINPV TK      DKRNYW  ANDTLPLHAEEWLESAESRPGSWWKDWDAMLAPQSGKQVAAPKSLGN  KEFPPLLAAPGSY 554  Chromobacterium violaceum
503  DSGHVAGVVNHP DN      AKYNYRL  NYDLSLSSNEWFMQATEYKGSWWNYWIDWLIKNNDTKMLVDSL  DYQNLDVIESAPGSY 581  Rickettsia prowazekii (PhaC2)
327 FPGGHIGIYVSG         KAQ                 KEVTPAIGKW    LNER                                  355  Chromatium vinosum D (PhaC)
326 FPGGHIGIYVSG         KAQ                 KEVTPAIGKW    LNERS                                 355  Thiocystis violaceae 2311 (PhaC)
328 FPGGHIGIYVSG         KAQ                 EGVTPAIGRW    LNERG                                 357  Thiocapsa pfennigii 9111 (PhaC)
349 FPVGHIGMYVSG         KVQ                 RDLPPAIAHW    LSERQ                                 379  Synechocystis sp. PCC6803 (PhaC)
332 LQTGHVSVVFGP         KAV                 KETYPSIGDW    LEKRSK                                362  Bacillus megaterium
318  KGGHISYLIND         KLD                 KLLKEYT       L                                     339  Rickettsia prowazekii (PhaC1)
      SGHIAG NPP         K              W  A   GSWW  W  WL  R     A   G       E APG Y              CONSENSUS
```

Fig. 2d (continued)

Ralstonia eutropha
35 -10
phaC (1767 bp)
phaB (738 bp)
ORF1 (551 bp)
ORF2 (968 bp)
ORF3 (1532 bp)
phaP (670 bp)
314 bp
84 bp
74 bp
450 bp
1033 bp
3 bp
74 bp
Alcaligenes latus
35 -10
phaC (1608 bp)
phaB (735 bp)
286 bp
56 bp
118 bp
Burkholderia cepacia
phaC (1875 bp)
phaB (738 bp)
(372 bp)
ORF4
99 bp
112 bp
121 bp
Chromobacterium violaceum
phaC (1704 bp)
phaA
71 bp
Acinetobacter sp. RA3849
phaB (746 bp)
phaP (344 bp)
phaC (1772 bp)
51 bp
125 bp
83 bp
Bacillus megaterium
phaP (513bp)
phaQ (441bp)
phaR (609 bp)
phaB (744 bp)
phaC (1089 bp)
169 bp
385 bp
82 bp
77 bp
Pseudomonas sp. 61-3
35 -10
phbR (1137 bp)
phaB (744 bp)
phaC (1701 bp)
268 bp
65 bp
161 bp
Comamonas acidovorans
phaC (1893 bp)
124 bp

sum [55]. Upon exchange of this cysteine residue by another amino acid, the activity of the enzyme is lost. Therefore, it is very likely that this cysteine residue provides one of the thiolate groups for covalent catalysis. However, this is the only cysteine residue which is conserved in any PHA synthase. How is the second thiolate group provided? The published studies have not yet revealed a clear answer to this question and different speculations and hypotheses were made in the past. All PHA synthases possess a highly conserved serine residue, which is, for example, the highly conserved Ser-260 in the PHA synthase of *R. eutropha*. It is known for the fatty acid synthases, polyketide synthases, and some other enzymes that contain 4-phosphopantetheine as a prosthetic group covalently linked to serine residues (see references cited in [71]). There was some experimental evidence that the PHA synthase of *R. eutropha* is posttranscriptionally modified by the transfer of a 4-phosphopantetheinyl moiety when it was expressed in an auxotrophic *panD* mutant of *E. coli* that was fed with [2-^{14}C]-β-alanine [71] although the authors did not identify a peptide carrying 4-phosphopantethein. However, in *panD* mutants generated from *R. eutropha* and fed with [U-^{14}C]-β-alanine no evidence for this covalent modification was obtained [72]. Nevertheless, it could be shown that Ser-260 and also Ser-546, which also occurs in many OHA synthases, are important for the catalytic activity, since mutants in which these serine residues were replaced by other amino acids exhibited only 0.8% or 3.7% of the wild type activity [72], respectively. Therefore, provision of the second thiolate group by a 4-phosphopantetheinyl moiety linked to a serine residue is unlikely. Based on kinetic studies investigating the lag phase of the enzyme reaction, it is now proposed that in type I- and type II-PHA synthases the two thiolate groups are provided by two Cys-319 residues located on two subunits and that a dimer of PhaC is therefore the minimum size of these PHA synthases. In this model Cys-319 alternates between the role of the loading thiol and the role of the elongation thiol as outlined at the beginning of this section [55]. This could of course also be true for the type-III PHA synthases. However, analysis of the primary structures of all so far sequenced type III-PHA synthases revealed a second highly conserved cysteine residue which is, for example, Cys-130 in PhaC of *C. vinosum* [53]. Sinskey and coworkers replaced this cysteine by an alanine, and they obtained an enzyme that exhibited only 0.003% of the activity of the wild type enzyme [55]. At present, no clear final conclusion can be drawn regarding the provision of the second thiolate group for the catalytic cycle of PHA synthases.

Fig. 3. Molecular organization of PHA synthases which are co-localized with other genes relevant for biosynthesis of PHA. The organization of PHA synthase genes encoding type I PHA synthases from various bacteria are presented. *PhaC*, gene encoding PHA synthase; *phaA*, gene encoding β-ketothiolase; *phaB*, gene encoding acetoacetyl-CoA reductase; *phbR*, gene encoding putative regulator protein; ORF, open reading frame with unknown function. See Table 1 for references. PHA synthase structural genes are shown in *black shading with white letters*, β-ketothiolase genes in *dark grey shading with white letters*, acetoacetyl-CoA reductase genes in *medium grey shading with black letters* and genes for phasins or other none catalytically active granule bound proteins in *light grey shading with black letters*

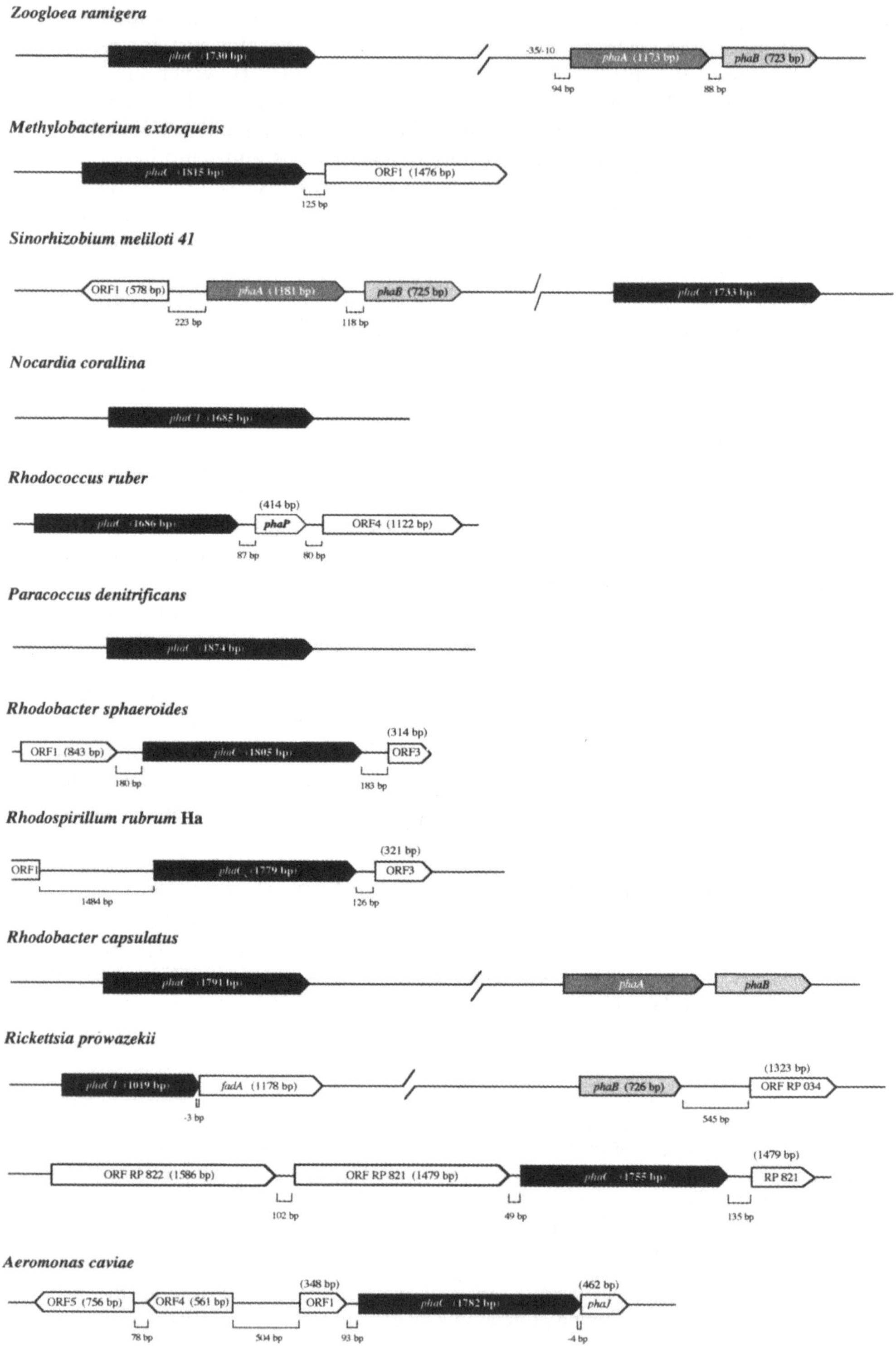

Zoogloea ramigera
phaC (1730 bp)
-35/-10
phaA (1173 bp)
phaB (723 bp)
94 bp
88 bp
Methylobacterium extorquens
phaC (1815 bp)
ORF1 (1476 bp)
125 bp
Sinorhizobium meliloti 41
ORF1 (578 bp)
phaA (1181 bp)
phaB (725 bp)
phaC (1733 bp)
223 bp
118 bp
Nocardia corallina
phaC1 (1685 bp)
Rhodococcus ruber
phaC (1686 bp)
(414 bp)
phaP
ORF4 (1122 bp)
87 bp
80 bp
Paracoccus denitrificans
phaC (1874 bp)
Rhodobacter sphaeroides
ORF1 (843 bp)
phaC (1805 bp)
(314 bp)
ORF3
180 bp
183 bp
Rhodospirillum rubrum Ha
ORF1
phaC (1779 bp)
(321 bp)
ORF3
1484 bp
126 bp
Rhodobacter capsulatus
phaC (1791 bp)
phaA
phaB
Rickettsia prowazekii
phaC1 (1019 bp)
fadA (1178 bp)
phaB (726 bp)
(1323 bp)
ORF RP 034
-3 bp
545 bp
ORF RP 822 (1586 bp)
ORF RP 821 (1479 bp)
phaC (1755 bp)
(1479 bp)
RP 821
102 bp
49 bp
135 bp
Aeromonas caviae
ORF5 (756 bp)
ORF4 (561 bp)
(348 bp)
ORF1
phaC (1782 bp)
(462 bp)
phaJ
78 bp
504 bp
93 bp
-4 bp

2.5 Inhibitors of PHA Synthases

Intermediates of the metabolism have so far not been identified as inhibitors or activators of PHA synthases. The only exception is coenzyme A which inhibits the PHA synthases of *R. eutropha*, *C. vinosum*, and *P. aeruginosa* at relatively low concentrations [73, 74]. It is not known whether this inhibition is physiologically relevant. The inhibition by coenzyme A has, however, to be taken into account during the design of *in vitro* PHA biosynthesis processes, if PHA is being prepared on a preparative scale; recycling of coenzyme is then recommended not only to reduce the costs but also to improve the kinetics of PHA formation.

On the other hand various nonphysiological chemicals were identified as inhibitors of PHA synthases such as, for example, phenazine methosulfonate. None of these chemicals specifically inhibit PHA synthases and they are therefore not suitable for application in strategies to enrich mutants with altered PHA metabolism or altered PHA synthase activity. These inhibitors are only interesting in that they reveal some biochemical features of the enzymes.

2.6 Localization of PHA Synthases in the Cell

Enzymatic studies showed that the PHA synthases from *Zoogloea ramigera* and *R. eutropha* occur in a soluble form in the cytoplasm if the cells do not accumulate PHA. When PHA synthesis starts, the enzyme activity can be sedimented together with the granules. Immunological methods employing gold-labeled antibodies clearly demonstrated that the PHA synthases from *R. eutropha* [75], *C. vinosum* [54], and *P. oleovorans* [76] are bound to the PHA granules and are localized at the surface of the granules.

3 Localization and Organization of PHA Biosynthesis Genes

At least 53 PHA synthases have meanwhile been cloned from 42 different bacterial strains, and the nucleotide sequences of 38 different PHA synthases from 31 different bacteria have been obtained (Table 1, Fig. 2). That the number of sequenced PHA synthases exceeds the number of bacteria from which they have been cloned is due to the occurrence of two or even more PHA synthase genes in some bacteria. The organization of PHA synthase genes and other genes of

Fig. 4. Molecular organization of PHA synthases which are not co-localized with other genes relevant for biosynthesis of PHA. *PhaC*, gene encoding PHA synthase; *phaA*, gene encoding *β*-ketothiolase; *phaB*, gene encoding acetoacetyl-CoA reductase; *phaJ*, gene encoding enoyl-CoA hydratase; ORF, open reading frame with unknown function. See Table 1 for references. PHA synthase structural genes are shown in *black shading with white letters*, *β*-ketothiolase genes in *dark grey shading with white letters*, acetoacetyl-CoA reductase genes in *medium grey shading with black letters* and genes for phasins or other none catalytically active granule bound proteins in *light grey shading with black letters*

PHA biosynthesis pathways, if the latter are mapping close to the PHA synthase genes in the genomes, are shown in Figs. 3–6. In several bacteria possessing a type I-PHA_{SCL} synthase, the structural gene form a clusters together with, for example, the structural genes for a β-ketothiolase and an acetoacetyl-CoA reductase (Fig. 3), whereas in other bacteria possessing this type of PHA synthase *phaC* is isolated from other genes related to PHA_{SCL} biosynthesis (Fig. 4). The organization of *phaC* in pseudomonads capable of PHA_{MCL} biosynthesis and possessing type II-PHA synthases is shown in Fig. 5. Figure 6 shows the organization of the genes in bacteria possessing a type III-PHA synthase.

In some PHA_{SCL} accumulating bacteria possessing type I-PHA synthase clusters of genes comprising the structural genes for PHA synthase, β-ketothiolase, and acetoacetyl-CoA reductase occur in this or a different order. Examples are *R. eutropha*, *A. latus*, *B. cepacia*, *Acinetobacter* sp. RA3849, and *Pseudomonas* sp. 61–3 (Fig. 3). In some bacteria these three genes constitute an operon, and in some bacteria additional genes relevant to PHA metabolism are inserted in these clusters like, for example, in *Acinetobacter* sp. RA3849. However, in most PHA_{SCL} accumulating bacteria possessing a type I-PHA synthase, the PHA synthase genes are isolated, and further PHA biosynthesis genes are definitely not located close to the PHA synthase structural genes (Fig. 4).

All pseudomonads *senso strictu* possess two different PHA synthase genes that are closely related in the genome and occur in the same orientation but are separated by a gene presumably encoding an intracellular PHA_{MCL} depolymerase as shown for *P. aeruginosa*, *P. oleovorans*, and *P. mendocina* in Fig. 5. An identical organization of the genes was recently found in *P. putida* strain U [37]. The genes were not included in Figs. 1, 2 and 5 nor in Table 2, because this study was published at a time when these complex figures and tables were already prepared. A similar cluster of genes also occurs in *Pseudomonas* sp. 61–3 (Fig. 5); in addition this bacterium possesses a third PHA synthase gene clustered with the structural genes for a β-ketothiolase and an acetoacetyl-CoA reductase (Fig. 3). The only other investigated bacterium, which possesses three different PHA synthases, is *Pseudomonas* sp. strain GP4BH1 [32]. Interestingly, *Rickettsia prowazekii* also possesses two PHA synthase structural genes; however, the organization of the genes in this bacterium is different from those occurring in the pseudomonads (Fig. 4).

Southern hybridization experiments, employing the cloned PHA synthase structural gene of *Acinetobacter* sp. and sucrose gradient fractions of DNA preparations separated in plasmid and chromosomal DNA fractions gave two hybridization signals and revealed some but not yet conclusive evidence for

Fig. 5. Molecular organization of genes encoding PHA synthases of type II, which are co-localized with PHA depolymerase genes. *PhaC1/C2*, genes encoding PHA synthase; *phaZ*, gene encoding PHA depolymerase; *phaD* and ORF, open reading frames with unknown function. See Table 1 for references. PHA synthase structural genes are shown in *black shading with white letters*, β-ketothiolase genes in *dark grey shading with white letters*, acetoacetyl-CoA reductase genes in *medium grey shading with black letters* and genes for phasins or other none catalytically active granule bound proteins in *light grey shading with black letters*

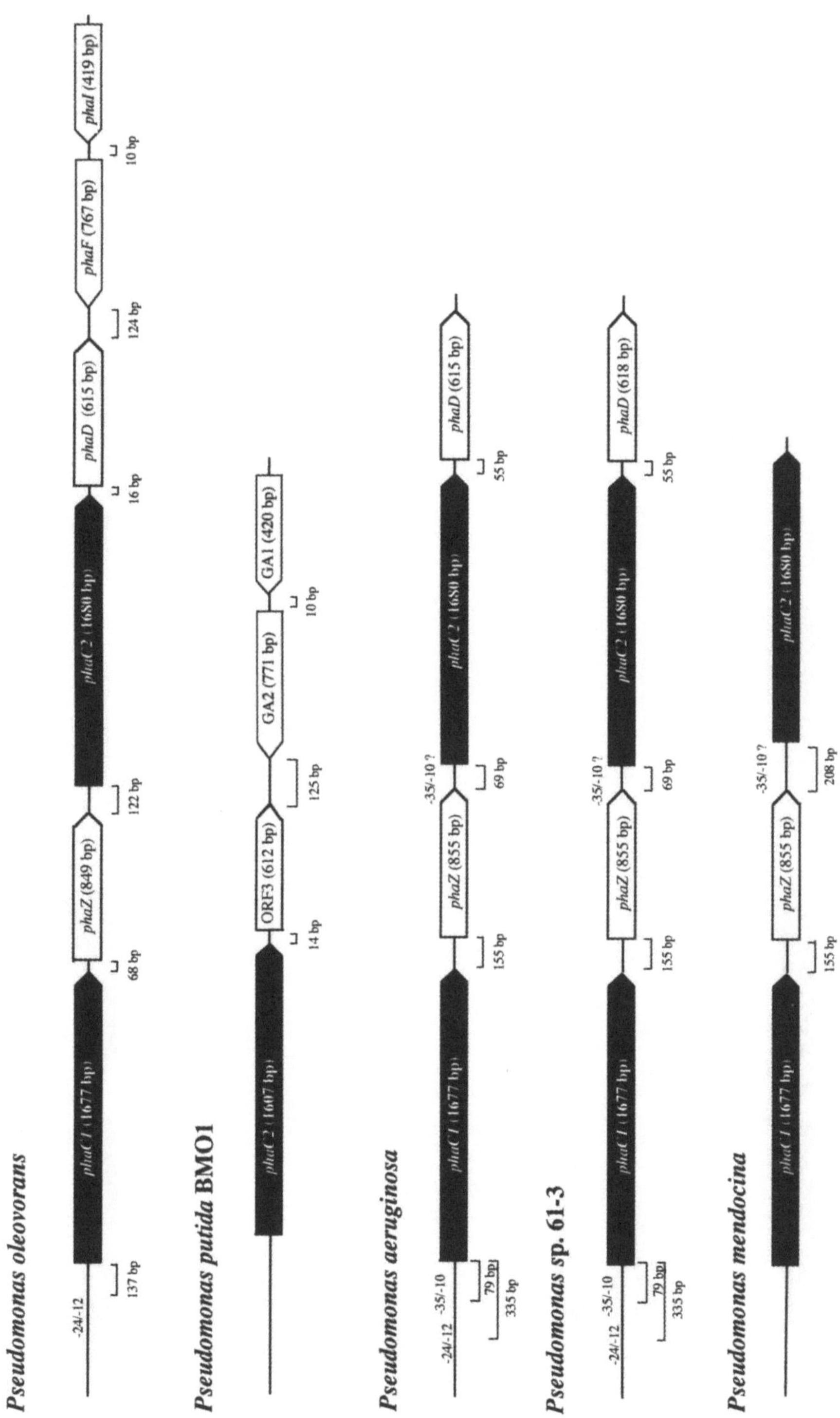

Pseudomonas oleovorans
-24/-12
137 bp
phaC1 (1677 bp)
68 bp
phaZ (849 bp)
122 bp
phaC2 (1680 bp)
16 bp
phaD (615 bp)
124 bp
phaF (767 bp)
10 bp
phaI (419 bp)
Pseudomonas putida BMO1
phaC2 (1607 bp)
14 bp
ORF3 (612 bp)
125 bp
GA2 (771 bp)
10 bp
GA1 (420 bp)
Pseudomonas aeruginosa
-24/-12
-35/-10
79 bp
335 bp
phaC1 (1677 bp)
155 bp
phaZ (855 bp)
-35/-10 ?
69 bp
phaC2 (1680 bp)
55 bp
phaD (615 bp)
Pseudomonas sp. 61-3
-24/-12
-35/-10
79 bp
335 bp
phaC1 (1677 bp)
155 bp
phaZ (855 bp)
-35/-10 ?
69 bp
phaC2 (1680 bp)
55 bp
phaD (618 bp)
Pseudomonas mendocina
phaC1 (1677 bp)
155 bp
phaZ (855 bp)
-35/-10 ?
208 bp
phaC2 (1680 bp)

Fig. 5

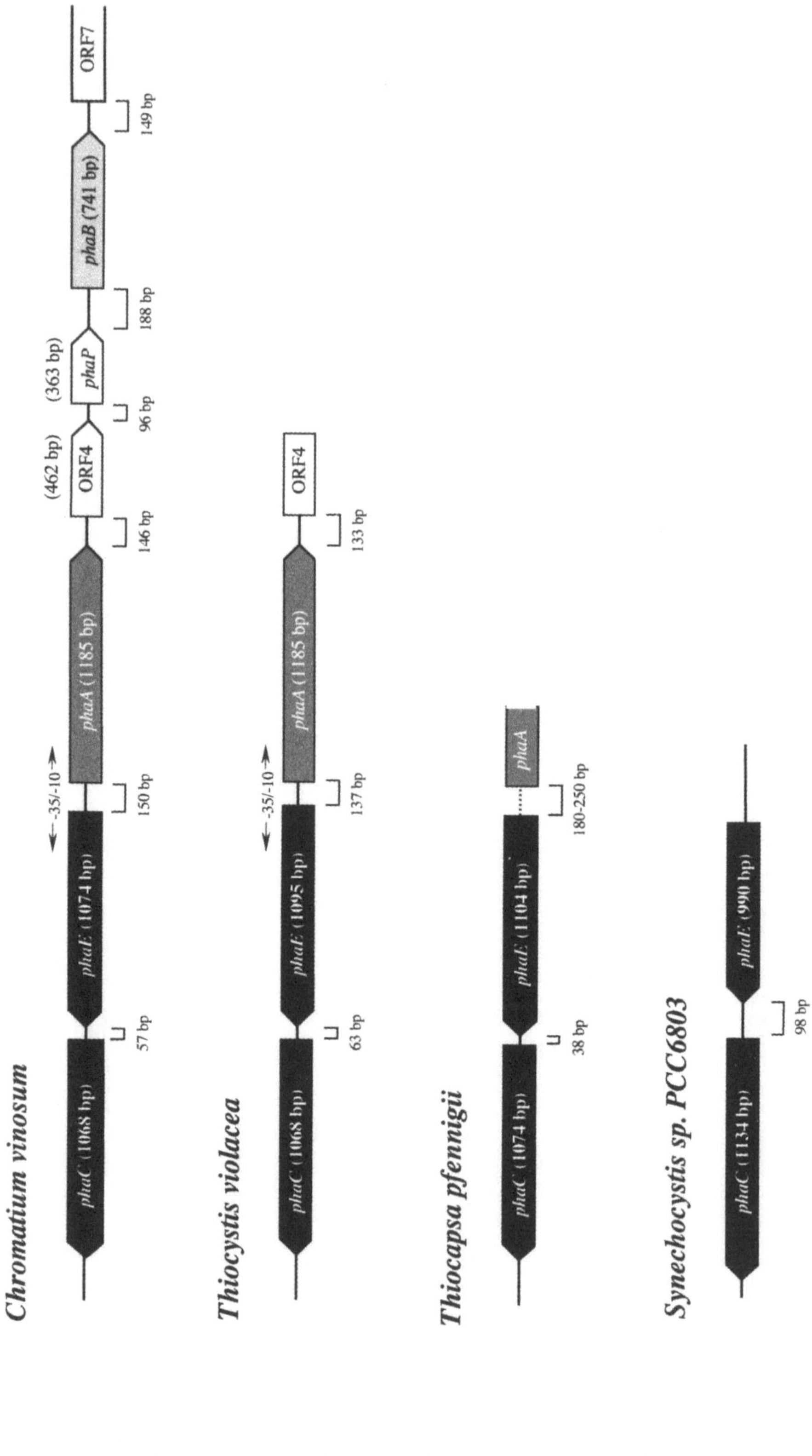
Chromatium vinosum
phaC (1068 bp)
57 bp
phaE (1074 bp)
-35/-10
150 bp
phaA (1185 bp)
146 bp
(462 bp)
ORF4
96 bp
(363 bp)
phaP
188 bp
phaB (741 bp)
149 bp
ORF7
Thiocystis violacea
phaC (1068 bp)
63 bp
phaE (1095 bp)
-35/-10
137 bp
phaA (1185 bp)
133 bp
ORF4
Thiocapsa pfennigii
phaC (1074 bp)
38 bp
phaE (1104 bp)
180-250 bp
phaA
Synechocystis sp. PCC6803
phaC (1134 bp)
98 bp
phaE (990 bp)

Fig. 6

two copies of *phaC* and that one is plasmid encoded [16]. This would be the first and so far only naturally occurring bacterium in which a gene involved in PHA biosynthesis is plasmid encoded.

4
Enzymology of the Formation of Hydroxyacyl-CoA Thioesters as Substrates for PHA Synthases

The presence of a PHA synthase alone is not sufficient to allow synthesis of PHAs. PHA biosynthesis will not occur if genes encoding enzymes required for the synthesis of hydroxyacyl-coenzyme A thioesters are absent or if the pathways constituted by these enzymes are for whatever reason not functionally active. This becomes evident, for example, when a PHA synthase gene is expressed in wild type or normal laboratory strains of *E. coli*; even if a functionally active PHA synthase is expressed, no or only traces of PHAs are accumulated.

4.1
Pathways Providing 3-Hydroxybutyryl-Coenzyme A

Starting from acetyl-CoA, two naturally occurring pathways and one genetically engineered pathway for the biosynthesis of D-(−)-3-hydroxybutyryl-CoA are known (for a review see [77]). First, the condensation of two molecules of acetyl-CoA to acetoacetyl-CoA by a *β*-ketothiolase and the subsequent reduction of acetoacetyl-CoA to D-(−)-3-hydroxybutyryl-CoA by a pyrimidine nucleotide-dependent acetoacetyl-CoA reductase, which is NADPH-dependent in most bacteria, is probably the most widespread pathway for the provision of D-(−)-3HB-CoA in bacteria [78]. Second a modified pathway was described in *Rhodospirillum rubrum*, where acetoacetyl-CoA is first reduced to L-(+)-3-hydroxybutyryl-CoA by an NADH-dependent acetoacetyl-CoA reductase, which is subsequently converted into D-(−)-3-hydroxybutyryl-CoA by stereospecific enoyl-CoA hydratases [79]. Third, a genetically modified rat fatty acid synthase and the *R. eutropha* PHA synthase constituted a novel genetically engineered pathway that was functionally expressed in *Spodoptera frugiperda* cells and involved intermediates of the fatty acid *de novo* synthesis bound to the acyl carrier protein (ACP) [80].

Fig. 6. Molecular organization of genes encoding PHA synthases of type III, which are composed of two different subunits. *PhaC*, gene encoding subunit of PHA synthase; *PhaE*, gene encoding subunit of PHA synthase; *phaA*, gene encoding *β*-ketothiolase; *phaB*, gene encoding acetoacetyl-CoA reductase; ORF, open reading frames with unknown function. See Table 1 for references. PHA synthase structural genes are shown in *black shading with white letters*, *β*-ketothiolase genes in *dark grey shading with white letters*, acetoacetyl-CoA reductase genes in *medium grey shading with black letters* and genes for phasins or other none catalytically active granule bound proteins in *light grey shading with black letters*

4.2 Pathways Providing Coenzyme A Thioesters of Other HA_{SCL}

Besides 3HB, several other HA_{SCL} are incorporated into PHAs. This includes 3-hydroxypropionate (3HP), 3-hydroxyvalerate (3HV), 4-hydroxybutyrate (4HB), 4-hydroxyvalerate (4HV), and 5-hydroxyvalerate (5HV). With only few exceptions these constituents are only incorporated into PHAs if precursor substrates are provided, and most of these polyesters cannot be synthesized from simple, structurally unrelated carbon sources.

In the simplest case the respective hydroxyalkanoic acid or a derivative of it is directly provided as carbon source and is subsequently converted into the corresponding coenzyme A thioester by a thiokinase. This was done with 3-hydroxypropionate, 4-hydroxybutyrate, and 5-chlorovalerate in order to synthesize PHA_{SCL} copolyesters containing 3HP [99], 4HB [100], or 5HV [101], respectively. In addition, other compounds can also be provided as carbon sources, such as α,ω-alkanediols, which are first oxidized to the corresponding α,ω-hydroxyalkanoic acid, then activated by a thiokinase to the corresponding α,ω-hydroxyacyl-coenzyme A thioester and are further oxidized via β-oxidation resulting in the formation of acetyl-CoA and shorter α,ω-hydroxyacyl-coenzyme A thioesters; the latter is subsequently incorporated into the polyester. From even chain-numbered α,ω-alkanediols such as 1,4-butandiol, 1,6-hexanediol, 1,8-octanediol, 1,10-decanediol, or 1,12-dodecanediol, PHA_{SCL} consisting of 4HB were obtained [99], whereas odd-numbered α,ω-alkanediols such as 1,3-propanediol, 1,5-pentanediol, 1,7-heptanediol, or 1,9-nonanediol were used to produce PHA_{SCL} containing 3HP [99]. Accordingly, various odd-numbered alkanoic acids were converted into $3HA_{SCL}$ containing PHAs. To mention only two examples, propionate or valerate can be used to produce copolyesters containing 3HV [101, 102]. Lactones are also suitable precursor substrates to produce PHA_{SCL}. From γ-butyrolactone or γ-valerolactone, PHA_{SCL} containing 4HB [99] or 4HV [103], respectively, were obtained. In addition, ketoacids can also be used as precursor substrates; levulinic acid has been intensively studied as a suitable carbon source for the production of 4HV containing PHAs [104].

Two new pathways were recently engineered *in vitro* employing various purified enzymes. One consisted of the PHA synthase from *C. vinosum* ($PhaEC_{Cv}$), the propionyl-CoA transferase from *Clostridium propionicum* (Pct) plus the acetyl-CoA synthetase from yeast and allowed ATP driven poly(3HB) biosynthesis from 3-hydroxybutyrate on a preparative scale with recycling of coenzyme A [73]. The other pathway also consisted of $PhaEC_{Cv}$, plus butyrate kinase (Buk) and phosphotransbutyrylase (Ptb) from *Clostridium acetobutylicum* also allowing ATP-driven poly(3HB) biosynthesis from 3-hydroxybutyrate [105]. Whereas in the first pathway AMP was formed from ATP, ADP was formed in the latter. The three enzymes of the latter pathway were also suitable *in vitro* to produce poly(3HB-*co*-4HB) copolyesters containing various molar fractions of 3HB and 4HB if these hydroxy fatty acids were provided as substrates. The latter pathway was also expressed from the genes cloned in pBR322 *in vivo* in recombinant strains of *E. coli*; however, the *C. vinosum* PHA synthase was replaced by the PHA synthase of *Thiocapsa pfennigii* ($PhaEC_{Tp}$) because the latter enzyme exhibited a broader substrate range [96]. The cells of the recombinant strain of *E. coli* accumulated homopolyesters of 3HB, 4HB, or 4HV when cultivated in a mineral salts medium containing glucose and the respective hydroxy fatty acids as carbon sources. Various copolyesters and terpolyesters of these hydroxy fatty acids were also obtained when two or three of these hydroxy fatty acids were, respectively, fed at equal amounts [96].

Copolyesters consisting of 3HV represent a rare example for PHA_{SCL} that can be synthesized from simple, unrelated carbon sources. Because copolyesters consisting of 3HB plus 3HV have more suitable technical properties than the poly(3HB) homopolyester, many attempts were made to obtain such copolyesters. The first approach was to feed propionic acid as a second carbon source in addition to glucose during the cultivation of *R. eutropha*. Processes based on this approach were already developed at an industrial scale by ICI and later by ZENECA to produce Biopol [102, 106]. In a second approach the metabolism of bacteria was engineered by isolating mutants or by expressing foreign genes. A spontaneous revertant to prototrophy of an isoleucine-auxotrophic mutant of *R. eutropha* accumulated poly(3HB-*co*-3HV) from various single unrelated carbon sources. The primary mutant excreted the amino acids valine, leucine, and isoleucine into the medium [107] due to a defective threonine dehydratase [108, 109]; the secondary mutant overproduced acetolactate synthase to compensate for this [109]. Valine and threonine or the respective methyl-branched 2-ketoalkanoates are degraded to propionyl-CoA or to propionyl-CoA plus acetyl-CoA; obviously these mutants were generating propionyl-CoA intracellularly due to a misregulated metabolism of these amino acids. Another strategy was applied to *E. coli*. A pathway was engineered utilizing threonine, which was converted into 2-ketobutyrate by a threonine deaminase and subsequently into propionyl-CoA by the *E. coli* pyruvate dehydrogenase multienzyme complex [110]. This pathway was also used later to establish poly(3HB-*co*-3HV) biosynthesis in transgenic plants (see below [111]). During these studies it was detected that the β-ketothiolase of *R. eutropha*, which is encoded by $phaA_{Re}$, cannot condense propionyl-CoA with acetyl-CoA. With respect to the substrate specificity $PhaA_{Re}$ is strictly a C4 (acetoacetyl coenzyme A) forming enzyme [112]. These findings were in contrast to previously published data [113] and resulted from studies of enzyme preparations which were obviously not homogenous. However, *R. eutropha* possesses at least two other β-ketothiolases [112]; one is encoded by $bktB_{Re}$ and is located a few thousand base pairs downstream of the *pha* operon in the genome (Fig. 3).

Poly(3HB-*co*-3HV) with a very high content of 3HV are also synthesized by various species belonging to the genera *Rhodococcus* or *Nocardia*. Synthesis of this copolyester occurs naturally in the wild type strains from many different substrates not structurally related to 3HV [114]. Experiments employing labeled succinate during poly(3HB-*co*-3HV) formation provided evidence that succinyl-CoA is synthesized from succinyl-CoA via the methylmalonyl-CoA pathway in *R. ruber* [115]. Employing labeled glucose in these experiments provided evidence that propionyl-CoA is formed by the carboxylation of pyruvate to oxaloacetate and subsequent conversion of the latter by the reverse citric acid cycle in *R. ruber* [116]. From the analysis of methylmalonyl-CoA mutase deficient mutants evidence was obtained that propionyl-CoA is synthesized from succinyl-CoA via methylmalonyl-CoA in *Nocardia corallina* [117].

PHAs consisting of 4HB are other examples which can be synthesized from simple, unrelated carbon sources. Recombinant strains of *E. coli* expressing *phaC*$_{Re}$ and orfZ from *Clostridium kluyveri* synthesized poly(3HB-*co*-4HB) or even poly(4HB) homopolyester if 4-hydroxybutyrate was used as carbon source [90]. When additional succinate degradation genes from *C. kluyveri* were expressed in the recombinant *E. coli*, poly(3HB-*co*-4HB) with a low content of 4HB was synthesized from glucose [118].

4.3 Pathways Providing Coenzyme A Thioesters of 3HA$_{MCL}$

Biosynthesis of poly(3HA$_{MCL}$) has been studied in great detail in pseudomonads and in recombinant strains of *E. coli* expressing PHA$_{MCL}$ synthases. Most pseudomonads *senso strictu* such as *P. oleovorans*, *P. putida*, or *P. aeruginosa* synthesize PHA$_{MCL}$ from fatty acids as well as from carbohydrates [119–121]. They are generally unable to synthesize poly(3HB) and other PHA$_{SCL}$. There are few other bacteria that incorporate 3HA$_{MCL}$ into PHAs such as, for example, *Aeromonas caviae*, which synthesizes copolyesters consisting of 3HB plus 3HHx.

Synthesis of PHA$_{MCL}$ from fatty acids such as octanoic acid or from the corresponding alkanes such as octane was first detected in *P. oleovorans* [119]. The alkanes are oxidized to the fatty acids; the latter are activated by thiokinases and then degraded via the fatty acid β-oxidation pathway. Obviously intermediates of this pathway accumulate under conditions favorable for the synthesis of PHA and are subsequently converted into substrates for the PHA synthase. Many reactions for the conversion of an intermediate of the β-oxidation cycle into R-(–)-3-hydroxyacyl-CoA were considered. These were:

1. An epimerase directly converting the L-(+) of 3-hydroxyacyl-CoA into the R-(–) stereoisomer
2. A ketoacyl-CoA reductase using 3-ketoacyl-CoA as substrate
3. An enoyl-CoA hydratase using the *trans*-2-enoyl-CoA as substrate

Which of these reactions is providing the missing link between fatty acid β-oxidation and PHA synthase remained unclear for a long time, and has not really been elucidated for the pseudomonads. In *A. caviae* an enoyl-CoA hydratase (encoded by *phaJ*) was identified; the gene is located in close proximity to the PHA synthase structural gene [81]. Two (R)-specific enoyl-CoA hydratase genes (*phaJ1* and *phaJ2*) were also recently identified by this laboratory in *P. aeruginosa* [122] indicating that pathway 3 represents the missing link between fatty acid β-oxidation and PHA synthase in wild type bacteria. Recombinant strains of *fadB* mutants of *E. coli* that express either PHA synthase of *P. aeruginosa* and that are defective in the dehydrogenase function of FadB utilize the epimerase function of FadB [91, 92]. The same pathway is probably also used in *E. coli* strains possessing wild type β-oxidation enzymes if the β-ketothiolase of this cycle is inhibited by the addition of acrylic acid [93].

Synthesis of PHA_{MCL} from glucose and other simple, non-related carbon sources occurs only in various species of the genus *Pseudomonas* and was published approximately 10 years ago [120, 121]. The main constituent of the polyesters accumulated from glucose is 3-hydroxydecanoate together with small amounts of 3-hydroxyoctanoate and 3-hydroxydocanoate. It seems to be a general feature of this group of bacteria, and only few members of the pseudomonads *senso strictu* lack this feature; one example is *P. oleovorans*. Analysis of PHA_{MCL} formed from labeled glucose and inhibitor studies done with *P. putida* showed that the fatty acid *de novo* synthesis pathway is the major pathway for the provision of the 3-hydroxyacyl moiety [123]. A 3-hydroxyacyl-acyl carrier protein-coenzyme A transferase was recently identified in *P. putida* as the missing link between fatty acid *de novo* synthesis and PHA synthase and transfers the 3-hydroxyacyl moiety from the acyl carrier protein to coenzyme A [124]. The transferase is referred to as $PhaG_{Pp}$, and the gene is not linked to the *pha* gene cluster. Mutants of *P. putida* defective in this gene accumulate only minor amounts of PHA_{MCL} [124], which are probably synthesized by two minor pathways also contributing to the PHA_{MCL} biosynthesis in the wild type [123]. On the other hand, $phaG_{Pp}$ conferred to *P. fragi* and *P. oleovorans* the capability to synthesize PHA_{MCL} from glucose [124, 125]. Highly homologous genes were recently also identified in *P. aeruginosa* [126] and *P. oleovorans* [127]. The occurrence of *phaG* in *P. oleovorans* was surprising, because this bacterium does not accumulate PHA_{MCL} from glucose or other carbon sources structurally not related to the constituent of the polyester indicating that $phaG_{Po}$ is probably a silent gene.

It is worth mentioning that metabolic engineering of *E. coli* recently provided recombinant strains which synthesized PHA_{MCL} from gluconate. For this, beside $phaC2_{Po}$ or $phaC1_{Pa}$, the thioesterase I from *E. coli* (TesA) [128] or the acyl-ACP thioesterase from *Umbellularia californica* [129], respectively, were expressed in *E. coli*. However, the amounts of PHA_{MCL} accumulated in the cells were rather low, and this artificial pathway was not very efficient.

A new pathway was recently engineered *in vitro* employing purified His_6-tagged PhaC1 and PhaC2 from *P. aeruginosa* plus a commercially available acyl-CoA synthetase allowing *in vitro* de novo PHA granules and *in vitro* synthesis of poly(3HD) from 3-hydroxydecanoate [74].

5 PHA-Negative Mutants and Mutated PHA Synthases

Mutants impaired in the synthesis and accumulation of PHAs have been obtained from conventional chemical mutagenesis or transposon mutagenesis from many bacteria. Mutant $PHB^{-}4$ from *R. eutropha* strain H16 [130] and mutant GpP104 from *P. putida* strain KT2442 [36] are most probably those mutants which were mostly used, although the genotypes of these mutants were never analyzed.

Site-directed mutants have so far only been obtained for the PHA synthases from *R. eutropha* and *C. vinosum*. These mutants were useful to identify particular amino acid residues that are essential for a catalytically active enzyme and have already described in detail in Sect. 2.4 of this review. Site-directed PHA synthase mutants that were altered with respect to the substrate specificity or the kinetic properties of a PHA synthase have so far not been identified, although most probably much effort has been made to generate such mutants.

It should be mentioned that from the PHA synthase of *R. eutropha*, truncated *phaC*$_{Re}$ genes were obtained, which were deleted for the first 108 or 300 bp. These genes encoded truncated PhaC$_{Re}$s lacking 36 or 100 amino acids from the amino terminal region, respectively. The PHA synthase lacking 36 amino acids still exhibited reasonable poly(3HB) synthase activity, and even with the PHA synthase lacking 100 amino acid low, but significant, activity was detected [131]. Although 6.1% or 16.7%, respectively, of the PhaC$_{Re}$ protein were deleted, these deletions did not cause a complete inactivation of the enzyme protein. This is consistent with the lack of highly conserved amino acids in this region of PHA synthases [53] (Fig. 2).

6
Overview of Heterologous Expression of PHA Synthases

When the structural genes for β-ketothiolase (*phaA*$_{Re}$), acetoacetyl-CoA reductase (*phaB*$_{Re}$), and PHA synthase (*phaC*$_{Re}$) constituting the PHA operon from *R. eutropha* (*phaCAB*$_{Re}$) were cloned as the first representatives of bacterial genes for a PHA biosynthesis pathway, the laboratories involved in these studies also succeeded rather easily in expressing these genes in recombinant strains of *E. coli* and established a functional active poly(3HB) biosynthesis pathway in this non-PHA accumulating bacterium. Expression of *phaCAB*$_{Re}$ in *E. coli* allowed the cells to synthesize and accumulate not only traces but large amounts contributing to more than 50% poly(3HB) of the cellular organic material and resulting in the formation of very opaque and easily recognized colonies if the cells were cultivated on mineral salts or complex media containing an excess of a carbon source such as, for example, glucose. Most probably none of the laboratories had expected that the heterologous expression of these genes in *E. coli* would proceed that easily; otherwise other cloning strategies would have most probably been applied at the very beginning. In any case, the interest in these genes was rather large, and the members of my laboratory and myself were astonished about the large number of requests for PHA biosynthesis genes which we received since then. Tables 3 and 4 show quite impressively in how many different organisms the *R. eutropha* PHA biosynthesis genes were expressed; in the examples listed, heterologous expression always yielded enzymatically active PHA synthase protein and in most cases also functional active PHA biosynthesis pathways.

6.1
Rationale for Heterologous Expression of PHA Biosynthesis Genes

There are several good reasons to express PHA biosynthesis genes in a different organism, and they are summarized in Tables 3 and 4:

1. Genes for PHA synthases or entire PHA biosynthesis pathways were identified during cloning by phenotypic complementation.

Table 3. Heterologous expression of PHA biosynthesis genes and related genes

Host organism	Foreign genes	PHA accumulated	Ref.
Aim: Identification of PHA synthase genes			
A. caviae PHA-	$phaC_{Ac}$, $phaJ_{Ac}$	Poly(3HB-*co*-3HHx) or PHA	[81]
E. coli	$phaCAB_{Re}$	Poly(3HB)	[40]
E. coli	phaC_{Sw}	Poly(3HB)	[50]
P. putida GPp104	$phaC1_{Pp}$, $phaZ_{Pp}$, $phaC2_{Pp}$	Poly($3HA_{MCL}$)	[36]
P. putida GPp104	$phaC1_{Po}$, $phaZ_{Po}$, $phaC2_{Po}$	Poly($3HA_{MCL}$)	[36]
R. eutropha PHB$^-$4	$phaCAB_{Re}$	Poly(3HB)	[40]
R. eutropha PHB$^-$4	$phaC_{Rr}$, $phaP_{Rr}$	Poly(3HB)	[45]
R. eutropha PHB$^-$4	$phaC_{Cv}$	Poly(3HB)	[23]
R. eutropha PHB$^-$4	$phaC_{RrHa}$	Poly(3HB)	[82]
R. eutropha PHB$^-$4	$phaC_{Rs}$	Poly(3HB)	[82]
R. eutropha PHB$^-$4	$phaC_{Me}$	Poly(3HB)	[27]
R. eutropha PHB$^-$4	$phaC_{Tv}$	Poly(3HB)	[51]
R. eutropha PHB$^-$4	$phaC_{Lr}$	Poly(3HB)	[51]
R. eutropha PHB$^-$4	$phaC_{As}$	Poly(3HB)	[16]
R. eutropha PHB$^-$4	$phaC_{Al}$	Poly(3HB)	[18]
R. eutropha PHB$^-$4	$phaC_{Pa}$	Poly(3HB)	[30]
R. eutropha PHB$^-$4	$phaC_{Bc}$	Poly(3HB)	[22]
R. eutropha PHB$^-$4	$phaC_{RrATCC}$	Poly(3HB)	[46]
Aim: Overexpression of PHA synthesis genes for purification			
Trichoplusia ni	$phaC_{Re}$	PHA synthase protein	[83]
Aim: Large scale production of PHAs			
E. coli	$phaCAB_{Re}$	Poly(3HB)	[84]
E. coli	$phaCAB_{Re}$	Poly(3HB-*co*-3HV)	[85]
E. coli	$phaC_{Re}$ + ORFZ_{Ck}	Poly(4HB)	[86]
Aim: Novel PHAs			
P. putida GPp104	$phaC_{Tp}$	Poly(3HB-*co*-3HHx-*co*-3HO)	[26]
P. putida GPp104	$phaC_{Tp}$	Poly(3HB-*co*-3HHx-*co*-4HHx)	[87]
P. putida GPp104	$phaC_{Tp}$	PHAs containing 5HHx, 4HHp, 4HO	[88]
R. eutropha PHB$^-$4	$phaC_{Tp}$	Poly(3HB-*co*-3HHx-*co*-4HHx)	[87]
P. oleovorans	$phaCAB_{Re}$	Poly(3HB) and poly (3HO-*co*-3HHx)	[89]
R. eutropha PHB$^-$4	$phaCB_{Re}$	Poly(3HB-*co*-3HHx)	[60]
K. aerogenes	$phaCB_{Re}$	Poly(3HB-*co*-3HHx)	[60]
P. putida GPp104	$phaCB_{Re}$	Poly(3HB-*co*-3HHx)	[60]
E. coli (*fadB*)	$phaCAB_{Re}$	Poly(3HB-*co*-3HD-*co*-3HDD)	[61]
R. eutropha PHB$^-$4	$phaCA_{Cviol}$	Poly(3HB-co-3HHx), poly(3HB-*co*-3HV-*co*-3HHp)	[24]
K. aerogenes	$phaCA_{Cviol}$	Poly(3HB-co-3HHx)	[24]
Aim: Metabolic engineering			
E. coli	$phaC_{Re}$ + ORFZ_{Ck}	Poly(4HB)	[90]
E. coli (*fadB*)	$phaC1_{Pa}$	various poly($3HA_{MCL}$)	[91]
E. coli (*fadB*)	$phaC1_{Pa}$ and/or $phaC2_{Pa}$	various poly($3HA_{MCL}$)	[92]
E. coli	$phaC1_{Pa}$ and/or $phaC2_{Pa}$	various poly($3HA_{MCL}$)	[93]
E. coli	$phaC_{Ac}$ + $fabG_{Ec}$	Poly(3HB-*co*-3HHx)	[94]
E. coli	$phaC_{Ps}$ + $fabG_{Ec}$	Poly(3HHx-*co*-3HO-*co*-3HD)	[94]
E. coli	$phaC_{Ac}$ + $fabD_{Ps}$	Poly(3HB)	[95]

Table 3 (continued)

Host organism	Foreign genes	PHA accumulated	Ref.
E. coli	$phaC_{Ac} + fabD_{Ec}$	Poly(3HB)	[95]
E. coli	$phaC_{Ac} + fabH_{Ec}$	Poly(3HB)	[95]
E. coli	$phaEC_{Tp} + buk_{Ca} + ptb_{Ca}$	Poly(3HB-*co*-4HB), poly (3HB-*co*-4HV), poly(3HB-*co*-4HB-*co*-4HV)	[96]
R. eutropha JMP222	pha_{Re}	Poly(4HB)	[97]
R. eutropha H16–4HV-	pha_{Re}	Poly(3HB-*co*-3HV-*co*-4HV)	[98]

Table 4. Overview of successful heterologous expression of the *R. eutropha* PHA synthase in various organisms

Organisms	Refs.
Bacteria	
Escherichia coli	[40]
Klebsiella aerogenes	[60, 132]
Klebsiella oxytoca	[132]
Mycoplana rubra	[133]
Pseudomonas oleovorans	[134]
Synechococcus sp. PCC7002	[135, 136]
and many further examples	
Eukaryotic microorganisms	
Saccharomyces cerevisiae	[137]
Animal cells	
Spodoptera frugiperda	[80]
Trichoplusia ni	[83]
Plants	
Arabidopsis thaliana	[138]
Brassica napus	[139]
Gossypium hirsutum	[140]
Nicotiana tabacum[a]	[141]
Solanum tuberosum	[146]
Zea mays	[142]

[a] $phaB_{Re}$ and $phaC_{Ac}$ were expressed.

2. Expression in *E. coli* will make PHA synthase proteins more readily accessible for purification and therefore available in large quantities for further studies such as detailed biochemical analysis, generation of antibodies and *in vitro* PHA biosynthesis.
3. Heterologous expression allows one to prove the functionality of single enzymes or of entire pathways.
4. Genetic engineering of PHA synthase proteins in *E. coli* in combination with a suitable viable-colony staining method for the screening of PHA content [143] will make many methodological tools applicable to PHA biosynthesis genes and proteins and will allow studies on structure to function relation-

ships. Such studies might also reveal for example mutated PHA synthases exhibiting altered properties for example regarding the substrate range.

5. Metabolic engineering may reveal novel pathways for synthesis and production of PHAs.
6. Other organisms may, for several reasons, be much more suitable for the technical production of PHAs than the original host of the gene(s). Therefore, organisms which grow faster accumulate PHAs at a higher rate, grow to higher cell densities, or allow a higher yield per volume and time, and organisms using cheaper and more readily available carbon sources or CO_2, or which are more suitable for downstream processes or organisms having a different safety assessment, may become accessible for PHA production.

6.2
Expression in Bacteria

There are an overwhelming number of studies which successfully demonstrated heterologous expression of PHA synthesis genes in bacteria; it will therefore not be possible to mention them all. Establishment of functional active PHA biosynthesis pathways in *E. coli* requires not only a PHA synthase but also enzymes that allow the conversion of metabolites, which derive from the provided carbon source, into the *R* isomers of hydroxyacyl-coenzyme A thioesters that are used as a substrate by the respective PHA synthase. Otherwise, no or only marginal amounts of PHAs are accumulated.

PHA-negative mutants of *R. eutropha* and *P. putida*, which are defective in PHA synthases such as the mutants $PHB^{-}4$ [130] and GPp104 [36], respectively, were rather useful in the past, and they were often used for physiological studies and cloning experiments. These mutants require only an active PHA synthase and no further genes for phenotypic complementation to a PHA-accumulating recombinant strain, and such strains are easily recognized on agar plates by the enhanced opacity of the colonies [130], by staining the colonies with the lipophilic dye Sudan black [130] or by enhanced fluorescence of colonies in the presence of Nile red [143].

Much effort has been made to use recombinant strains of *E. coli*, which expressed the PHA biosynthesis genes from *R. eutropha*, or other combinations of a suitable host organism and foreign PHA biosynthesis genes for high cell density cultivation and production of bulk amounts of PHAs. The interested reader is referred to a different chapter of this book [13] and the references cited therein in which these systems are described in detail.

6.3
Expression in Eukaryotic Microorganisms

From this group of microorganisms, so far only *Saccharomyces cerevisiae* has been transformed to a poly(3HB) accumulating organism by expressing solely $phaC_{Re}$ in the cytoplasm ([137], Table 4). In contrast to *E. coli* and plants, which synthesize poly(3HB) only if a β-ketothiolase and an acetoacetyl-CoA reductase are also expressed, expression of foreign genes encoding these enzymes was not required in yeast for poly(3HB) synthesis. The authors said that these steps are being catalyzed by the native Erg10 and Fox2 proteins, respectively, which are enzymes involved in the β-oxidation and are exhibiting β-ketothiolase or acetoacetyl-CoA reductase activity [137]. The results from experiments studying the effect of cerulenin, which is an inhibitor of the condensing enzyme domain of the fatty acid synthetase, and compactin, which is an inhibitor of the 3-hydroxy-3-methylglutaryl-CoA reductase, on poly(3HB) accumulation in the recombinant yeast cells made it unlikely that the fatty acid *de novo* synthesis pathway or the mevalonate biosynthesis pathways are involved in the provision of D-(−)-3HB-CoA [137].

6.4 Expression in Animal Cells

There are two examples of the successful expression of bacterial PHA biosynthesis genes in animal cells (Table 4). Cells of the insect *Spodoptera frugiperda* expressing a dehydratase-domain mutant rat fatty acid synthase cDNA and $PhaC_{Re}$ provided R-(−)-3HBCoA as a premature termination product of fatty acid *de novo* synthesis. Since R-(−)-3HBCoA was subsequently polymerized to poly(3HB), an alternative pathway for the biosynthesis of this polyester was established [80]. $PhaC_{Re}$ was also expressed in *Trichoplusia ni* cells using the baculovirus *Autographa californica* nuclear polyhedrosis virus system. In the transgenic cells the PHA synthase from *R. eutropha* constituted approximately 50% of the total protein. This system was used to purify large amounts of highly pure PHA synthase protein in only one single liquid chromatography step for further biochemical characterization [83].

6.5 Expression in Plants

Since expression of bacterial PHA biosynthesis genes and production of PHAs by transgenic plants will be the subject of a separate chapter of this book [14], only a short summary will be provided here. Plants in which bacterial PHA biosynthesis genes have been successfully expressed, resulting in PHA biosynthesis, include *Arabidopsis thaliana* [138, 144, 145], *Brassica napus* (canola) [139], *Gossypium hirsutum* (cotton) [140], *Nicotiana tabacum* (tobacco) [141], *Solanum tuberosum* (potato) [146], and *Zea mays* (corn) [142], and are listed in Table 4. However, so far the PHAs accumulated in the transgenic plants contributed to only a few percent of the cellular dry matter in most studies. Biosynthesis pathways for poly(3HB), poly(3HB-*co*-3HV), and poly($3HA_{MCL}$) have been established in various compartments of plants cells and tissues, and different expression systems have been used relying on the PHA biosynthesis genes from *R. eutropha* and *P. aeruginosa*. To establish poly(3HB) biosynthesis in plants, $phaC_{Re}$ and $phaB_{Re}$ alone or together with $phaA_{Re}$ were used [138, 140, 142, 144]. For poly(3HB) synthesis, $phaA_{Re}$ could also be replaced by $bktB_{Re}$, which encodes a second β-ketothiolase in *R. eutropha* [112, 139]. The latter provides a much greater flexibility for the type of polyester that can be synthesized, since $BktB_{Re}$ uses, in contrast to $PhaA_{Re}$, also propionyl-CoA as substrate. If *bktB*$_{Re}$, *phaB*$_{Re}$, and *phaC*$_{Re}$ were coexpressed together with the gene from a feedback inhibition-impaired threonine deaminase mutant of *E. coli* (*ilvA*466), transgenic plants *of A. thaliana* and *B. napus* synthesized poly(3HB-*co*-3HV) with 3HV contributing up to 17 mol% of the constituents [111]. Transgenic plants of *A. thaliana* expressing *phaC1*$_{Pa}$ in the peroxisomes, synthesized PHA consisting of various $3HA_{MCL}$s [145]. The accumulation of PHA_{MCL} increased if the caproyl-ACP thioesterase from *Cuphea lanceolata* was also expressed in these plants [147].

7 Structure and Formation of PHA Granules

During the last decade our knowledge of the structure of PHA granules has been significantly increased. In any naturally occurring PHA accumulating bacterium investigated so far, the granules are covered by a surface layer consisting mainly of specific proteins and phospholipids. These surface layers cover the cores of the granules that consist of PHA in the amorphous state; a granule with a diameter of 350 nm will, for example, contain approximately 40,000 poly(3HB) molecules if the molecular weight is approximately 3×10^6 g/mol [148]. It became evident that PHA accumulating wild types of bacteria are principally possessing specific PHA granule bound proteins. At present, two different models for the structure of PHA granules occurring in bacteria accumulating PHA_{SCL} with poly(3HB) and in bacteria accumulating PHA_{MCL} such as poly(3HO) as the preferred storage polyesters were proposed, respectively.

The model for PHA_{SCL} accumulating bacteria predicts a monolayer consisting of proteins and phospholipids as revealed by electron microscopic analysis of thin sections or freeze-fractures showing one electron-dense borderline after staining with osmium tetroxide and uranyl acetate [56, 149]. The thickness of this surface layer was estimated to be approximately 3–4 nm [150], which would be approximately half of the diameter of the cytoplasmic unit membrane consisting of two different electron-dense layers according to two layers of phospholipids. Chemical analysis of the PHA granules isolated from *Bacillus megaterium* showed that they consisted of approximately 98% poly(3HB), 0.45% lipids, and 2% proteins [66]. In addition to the PHA synthase and PHA depolymerase proteins that are bound to the PHA granules, one or a few additional proteins, which contribute to the major fraction of granule-associated proteins, are bound to the surface of the granules. These are small, amphiphilic proteins mostly exhibiting molecular weights between 10 kDa and 30 kDa. Most likely these proteins, which were referred to as phasins in analogy to oleosin proteins at the surface of lipid inclusions in plant cells, have a structural function [148] homologous to that of oleosins occurring in oil seed bodies of plant cells (for a review see [151]). Most detailed studies were done on the phasins of *R. ruber* ($PhaP_{Rr}$) [152], and *R. eutropha* ($PhaP_{Re}$) [153, 154].

In contrast to the analogous oleosins, which are phylogenetically related and share a common ancestor, phasins are functionally but not phylogenetically related [154]. For the binding of the phasins to the PHA granules different models were proposed. In the *R. ruber* phasins two stretches of hydrophobic/amphiphilic amino acids (HD1 and HD2) were identified, which were essential for the binding of the protein to the surface of the granules [152]. It was proposed that HD1 and HD2 are domains spanning the membrane provided by the phospholipids. However, the *R. ruber* phasin is so far the only phasin that contains two of these membrane spanning domains, and also in the revised phasin sequence of *R. eutropha* two such regions were absent [154]. But since all phasins contain at least one markedly hydrophobic region, an alternative model for the binding of phasins to the granule surface was proposed postulating a direct binding of this region to PHA that is responsible for the anchoring of the phasin to the granules [154].

Phasins provide an amphiphilic layer between the hydrophobic core consisting of PHAs and the mostly hydrophilic components of the cytoplasm. Mutants of *R. eutropha*, which lacked the phasin protein, accumulated less PHA than the wild type and formed only one single large granule, whereas several smaller granules occur in the wild type cells. Recombinant strains of *E. coli*, which expressed a poly(3HB) biosynthesis pathway in the absence of a phasin protein or in the presence of a truncated phasin protein derived from $PhaP_{Rr}$, that is lacking the region probably responsible for the binding of the phasin to the granule surface, formed much larger but less PHA granules than recombinant strains of *E. coli* expressing functional phasin proteins $PhaP_{Rr}$ or $PhaP_{Re}$.

Whereas in *R. eutropha* the structural gene for the phasin ($phaP_{Re}$) was not related to the PHA biosynthesis genes in the genome (Fig. 3), the *R. ruber* phasin gene ($phaP_{Rr}$) mapped downstream from a gene adjacent to the PHA synthase gene (Fig. 4). In *Chromatium vinosum* the phasin gene is located between the structural genes for the β-ketothiolase ($phaA_{Cv}$) and the acetoacetyl-CoA reductase ($phaB_{Cv}$) as shown in Fig. 6 [23, 155, 156]. Also, in *Acinetobacter* sp. strain RA3849 and *Bacillus megaterium*, the phasin genes mapped in direct or close neighborhood of the PHA synthase structural genes, respectively [16, 21] as shown in Fig. 3. Yamane and coworkers [157] identified the phasin structural gene of *Paracoccus denitrificans* ($phaP_{Pd}$) downstream of $phaC_{Pd}$. The pseudomonads represent further examples of genes of phasins and granule-associated proteins that are linked to the PHA synthase genes in the genomes (Fig. 5, see also below). Coexpression of $phaP_{Pd}$ and the PHA biosynthesis genes in *E. coli* yielded an approximately 65% higher poly(3HB) content of the cells with a higher number of granules, whereas the size of the granules was smaller and the molecular weight of the polyester was lower compared to recombinant strains of *E. coli* expressing the PHA biosynthesis genes in the absence of $phaP_{Pd}$ [157].

In summary, these studies indicated an important, but not essential function of phasins for the metabolism of PHA. If intact phasins are lacking in PHA accumulating cells, the PHA granules tend to coalesce. This has, for example, significant effects on the surface area of PHA granules and therefore on the area that is available for the binding of the catalytically active proteins such as PHA synthases and PHA depolymerases, which catalyze the polymerization or depolymerization of PHAs, respectively. Whether the presence or absence of phasins at the PHA granule surface also affects the functionality of these enzymes or even contributes to the regulation of either process, is not yet known.

In contrast to this, a much more complex model of the surface of PHA granules was proposed for bacteria accumulating PHA_{MCL}. Electron microscopical studies suggested that the surface consists of two distinct crystalline protein lattice layers which are presumably separated by a phospholipid layer [76]. These studies also revealed the presence of two major proteins of 20 kDa and 43 kDa as grana-associated proteins in *P. oleovorans* [76] that were also found in our laboratory [148]. The larger protein was localized in the outer protein layer [76]. In *P. aeruginosa* we identified proteins of 15 kDa and 41 kDa as the major grana-associated proteins [148].

Molecular studies revealed further interesting information. In *P. oleovorans* two major grana-associated proteins of 18 kDa (PhaI) and 36 kDa (PhaF) were identified [158]. In *P. putida* strain BMO1 two major grana-associated proteins of 18 kDa (GA1) and 43 kDa (GA2) were also identified [35]. In both pseudomonads these proteins were encoded by two open reading frames located on the opposite strand and downstream of a gene homologous to *phaD* of *P. aeruginosa*, which is located immediately downstream of phaC2 [35, 158, 159]. The apparent molecular weights of GA1 and GA2 as revealed by SDS polyacrylamide gel electrophoresis were approximately 17% and 67% higher, respectively, than the molecular weights calculated from the nucleotide sequences of the corresponding genes. The aberrant migration behavior of GA2 was explained by the occurrence of eight repeats of a sequence of nine amino acids (AAAKPAAKT) located close to the C-terminus of the protein [35].

In *Bacillus megaterium* a different type of granule-associated protein ($PhaP_{Bm}$) was identified [21]. The gene of this protein was mapped together with other *pha* genes close to *$phaC_{Bm}$*. According to the primary structure, $PhaP_{Bm}$ was an extremely hydrophilic protein, thus being clearly distinguished from the phasins. It was speculated that $PhaP_{Bm}$ is a storage protein and that PHA granules are a source of amino acids [21].

7.1
Occurrence of Two Types of Granules in one Cell

When the *$phaCAB_{Re}$* operon from *R. eutropha* was cloned, and when it became evident that the PHA_{SCL} synthase from *R. eutropha* and the PHA_{MCL} synthase from *P. oleovorans* represent obviously two different classes of PHA synthases with respect to the substrate range, it was decided to express *$phaCAB_{Re}$* in *P. oleovorans*, to cultivate the cells under conditions conducive to the accumulation of PHAs on octanoic acid as carbon source, and to analyze the accumulated PHAs. The results were quite clear: the cells of the recombinant *P. oleovorans* strain accumulated a blend of two different polymers, poly(3HB) and poly (3HO-*co*-3HHx), just as the wild types of *R. eutropha* and *P. oleovorans* would accumulate, respectively; a copolyester consisting of 3HB plus 3HO was not accumulated [89]. According to the mechanism of PHA synthases, which requires that the growing PHA chain remains covalently bound to the enzyme until its synthesis is terminated, it is, in retrospect not surprising that a blend rather than a copolyester was synthesized. A copolyester of, e.g., 3HB, 3HHx, and 3HO can only be synthesized if the PHA synthase exhibits a broad substrate range and accepts the coenzyme A thioesters of either hydroxyalkanoic acid as substrates. Such an enzyme is represented by the PHA synthase from *T. pfennigii*, which exactly synthesizes such a copolyester (see below) [26]. Another such enzyme is the PHA synthase from *A. caviae* [59] as already mentioned in Sect. 2.2 of this review. Surprisingly, both types of PHAs synthesized by the recombinant strain of *P. oleovorans* did not occur in the same granules but were deposited in separate granules as revealed by separation of the granules in sucrose gradients and freeze-fracture electron microscopy [160].

It was recently found that *Burkholderia* sp. strain IPT77B simultaneously accumulated the two homopolyesters poly(3HB) and poly(3-hydroxy-4-pentenoic acid), poly(3HPE), during cultivation on sucrose [161], rather than a copolyester consisting of 3HB plus 3HPE as described before [162]. Mainly due to the relatively low poly(3HPE) content of the cells in comparison to the high poly(3HB) content, it could not be found whether or not these two homopolyesters were separated in two different granules. However, evidence was obtained that this bacterium possesses two different PHA synthases. One PHA synthase (PhaC1) has been already cloned and characterized at a molecular level [22]. Mutants, which contained an inactive PhaC1, still expressed some poly(3HB) synthase activity and accumulated small amounts of PHA [163].

Separation of PHA_{SCL} and PHA_{MCL} into two separate granules within the same cell often seems to occur if two different PHA synthases with non-overlapping substrate ranges are present in the cell. Obviously, each type of PHA synthase participates separately in the initiation of micelle formation of nascent PHA granules, because only PHA molecules of similar structure, which remain bound to the PHA synthase protein, can contribute to the formation of a micelle. When the granules become larger, further PHA synthase molecules can only bind to the surface of the granules if they represent the same type of PHA synthase.

7.2
Occurrence of Layered PHA Granules

P. oleovorans synthesized, during cultivation on a mixture of 5-phenylvaleric acid and nonanoic acid, a homopolyester of poly(3-hydroxy-5-phenylvaleric acid), poly(3HPV), and a copolyester of 3-hydroxynonanoic acid and some related $3HA_{MCL}$, poly(3HN-*co*-$3HA_{MCL}$) [164]; poly(3HPV) was also synthesized if 5-phenylvaleric acid was provided as sole carbon source. When *P. oleovorans* was fed with 5-phenylvaleric acid plus nonanoic acid as carbon sources, layered PHA granules occurred in the cells consisting of a core of poly(3HN-*co*-$3HA_{MCL}$) that was surrounded by a shell of poly(3HPV), which is synthesized upon exhaustion of nonanoic acid as revealed by transmission electron microscopy imaging of ruthenium tetroxide-stained thin sections obtained from these cells [165]. By controlling the substrate available to *R. eutropha*, layered PHA granules were obtained when the cells were cultivated on a mixture of excess fructose and limited quantities of valeric acid. Again, transmission electron microscopy imaging of ruthenium tetroxide-stained thin sections obtained from these cells and also differential scanning calorimetry revealed conclusive evidence of layered granules consisting of a core of poly(3HB-*co*-3HV) that was surrounded by a shell of poly(3HB), which is synthesized upon exhaustion of valeric acid [166]. That only 30–40% of the granules were layered, was explained by the occurrence of granule coalescence and by the fact that the formation of the granules was not initiated at the same time.

8
Regulation of PHA Formation at the Molecular Level

Biosynthesis pathways and the accumulation of the products of these pathways may be regulated in quite different ways. A simple although not a very efficient strategy is to express all necessary enzymes constitutively. The enzymes are always present in a functional active state at a more or less constant specific activity, and the availability of the first metabolite at the beginning of the respective PHA biosynthesis pathway triggers PHA synthesis and subsequent accumulation of the polyester. This allows the cells to react immediately on changes in the environment and to proceed with the synthesis of PHA as a storage compound. Modifications of this theme may be achieved if the activity of one of the enzymes, which is preferentially the first enzyme of the pathway, is regulated at the protein level. Such a situation seems to occur in *R. eutropha* with the *phaA* product encoding an allosteric β-ketothiolase that is inhibited by coenzyme A [113, 167]. However, the data from these two studies have to be carefully analyzed because the preparations of the enzyme were obviously not homogenous as revealed recently [112]. The relevance of the inhibition of the β-ketothiolase by coenzyme for PHA accumulation was discussed in detail in a previous review [168].

The PHA synthase is a constituent enzyme and is soluble in the cytoplasm in non-PHA accumulating cells. During synthesis and accumulation of PHA, the enzymes becomes particulate and is bound to the granules [169]. Whether this occurs in general in all PHA accumulating bacteria or whether there are also examples for a regulation at a transcriptional or translational level is hitherto unknown, mainly due to the lack of detailed studies. The reason for this is that, in the past, most laboratories active in the PHA area focussed on other aspects such as identification of novel polyesters, analysis of PHA biosynthesis routes, cloning of the genes, or metabolic engineering of PHA biosynthesis pathways.

In addition to the aspects mentioned above, the extent of PHA accumulation might depend on quite different features. The only physiological inhibitor of PHA synthase that has been identified is coenzyme A as pointed out in an earlier section. Since the concentration of coenzyme A in the cytoplasma will probably never rise to very high levels, it may be questionable whether this inhibition is physiologically relevant at all. In our opinion, the availability of a hydroxyacyl coenzyme A thioester provided by the biosynthesis pathway is most important to initiate PHA biosynthesis (compare also [8] – this book).

8.1
Competition Between PHA Synthesis and Degradation

Practically nothing is known about how PHA synthases and intracellular PHA depolymerases, which are both bound to the granules, interact and whether or not the phasins somehow mediate a putative interaction between these two counteracting enzymes. Some contrary evidence was obtained as to whether or not PHA synthesis and degradation can occur in parallel [114, 170]. If this occurs, a futile cycle would be generated. Some evidence was also obtained that proteins exhibiting significant homologies to the carbohydrate:phosphoenolpyruvate:phosphotransferase system are relevant for PHA accumulation in at least two different bacteria [171, 172]. In *R. eutropha phaH* and *phaI*, identified as putatively occurring in one single transcriptional unit, the translational products showed 34.1% and 38.9% sequence homology with PtsH and PtsI from *E. coli*, respectively, and also significant homologies to the corresponding proteins from several other bacteria. Since the only detected phenotype was less accumulation of PHA, and since the cells showed normal growth on any tested carbon source but degraded the accumulated PHA at a significantly higher rate, it was speculated whether in these PHA-leaky mutants the regulation of PHA degradation was altered, resulting in PHA degradation concomitantly with PHA synthesis. Interestingly, Espin and coworkers isolated a mutant from *Azotobacter vinelandii* that was defective in a gene homologous to *ptsP* from *E. coli* and accumulated 90% less PHA than the wild type. This mutant was partially complemented upon transfer of the *R. eutropha phaHI* genes [173].

8.2 Are PHA Synthesis and Accumulation Affected by Global Regulators?

A detailed analysis of the PHA gene loci from more than 30 different bacteria revealed in six PHA_{SCL} accumulating bacteria the presence of open reading frames of approximately 150–200 bp, the putative translational products of which shared significant sequence homologies to each other but not to other proteins from data banks with one exception. Such ORFs occurred in *R. eutropha* [112], *Burkholderia* sp. DSM 9243 ([22], 84% similarity of the putative translational product of the ORF in *R. eutropha*), *Thiocystis violacea* ([51], 76%), *Chromatium vinosum* ([23], 76%), *Sinorhizobium meliloti* ([64], 75%), and *Paracoccus denitrificans* ([157], 57%). These ORFs were, in the mentioned bacteria, closely related to the PHA loci or even integral parts of the PHA locus (Fig. 3, 4 and 6), and they were always proceeded by a motif perfectly matching the σs^{54} promoter binding sequence. The only other protein to which the translational products showed sequence homology (73% in the case of the *R. eutropha* protein) was a hypothetical, sterol inducible protein encoded by *stcC* of *Comamonas testosteroni* [174].

Knowledge of the function of these genes is rather preliminary, but there are some indications that these proteins affect PHA synthesis and that they might have a regulatory function. These highly homologous genes are certainly not accidentally located in these regions of the genomes, and they are most likely related to PHA metabolism. The respective gene ($phaR_{Pd}$) is in the *P. denitrificans* genome localized downstream of $phaP_{Pd}$, and in *E. coli* it seems to inhibit the expression of the latter. In *Sinorhizobium meliloti* the respective gene (ORF1) is located upstream and on the opposite strand of the gene for the β-ketothiolase [64]. If this gene was disrupted by mutagenesis, the mutants accumulated less poly(3HB) and glycogen and produced, instead of these storage compounds, more exopolysaccharides [175].

Therefore, some genes seem to exist in PHA accumulating bacteria, which regulate PHA synthesis and accumulation not simply by controlling the expression of PHA biosynthesis or phasin genes but by exerting regulation at a higher rank. These genes might affect the flow of carbon in the cells and might have an important function during the shift of the cells from the growth phase to the starvation phase. However, these studies are just beginning, and a clear answer to the question raised in the heading of this section cannot be provided, yet.

References

1. Steinbüchel A (1991) In: Byrom D (ed) Biomaterials. Macmillan, Basingstoke, pp 123–213
2. Steinbüchel A, Valentin HE (1995) FEMS Microbiol Lett 128:219
3. Kim YB, Lenz RW (2000) Adv Biochem Eng Biotechnol (this volume)
4. Holler E, Angerer B, Achhammer G, Miller S, Windisch C (1992) FEMS Microbiol Rev 103:109
5. Liu S, Steinbüchel A (1996) Appl Microbiol Biotechnol 46:273
6. Kolattukudy PE (2000) Adv Biochem Eng Biotechnol (this volume)
7. Anderson AJ, Dawes EA (1990) Microbiol Rev 54:450
8. Babel W, Ackermann JU, Breuer U (2000) Adv Biochem Eng Biotechnol (this volume)
9. Jendrossek D, Schirmer A, Schlegel HG (1996) Appl Microbiol Biotechnol 46:451
10. Jendrossek D (2000) Adv Biochem Eng Biotechnol (this volume)
11. Van der Walle GAM, De Koning GJM, Weusthuis RA, Eggink G (2000) Adv Biochem Eng Biotechnol (this volume)
12. Kessler B, Weusthuis R, Witholt B, Eggink G (2000) Adv Biochem Eng Biotechnol (this volume)
13. Lee SY, Choi JI (2000) Adv Biochem Eng Biotechnol (this volume)
14. Poirier Y (2000) Adv Biochem Eng Biotechnol (this volume)
15. Kobayashi S, Uyama H (2000) Adv Biochem Eng Biotechnol (this volume)
16. Schembri MA, Bayly RC, Davies JK (1994) FEMS Microbiol Lett 118:145
17. Fukui T, Doi Y (1997) J Bacteriol 179:4821
18. Choi J-I, Lee SY, Han K (1998) Appl Environ Microbiol 64:4897
19. Lee SP, Do V, Huisman GW, Peoples OP (1996) Direct submission Accession No U78047
20. Mandon K, Michel-Reydellet N, Encarnacion S, Kaminski PA, Leija A, Cevallos MA, Elmerich C, Mora J (1998) J Bacteriol 180:5070
21. McCool GJ, Cannon MC (1999) J Bacteriol 181:585
22. Rodrigues MFA, Valentin HE, Berger PA, Tran M, Asrar J, Gruys KJ, Steinbüchel A (2000) Appl Microbiol Biotechnol 53:453
23. Liebergesell M, Steinbüchel A (1992) Eur J Biochem 209:135
24. Kolibachuk D, Miller A, Dennis D (1999) Appl Environ Microbiol 65:3561
25. Sudesh K, Fukui T, Doi Y (1998) Appl Environ Microbiol 64:3437
26. Liebergesell M, Mayer F, Steinbüchel A (1993) Appl Microbiol Biotechnol 40:292
27. Valentin HE, Steinbüchel A (1993) Appl Microbiol Biotechnol 39:309
28. Hall B, Baldwin J, Rhie HG, Dennis D (1997) Direct submission Accession No AFO19964
29. Ueda S, Yabutani T, Maehara A, Yamane T (1996) J Bacteriol 178:774
30. Umeda F, Kitano Y, Murakami Y, Yagi K, Miura Y, Mizoguchi T (1998) Appl Biochem Biotechnol 70:341
31. Timm A, Steinbüchel A (1992) Eur J Biochem 209:15
32. Timm A, Wiese S, Steinbüchel A (1994) Appl Microbiol Biotechnol 40:669
33. Dennis D, unpublished data

34. Hein S, Steinbüchel A, unpublished results
35. Valentin HE, Stuart ES, Fuller RC, Lenz RW, Dennis D (1998) J Biotechnol 64:145
36. Huisman GW, Wonink E, Meinma R, Kazemier B, Perpstra P, Witholt B (1991) J Biol Chem 266:2191
37. Garcia B, Olivera ER, Minambres B, Fernandez-Valverde M, Canedo LM, Prieto MA, Garcia JL, Martinez M, Luengo JM (1999) J Biol Chem 274:29,228
38. Matsusaki H, Manji S, Taguchi K, Kato M, Fukui T, Doi Y (1998) J Bacteriol 180:6459
39. Ja Shin et al., unpublished results
40. Schubert P, Steinbüchel A, Schlegel HG (1988) J Bacteriol 170:5837
41. Peoples OP, Sinskey AJ (1989) J Biol Chem 264:15,298
42. Cavellos MA, Encarnacion S, Leija A, Mora Y, Mora J (1996) J Bacteriol 178:1646
43. Kranz RG, Gabbert KK, Locke TA, Madigan MT (1997) Appl Environ Microbiol 63:3003
44. Hustede E, Steinbüchel A (1993) Biotechnol Lett 15:709
45. Pieper U, Steinbüchel A (1992) FEMS Microbiol Lett 96:73
46. Clemente T, Shah D, Tran M, Stark D, Padgette S, Dennis D, Brückener K, Steinbüchel A, Mitsky T (2000) App Microbiol Biotechnol 53:420
47. Andersson SGE, Zomorodipour A, Andersson JO, Sicheritz-Ponten T, Alsmark UCM, Podowski RM, Naeslund AK, Eriksson AS, Winkler HH, Kurland CG (1998) Nature 396:133
48. Tombolini R, Povolo S, Buson A, Squartini A, Nuti MP (1994) In: Abstracts of the 4th International Symposium on Bacterial Polyhydroxyalkanoates, Montreal, Quebec, Abstract No 37
49. Hein S, Hai T, Steinbüchel A (1998) Arch Microbiol 170:162
50. McInnerney MJ, Amos DA, Kealy KS, Palmer JA (1992) FEMS Microbiol Rev 103:195
51. Liebergesell M, Steinbüchel A (1993) Appl Microbiol Biotechnol 38:493
52. Lee I, Rhee YH, Kim J-Y (1996) Direct submission. Accesssion No U78047
53. Rehm BHA, Steinbüchel A (1999) Int J Biol Macromol 25:3
54. Liebergesell M, Sonomoto K, Madkour M, Mayer F, Steinbüchel A (1994) Eur J Biochem 226:71
55. Müh U, Sinskey AJ, Kirby DP, Lane WS, Stubbe J (1999) Biochemistry 38:826
56. Mayer F, Madkour MH, Pieper-Fürst U, Wieczorek R, Liebergesell M, Steinbüchel A (1996) J Gen Appl. Microbiol 42:445
57. Liebergesell M, Rahalkar S, Steinbüchel A (2000) Appl Microbiol Biotechnol 54:186
58. Hai T, Steinbüchel A, unpublished results
59. Doi Y, Kitamura S, Abe H (1995) Macromolecules 28:482
60. Dennis D, McCoy M, Stangl A, Valentin HE, Wu Z (1998) J Biotechnol 64:177
61. Antonio RV, Steinbüchel A, Rehm BHA (2000) FEMS Microbiol Lett 182:111
62. Olsen GJ, Woese CR, Overbeck R (1994) J Bacteriol 176:1
63. Feng D-F, Doolittle RF (1987) J Mol Evol 25:351
64. Tombolini R, Povolo S, Buson A, Squartini A, Nuti MP (1995) Microbiology 141:2553
65. Kaneko T, Sato S, Kotani H, Tanaka A, Asamizu E, Nakamura Y, Miyajima N, Hirosawa M, Suigura M, Sasamoto S, Kimura T, Hosouchi T, Matsuno A, Muraki A, Nakazaki N, Naruo K, Okumura s, Shimpo S, Takeuchi C, Wada T, Tatnabe A, Yamada M, Yasuda M, Tabata S (1996) DNA Res 3:109
66. Griebel R, Smith Z, Merrick JM (1968) Biochemistry 7:3676
67. Ballard DGH, Holmes PA, Senior PJ (1987) In: Fontanille M, Guyot A (eds) Recent advances in mechanistic and synthetic aspects of polymerization. Reidel, Lancaster, pp 293–313
68. Kawaguchi Y, Doi Y (1992) Macromolecules 25:2324
69. Madden LA, Anderson AJ, Shah DT, Asrar J (1999) Int J Biol Macromol 25:43
70. Shah DT, Tran M, Berger PA, Aggarwal P, Asrar J, Madden LA, Anderson AJ (2000) Macromolecules 33:2875
71. Gerngross TU, Snell KD, Peoples OP, Sinskey AJ, Cushai E, Masamune S, Stubbe J (1994) Biochemistry 33:9311
72. Hoppensack A, Rehm BHA, Steinbüchel A (1999) J Bacteriol 181:1429

73. Jossek R, Steinbüchel A (1998) FEMS Microbiol Lett 168:319
74. Qi Q, Steinbüchel A, Rehm BHA (2000) Appl Microbiol Biotechnol 54:37
75. Gerngross TU, Reilly P, Stubbe J, Sinskey AJ, Peoples OP (1993) J Bacteriol 175:5289
76. Stuart ES, Lenz RW, Fuller RC (1995) Can J Microbiol 41 (Suppl 1):84
77. Steinbüchel A, Füchtenbusch B (1998) TIBTECH 16:419
78. Gottschalk G (1964) Arch Mikrobiol 47:236
79. Moskowitz GJ, Merrick JM (1969) Biochemistry 8:2748–2755
80. Williams, MD, Rahn JA, Sherman DH (1996) Appl Environ Microbiol 62:2540
81. Fukui T, Shiomi N, Doi Y (1998) J Bacteriol 180:667
82. Hustede E, Steinbüchel A, Schlegel HG (1992) FEMS Microbiol Lett 93:285
83. Williams MD, Fieno AM, Grant RA, Sherman DH (1996) Protein Expression Purif 7:203
84. Choi J, Lee SY (1998) Appl Environ Microbiol 65:4363
85. Choi J, Lee SY, Han K (1999) Appl Environ Microbiol 64:4897
86. Song S, Hein S, Steinbüchel A (1999) Biotechnol Lett 21:193
87. Valentin HE, Lee EY, Choi CY, Steinbüchel A (1994) Appl Microbiol Biotechnol 40:710
88. Valentin HE, Schönebaum A, Steinbüchel A (1996) Appl Microbiol Biotechnol 46:261
89. Timm A, Byrom D, Steinbüchel A (1990) Appl Microbiol Biotechnol 33:296
90. Hein S, Söhling B, Gottschalk G, Steinbüchel A (1997) FEMS Microbiol Lett 153:411
91. Langenbach S, Rehm BHA, Steinbüchel A (1997) FEMS Microbiol Lett 150:303
92. Qi Q, Rehm BHA, Steinbüchel A (1997) FEMS Microbiol Lett 157:155
93. Qi Q, Steinbüchel A, Rehm BHA (1998) FEMS Microbiol Lett 167:89
94. Taguchi K, Aoyagi Y, Matsusaki H, Fukui T, Doi Y (1999) FEMS Microbiol Lett 176:183
95. Taguchi K, Aoyagi Y, Matsusaki H, Fukui T, Doi Y (1999) Biotechnol Lett 21:579
96. Liu SJ, Steinbüchel A (2000) Appl Environ Microbiol 66:739
97. Steinbüchel A, Valentin HE, Schönebaum A (1994) J Environ Polym Degrad 2:67
98. Valentin HE, Steinbüchel A (1995) J Environ Polym Degrad 3:169
99. Doi Y, Segawa A, Nakamura S, Kunioka M (1990) In: Dawes EA (ed) Novel biodegradable microbial polymers. Kluwer Press, Dordrecht, pp 37–48
100. Kunioka M, Nakamura Y, Doi Y (1988) Polym Commun 29:174
101. Doi Y, Tamaki A, Kunioka M, Soga K (1987) Macromol Chem, Rapid Commun 8:631
102. Holmes PA, Wright LF, Collins SH (1981) Eur Pat Appl EP 052,459
103. Valentin HE, Schönebaum A, Steinbüchel A (1992) Appl Microbiol Biotechnol 36:507
104. Schmack G, Gorenflo V, Steinbüchel A (1998) Macromolecules 31:644
105. Liu SJ, Steinbüchel A (2000) Appl Microbiol Biotechnol 53:545
106. Byrom D (1992) FEMS Microbiol Lett 103:247–250
107. Reh M (1970) In: Dellweg H (ed) Proceedings Zweites Symposium Technische Mikrobiologie. Verlag Versuchs- und Lehranstalt für Spiritusfabrikation im Institut für Gärungsgewerbe, Berlin, pp 151–158
108. Reh M, Schlegel HG (1969) Arch Mikrobiol 67:99
109. Reh M, Schlegel HG (1969) Arch Mikrobiol 67:110
110. Eschenlauer AC, Stoup SK, Srienc F, Somers DA (1996) Int J Biol Macromol 19:121
111. Slater S, Mitsky TA, Houmiel KL, Hao M, Reiser SE, Taylor NB, Tran M, Valentin HE, Rodriguez DJ, Stone DA, Padgette SR, Kishore G, Gruys KJ (1999) Nature Biotechnol 17:1011
112. Slater S, Houmiel KL, Tran H, Mitsky TA, Taylor NB, Padgette SR, Gruys K (1998) J Bacteriol 180:1979
113. Haywood GW, Anderson AJ, Chu L, Dawes EA (1988) FEMS Microbiol Lett 52:91
114. Haywood G, Anderson AJ, Williams DR, Dawes EA, Ewing D (1991) Int J Biol Macromol 13:83
115. Williams DR, Anderson AJ, Dawes EA, Ewing DF (1994) Appl Microbiol Biotechnol 40:717
116. Anderson AJ, Williams DR, Dawes EA, Ewing DF (1995) Can J Microbiol 41(Suppl 1):4
117. Valentin HE, Dennis D (1996) Appl Environ Microbiol 62:372
118. Valentin HE, Dennis D (1997) J Biotechnol 58:33
119. De Smet MJ, Eggink G, Witholt B, Kingma J, Wynberg H (1983) J Bacteriol 154:870

120. Haywood GW, Anderson AJ, Ewing DF, Dawes EA (1990) Appl Environ Microbiol 56:3354
121. Timm A, Steinbüchel A (1990) Appl Environ Microbiol 56:3360
122. Tsuge T, Fukui T, Matsusaki H, Taguchi S, Kobayashi G, Ishizaki A, Doi Y (2000) FEMS Microbiol Lett 184:193
123. Huijberts GNM, De Rijk T, De Waard P, Eggink G (1994) J Bacteriol 176:1661
124. Rehm BHA, Krüger N, Steinbüchel A (1998) J Biol Chem 273:24,044
125. Fiedler S, Steinbüchel A, Rehm BHA (2000) Appl Environ Microbiol 66:2117
126. Hoffmann N, Steinbüchel A, Rehm BHA (2000) FEMS Microbiol Lett 184:253
127. Hoffmann N, Steinbüchel A, Rehm BHA (2000) Appl Microbiol Biotechnol (in press)
128. Klinke S, Ren Q, Witholt B, Kessler B (1999) Appl Environ Microbiol 65:540
129. Rehm BHA, Steinbüchel A (2000) Appl Microbiol Biotechnol (in press)
130. Schlegel HG, Lafferty R, Krauss I (1970) Arch Mikrobiol 71:283
131. Schubert P, Krüger N, Steinbüchel A (1991) J Bacteriol 173:168
132. Zhang H, Obias V, Gonyer K, Dennis D (1994) Appl Environ Microbiol 60:1198
133. Föllner CG, Müller S, Steinbüchel A, Babel W (1995) J Basic Microbiol 35:179
134. Steinbüchel A, Schubert P (1989) Arch Microbiol 153:101
135. Suzuki T, Miyake M, Tokiwa Y, Saegusa H, Saito T, Asada Y (1996) Biotechnol Lett 18:1047
136. Takahashi H, Miyake M, Tokiwa Y, Asada Y (1998) Biotechnol Lett 20:183
137. Leaf TA, Peterson MS, Stoup SK, Somers D, Srienc F (1996) Microbiology 142:1169
138. Poirier Y, Dennis DE, Klomparens K, Somerville CR (1992) Science 256:520
139. Valentin HE, Broyles DL, Casagrande LA, Colburn SM, Creely WL, DeLaquil PA, Felton HM, Gonzalez KA, Houmiel KL, Lutke K, Mahadeo DA, Mitsky TA, Padgette SR, Reiser SE, Slater S, Stark DM, Stock RT, Stone DA, Taylor NB, Thorne GM, Tran M, Gruys KJ (1999) Int J Biol Macromol 25:303
140. John ME, Keller G (1996) Proc Natl Acad Sci USA 93:12,768
141. Nakashita H, Arai Y, Yoshioka K, Fukui T, Doi Y, Usami R, Horikoshi K, Yamaguchi I (1999) Biosci Biotechnol Biochem 63:870
142. Hahn JJ, Eschenlauer AC, Narrol MH, Somers DA, Srienc F (1997) Biotechnol Prog 13:347
143. Spiekermann P, Rehm BHA, Kalscheuer R, Baumeister D, Steinbüchel A (1999) Arch Microbiol 171:73
144. Nawrath C, Poirier Y, Somerville (1994) Proc Natl Acad Sci USA 91:12,760
145. Poirier Y, Nawrath C, Somerville CR (1995) Biotechnology 13:142
146. Mittendorf V, Robertson EJ, Leech RM, Krüger N, Steinbüchel A, Poirier Y (1998) Proc Natl Acad Sci USA 95:13,397
147. Willmitzer L, unpublished results; Mittendorf V, Bongcam V, Allenbach L, Coullerez G, Martini N, Poirier Y (1999) Plant J 20:45
148. Steinbüchel A, Aerts K, Babel W, Föllner C, Liebergesell M, Madkour MH, Mayer F, Pieper-Fürst U, Pries A, Valentin HE, Wieczorek R (1995) Can J Microbiol 41(Suppl 1):94
149. Boatman ES (1964) J Gen Microbiol 34:441
150. Mayer F, Hoppert M (1997) J Basic Microbiol 37:45
151. Murphy DJ (1993) Progr. Lipid Res 32:247
152. Pieper-Fürst U, Madkour MH, Mayer F, Steinbüchel A (1995) J Bacteriol 177:2513
153. Wieczorek R, Pries A, Steinbüchel A, Mayer F (1995) J Bacteriol 177:2425
154. Hanley SZ, Pappin DJC, Rahman D, White AJ, Elborough KM, Slabas AR (1999) FEBS Lett 447:99
155. Liebergesell M, Schmidt B, Steinbüchel A (1992) FEMS Microbiol Lett 99:227
156. Liebergesell M, Steinbüchel A (1996) Biotechnol Lett 18:719
157. Maehara A, Ueda S, Nakano H, Yamane T (1999) J Bacteriol 181:2941
158. Prieto MA, Bühler B, Jung K, Witholt B, Kessler B (1999) J Bacteriol 181:858
159. Stuart ES, Tehrani A, Valentin HE, Dennis D, Lenz RW, Fuller RC (1998) J Biotechnol 64:137
160. Preusting H, Kingma J, Huisman GW, Steinbüchel A, Witholt B (1993) J Environ Polym Degrad 1:11
161. Valentin HE, Berger PA, Gruys KJ, Rodrigues MFA, Steinbüchel A, Tran M, Asrar J (1999) Macromolecules 32:7389

162. Rodrigues MFA, Silva LF, Gomez JGC, Valentin HE, Steinbüchel A (1995) Appl Microbiol Biotechnol 43:880
163. Rodrigues FM, Steinbüchel A, unpublished results
164. Kim YB, Lenz RW, Fuller RC (1991) Macromolecules 24:5256
165. Curley JM, Lenz RW, Fuller RC (1996) Int J Biol Macromol 19:247
166. Kelley AS, Srienc F (1999) Int J Biol Macromol 25:61
167. Oeding V, Schlegel HG (1973) Biochem J 134:239
168. Steinbüchel A, Schlegel HG (1991) Molecular Microbiology 5:535–542
169. Haywood GW, Anderson AJ, Dawes EA (1989) FEMS Microbiol Lett 57:1
170. Doi Y, Segawara A, Kawaguchi Y, Kunioka M (1990) FEMS Microbiol Lett 67:165
171. Pries A, Priefert H, Krüger N, Steinbüchel A (1991) J Bacteriol 173:5843
172. Segura D, Espin G (1998) J Bacteriol 180:4790
173. Grote S, Steinbüchel A, unpublished results
174. Cabrera JE, Panzetta-Dutari G, Pruneda JL, Genti-Raimondi S (1997) J Steroid Biochem Molec Biol 63:91
175. Povolo S, Casella S (1998) Poster presented at the International Symposium on Biological Polyhydroxyalkanoates – ISBP '98, Saitama (Japan)

Received: June 2000

Physiology, Regulation, and Limits of the Synthesis of Poly(3HB)

Wolfgang Babel[1], Jörg-Uwe Ackermann[2], Uta Breuer[1]

[1] UFZ Umweltforschungszentrum Leipzig-Halle, Sektion Umweltmikrobiologie, Permoserstraße 15, D-04318 Leipzig, Germany
E-mail: babel@umb.ufz.de

[2] SIAB Sächsisches Institut für Angewandte Biotechnologie e.V. an der Universität Leipzig, Permoserstraße 15, 04318 D-Leipzig, Germany

The properties of poly(3-hydroxybutyrate) combined with the fact that it can be produced easily by numerous prokaryotes from renewable resources and even from potentially toxic waste products using well-known fermentation processes have generated keen interest in this biopolyester as a substitute for chemo-synthetic petroleum-derived polymers in many applications. However, the high price of poly(3HB) compared with the conventional synthetic materials currently in use has restricted its availability in a wide range of applications. If the economic viability of poly(3HB) production and its competitiveness are to be improved, more must be found out about the phenotypic optimization and the upper limits of bacterial systems as the factory of poly(3HB). In this chapter, two aspects of poly(3HB) are reviewed – poly(3HB) formation as a physiological response to external limitations and overcoming internal bottlenecks, and poly(3HB) as a commercially attractive polyester. From a physiological viewpoint, the ability to synthesize and degrade poly(3HB) is considered an investment in the future and provides organisms with a selective advantage. Poly(3HB) is presented as a strategic survival polymer, and it is shown that growth-associated synthesis is not as rare as reported. The influence of the efficiency and velocity of cell multiplication and product formation, of poly(3HB) content and of productivity on the overall yield, and finally on the economics of the whole process are discussed and evaluated from the technological or consumer's point of view. The specific production rate and poly(3HB) content appear to be more important than the yield coefficients.

Keywords. Poly(3-hydroxybutyrate), Metabolic sequences, "Fine" regulation, Poly(3HB) cycle, Strategic survival polymer, Growth-associated synthesis, Energy-generating and -consuming synthesis, Optimization

Advances in Biochemical Engineering/
Biotechnology, Vol. 71
Managing Editor: Th. Scheper

List of Abbreviations and Symbols

AcAcCoA	acetoacetyl-CoA
AcCoA	acetyl-CoA
CCE	carbon conversion efficiency
$C_aH_bO_c$	carbon substrate
$C_\alpha H_\beta O_\gamma N_\delta$	hypothetical "average" cell molecule
CMD	carbon metabolism determined
CoA	CoASH
E	energy and reducing power necessary for the synthesis of cell substance
E^a	"assimilatory" energy
E^d	"dissimilatory" energy
2/H/	reducing power, reduction equivalents, e.g., NADPH
K_i	inhibitor constant
K_m	Michaelis-Menten constant
μ	specific growth rate
p	poly(3HB) content of the biomass
PGA	3-phosphoglycerate
poly(3HB) or PHB	poly(3-hydroxybutyrate)
q_p	specific poly(3HB) formation rate
TCA cycle	tricarboxylic acid cycle
X_0	initial biomass concentration
Y_{ov}	overall yield coefficient
Y_P	yield coefficients of the poly(3HB) synthesis
$Y_{X'}$	yield coefficients of the poly(3HB)-free biomass

1 Introduction

Poly(3-hydroxybutyrate), poly(3HB), or PHB is a bacterially produced polyester which appears in bacterial cells as discrete granules. The capability of synthesizing poly(3HB) is widespread in prokaryotes [1–3]. This capability is neither a taxon-related property nor an ecophysiological peculiarity. It is largely independent of both the carbon source and the type of nutrition [4, 5]. Poly(3HB)-forming organisms are found among Gram-positives and Gram-negatives, aerobic and anaerobic chemo-organo-heterotrophs, chemo-litho-autotrophs, as well as aerobic and anaerobic phototrophs. Because the synthesis of poly(3HB) is started from acetyl-CoA and the activated acetic acid is a central intermediate in all living systems, it is not surprising that Archaebacteria can also synthesize and accumulate poly(3HB). Since poly(3HB) formation is a ubiquitously occurring performance, and a range of environmental factors stimulate its synthesis, it seems logical to conclude that (over-)production of poly(3HB) should play a physiologically essential role in the life of prokaryotes.

Poly(3HB) was first discovered by M. Lemoigne in 1925, and it has been the subject of academic research for many years. Since we have come to know the properties of this polyester, and that it can be accumulated to levels representing more than 80% of cellular dry mass, it has become increasingly commercially attractive [1, 2]; cf. Chap. 9 of this book.

2 Metabolism of Poly(3HB) and its Regulation

Poly(3HB) is not an essential, constitutive component of bacteria or other living beings. Thus, it is not formed under all circumstances. As a rule, its synthesis is switched on as a response to sub-optimal conditions for growth and multiplication, e.g., imbalanced supply of nutrients or deviations from optimum levels of physical factors. In an early study of its regulation, Macrae and Wilkinson [6] found that the rate of poly(3HB) accumulation in *Bacillus megaterium* increased as the ratio of glucose to ammonia increased. However, its synthesis can also be triggered by factors such as limitations of oxygen and phosphate. Such a physiological response in microorganisms to limitations is called overflow metabolism [5, 7]. Of course, the capability of synthesizing poly(3HB) is genotypically determined, but not carbon substrate-related. As a rule, the synthesis of poly(3HB) is initiated from the central intermediate of carbon metabolism, i.e., acetyl-CoA (Fig. 1). The monomers in poly(3HB) are all in the D-(–) configuration, implying that the biosynthetic steps are stereospecific.

2.1 Biosynthesis of Poly(3HB)

2.1.1 *Metabolic Pathways*

In the synthesis route from acetyl-CoA to poly(3HB), at least three steps and three enzymes are involved (Fig. 1). The first step is catalyzed by the 3-ketothiolase (EC 2.3.1.9) which reversibly links two acetyl-CoA moieties to acetoacetyl-CoA in a Claisen-condensation. The conversion of acetoacetyl-CoA into D-(–)-3-hydroxybutyryl-CoA can be mediated by a reductase (step 2) or via a sequence catalyzed by a reductase (step 4) and two hydratases (steps 5, 6). The last step, i.e., the polymerization, is catalyzed by a polymerase (step 3). This

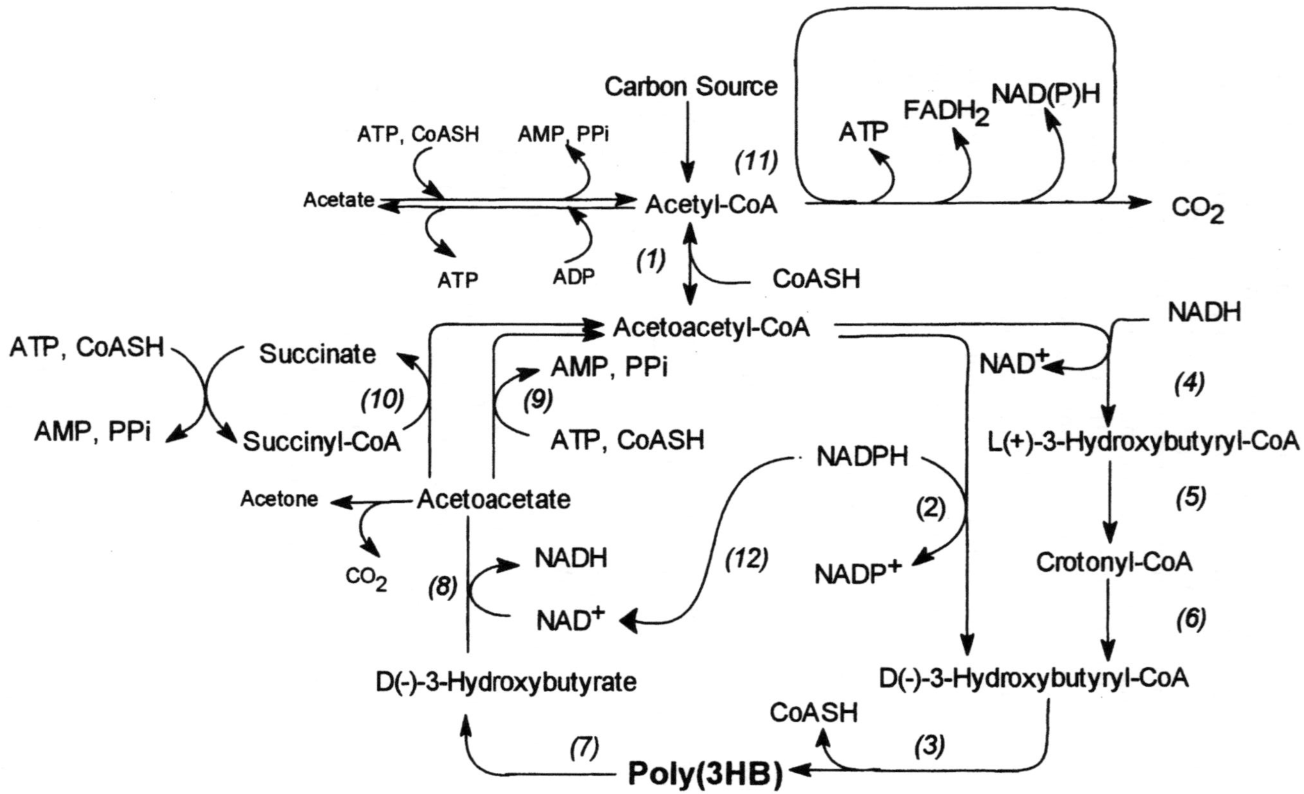

Fig. 1. The metabolic cycle for the synthesis and degradation of poly(3HB). *(1)* 3-ketothiolase; *(2)* NADPH-dependent acetoacetyl-CoA reductase; *(3)* poly(3HB) synthase; *(4)* NADH-dependent acetoacetyl-CoA reductase; *(5)*, *(6)* enolases; *(7)* depolymerase; *(8)* D-(−)-3-hydroxybutyrate dehydrogenase; *(9)* acetoacetyl-CoA synthetase; *(10)* succinyl-CoA transferase; *(11)* citrate synthase; *(12)* see Sect. 3

polymerase or synthase, which is associated with the polyhydroxyalkanoates granule [8] (for review see [9]) adds the hydroxybutyryl monomer to the growing chain.

The 3-ketothiolase has been purified and investigated from several poly(3HB)-synthesizing bacteria including *Azotobacter beijerinckii* [10], *Ralstonia eutropha* [11], *Zoogloea ramigera* [12], *Rhodococcus ruber* [13], and *Methylobacterium rhodesianum* [14]. In *R. eutropha* the 3-ketothiolase occurs in two different forms, called A and B, which have different substrate specificities [11, 15]. In the thiolytic reaction, enzyme A is only active with C_4 and C_5 3-ketoacyl-CoA whereas the substrate spectrum of enzyme B is much broader, since it is active with C_4 to C_{10} substrates [11]. Enzyme A seems to be the main biosynthetic enzyme acting in the poly(3HB) synthesis pathway, while enzyme B should rather have a catabolic function in fatty-acid metabolism. However, *in vitro* studies with reconstituted purified enzyme systems have demonstrated that enzyme B can also contribute to poly(3HB) synthesis [15].

In studies of *R. ruber*, only one active enzyme has been found, although the possibility that a second unstable enzyme exists has not been excluded. Its activity was greatest with acetoacetyl-CoA, and two-thirds lower with 3-ketovaleryl-CoA. Activity was also found with C_6 to C_8 3-ketoacyl-CoAs [13].

In *M. rhodesianum* no isoenzymes of the 3-ketothiolase have been found [14]. The enzyme's capability to react with long-chain acyl-CoAs has not been tested. The enzyme was found to be very similar to the 3-ketothiolases from *R. eutropha*, *A. beijerinckii*, *Z. ramigera*, and *R. ruber* with respect to molecular weight, optimum pH, and kinetic properties [14].

The kind of enantiomer [D-(–)- or L-(+)-] synthesized in the formation of the C_4 intermediate varies. The acetoacetyl-CoA reductase (EC 1.1.1.36), which is NADPH-dependent, stereoselectively reduces acetoacetyl-CoA to D-(–)-3-hydroxybutyryl-CoA (*R. eutropha* [15]). The NADH-dependent reductase catalyzes the reduction of acetoacetyl-CoA to L-(+)-3-hydroxybutyryl-CoA. Afterwards two stereospecific crotonyl-CoA hydratases, L-(+)- and D-(–)-specific, convert the L-(+)-3-hydroxybutyryl-CoA into the D-(–)-isomer (*Rhodospirillum rubrum* [16]).

The NADPH-dependent reductase is active with C_4 to C_6 D-(–)-3-hydroxyacyl-CoAs, it has no activity with L-(+)-substrates, and the reduction of acetoacetyl-CoA yields only D-(–)-3-hydroxybutyryl-CoA. The NADH-dependent reductase can use the L-(+)-enantiomers of these compounds and, in addition, C_7, C_8, and C_{10} L-(+)-3-hydroxyacyl-CoAs as substrates. From acetoacetyl-CoA the NADH-dependent reductase produces only L-(+)-3-hydroxybutyryl-CoA, but in the reverse direction it is active with both substrates [15].

Other bacteria producing the polymer via enoyl-CoA hydratases include *Methylobacterium rhodesianum* [17], a pink pigmented facultatively methylotrophic bacterium, that uses the serine pathway for assimilation of the reduced C_1 compounds, and *Aeromonas caviae*, a facultatively aerobic enteric bacterium that has a fermentative metabolism on sugar-containing media [18]. In *Z. ramigera* [19], *R. eutropha* [15], and *M. rhodesianum* [20], multiple reductase isoenzymes, with differing coenzyme specificities, have been found in each organism. In *Rhodopseudomonas spheroides, Rhodomicrobium vanielii* [21], *A. beijerinckii* [22], and *Syntrophomonas wolfei* [23] only single forms have been found. However, the enzymes in the latter group of organisms show activities with NADPH as well as NADH as coenzymes. In *Chromatium vinosum*, an anaerobic photosynthetic purple sulfur bacterium, an NADH-linked acetoacetyl-CoA reductase catalyzes the reduction of the C_4 compound within the poly(3HB) synthesis pathway [24].

In *Aeromonas caviae*, 3-ketothiolase and acetoacetyl-CoA reductase are absent. In this species, the synthesis of poly(3HB) proceeds via an enoyl-CoA hydratase from either crotonyl-CoA or hexenoyl-CoA. The enoyl-CoA derivatives stem from the fatty-acid oxidation pathway [18].

The polymerase can occur in either a soluble (hydrophilic) or a granule-bound (hydrophobic) form, depending on prevailing conditions of life. During growth under carbon substrate limitation, the enzyme exists mainly in the soluble state, and after transfer to nutrient-limiting conditions, e.g., ammonium limitation, the granule-associated polymerase appears (for review see [25]).

The polymerase is stereospecific. It accepts only the D-(–)-stereoisomer which is generally formed by the NADPH-linked reductase. With respect to chain length of the activated fatty acids the specificity of the polymerase varies in different organisms. It links not only C_4 3-acyl moieties but also C_5 compounds when forming the polyester molecule [26]. It also polymerizes 3-hydroxy-, 4-hydroxy-, and 5-hydroxyalkanoates from C_3 to C_5 monomers, but not C_6 or higher (e.g., in *R. eutropha*) [27–31]. In pseudomonads, in contrast, it links C_6 to C_{14} 3-hydroxyalkanoyl-CoA [32].

The concentration of the synthase or the number of enzyme copies has been assumed to have an influence on the molecular mass and molecular mass distribution of the synthesized polymer [33], but this has not been confirmed. The only variables found so far to control the molecular mass of the polymer are the initial ratio of substrate to enzyme levels, and the concentration of inducing factors in the culture medium [34–36]; cf. also Chap. 9 of this book.

Acetyl-CoA as a central intermediate in the metabolism of all carbon compounds can be dissimilated to generate biologically useful energy or assimilated and used for growth and multiplication. But the shortest and quickest way to store this carbon skeleton is synthesis of poly(3HB) via formation of acetoacetyl-CoA (Fig. 1). Since the enzymes involved in the metabolic route to poly(3HB) are unspecific, the synthesis of other homopolyesters and heteropolyesters is possible. Such analogues are formed if appropriate; prefabricated substrates (which merely need to be activated and incorporated) are offered. Compounds of this type are called related substrates.

The following examples, found in *R. eutropha*, illustrate the formation of copolymers (cf. [37]). With propionic acid as an additional carbon source, the 3-ketothiolase catalyzes the condensation of the propionyl-CoA unit with acetyl-CoA to form 3-ketovaleryl-CoA, which is reduced to 3-hydroxyvalerate moieties and polymerized by the synthase [27].

If the D-(–)-3-hydroxybutyryl moiety or D-(–)-3-hydroxyvaleryl moiety are derived from butyric acid or valeric acid via acetoacetyl-CoA [28, 29], a 3-ketothiolase is obviously not necessary.

4-Hydroxybutyric acid can also be directly incorporated [38], though some of it reacts with the corresponding enoyl-CoA. A more likely pathway proposed is via succinate semialdehyde, succinate, pyruvate, and acetyl-CoA, derived from 4-hydroxybutyrate, generally leading to a copolymer of 3-hydroxybutyryl and 4-hydroxybutyryl monomers from 4-hydroxybutyric acid [39].

In *Rhodococcus ruber* and *Nocardia corallina* the polymers composed of 3-hydroxybutyryl and 3-hydroxyvaleryl residues are synthesized from sugars by methyl-malonyl-CoA. Succinyl-CoA is decarboxylated via methyl-malonyl-CoA to propionyl-CoA as the precursor of 3-hydroxyvaleryl-CoA [40].

There is considerable interest in synthesizing copolymers. This is actually possible if organisms are confronted with mixtures of so-called related and unrelated substrates. Copolymers can also be synthesized from unrelated substrates, e.g., from glucose and gluconate. The 3-hydroxydecanoate involved in the polyester is formed by diversion of intermediates from *de novo* fatty-acid synthesis [41, 42]. "Related", in this context, refers to substrates for which the monomer in the polymer is always of equal carbon chain length to that of the substrate offered. Starting from related substrates, the synthesis pathway is closely connected to the fatty-acid β-oxidation cycle [43]. In *Pseudomonas oleovorans*, for example, cultivated on octane, octanol, or octanoic acid, the synthesized medium chain length polyester consists of a major fraction of 3-hydroxyoctanoic acid and a minor fraction of 3-hydroxyhexanoic acid. If *P. oleovorans* is cultivated on nonane, nonanol, or nonanoic acid, the accumulated polyester comprises mainly of 3-hydroxynonanoate [44].

2.1.2
"Fine" Regulation of Poly(3HB) Biosynthesis

Synthesis and intracellular accumulation of poly(3HB) are promoted when bacteria are subjected to limitations. Although the physiological role and affects of limiting factors (ammonium, oxygen, phosphate, sulfate, K^+, Mg^{2+}, or Fe^{2+}) differ, the bacteria respond to the different limitations in a qualitatively similar fashion. This gives rise to the suspicion that the synthesis control mechanisms are also similar. For synthesis, two substrates are necessary, i.e., acetyl-CoA and reducing power (2/H/). For kinetic reasons, the intracellular concentration of acetyl-CoA should be high, to favor the synthesis of acetoacetyl-CoA, since the equilibrium of the reaction lies towards hydrolysis, and reducing equivalents must be available to withdraw the acetoacetyl-CoA formed from the equilibrium.

If nitrogen (in the form of ammonia) is growth limiting, the potential applications of acetyl-CoA and NAD(P)H are restricted. Liberated NAD(P)H cannot be consumed for reductive syntheses, for instance of amino acids, it remains available and starts to inhibit citrate synthase [45, 46]. To the extent that the TCA cycle is thereby inhibited, acetyl-CoA should become available for the 3-ketothiolase, and could flow into poly(3HB) (Fig. 1, Table 1).

If oxygen limits growth and multiplication of aerobic bacteria, the adaptive value and inductive mechanism of poly(3HB) synthesis are quite similar to those outlined for nitrogen limitation. 2/H/ generated are not oxidized via elec-

Table 1. Kinetic parameters of some bacterial poly(3HB) synthesizing and degrading enzymes (values in mmol l^{-1})

Organism	3-ketothiolase		Citrate synthase	Acetoacetyl-CoA reductase		3-hydroxybutyryl-CoA dehydrogenase	References
	thiolytic	condensing		NADH-dependent	NADPH-dependent		
Azotobacter beijerinckii	K_m CoA 0.015	K_m AcCoA 0.9	K_m AcCoA 0.0018		K_m AcAcCoA 0.0019	K_m 3-hydroxy-butyrate 0.877; K_m NAD 0.07	10, 54, 64, 65
Hydrogeno-monas eutropha		K_m AcCoA 0.39					53
Ralstonia eutropha (formerly *Alcali-genes eutrophus*)		K_m AcCoA 1.1; K_i CoA 0.016		$K_{m\,Ac}$AcCoA 0.022; K_m 3-hydroxybutyryl-CoA 0.026; K_m NADH 0.013	K_m AcAcCoA 0.005; K_m 3-hydroxybutyryl-CoA 0.033; K_m NADH 0.4; K_m NADPH 0.019		66
Rhodospirillum rubrum						K_m 3-hydroxy-butyrate 0.84; K_m NAD 0.07; K_m Acetoacetate 0.07	65
Rhodopseudo-monas spheroides						K_m 3-hydroxy butyrate 0.41; K_m NAD 0.08; K_m NADH 0.054; K_m Acetoacetate 0.28	65

Pseudomonas lemoignei						K_m 3-hydroxy butyrate 0.6; K_m Acetoacetate 0.2	65
Zoogloea ramigera		K_m AcCoA 0.33		K_m AcAcCoA 0.009; K_m NADH 0.007	K_m AcAcCoA 0,008; K_m NADPH 0.021	K_m NADH 0.057; K_m 3-hydroxybutyrate 0.32	54, 19
Methylobacterium rhodesianum		K_m AcCoA 0.5; K_i CoA 0.02	K_m AcCoA 0.07; K_i NADH 0.15	K_m AcAcCoA, NADPH 0.03	K_m AcAcCoA 0.015; K_m NADPH 0.018		20, 67, 14
Methylobacterium extorquens			K_m CoA 0.084;		K_m AcAcCoA 0.011; K_m NADPH 0.041		51, 55
Azotobacter vinelandii	K_m AcAcCoA 0.053; K_m CoA 0.012	K_m AcCoA 0.83			K_m AcAcCoA 0.011; K_m NADPH 0.02	K_m NADH 0.042; K_m 3-hydroxybutyrate 1.06	68

tron transport phosphorylation. Consequently, the TCA cycle is stopped due both to metabolic control by 2/H/ and for thermodynamic reasons, since as it is well known the TCA cycle is endergonic in the absence of oxygen.

When phosphate is the limiting factor, the induction mechanism is more difficult to deduce. However, bacteria, like all other living systems, are not able to generate ATP by phosphorylating ADP in the absence of phosphate. It seems quite plausible that, under these conditions, 2/H/ do not flow away. As in the two previously mentioned limitations, 2/H/ remain and acetyl-CoA becomes available, thus both substrates could be assimilated and deposited intracellularly as poly(3HB).

As a rule, biochemical reactions and pathways cannot be stopped completely by metabolic regulation, i.e., inhibition of enzyme activities, but the specific reaction rates and flow rates can be considerably diminished. When the TCA cycle is inhibited, the rate of 2/H/ generation via the TCA cycle, and the amount of 2/H/ made available, are reduced. Thus, if the TCA cycle is switched off, reactions occurring before the TCA cycle must provide the 2/H/ required for the reduction of acetoacetyl-CoA to form 3-hydroxybutyryl-CoA instead. This applies to reactions involving glucose (1 glucose $\rightarrow$ 2 acetyl-CoA + $2CO_2$ + 4 NAD(P)H + 2 ATP), so if glucose is offered, the TCA cycle is dispensable for poly(3HB) synthesis. However, if acetate is supplied as sole source of carbon and energy, the TCA cycle is needed both for growth and poly(3HB) formation (Fig. 1). Therefore, when acetate is intended for use as sole source of both carbon and energy for the production of poly(3HB), it seems that it is not appropriate to limit growth by oxygen or phosphate.

At least two enzymes compete for acetyl-CoA – the citrate synthase and 3-ketothiolase. The affinities of these enzymes differ for acetyl-CoA (Table 1), and at low concentrations of it the citrate synthase reaction tends to dominate, provided that the concentration of 2/H/ is not inhibiting. The "fine"regulation of the citrate synthases of various poly(3HB) accumulating bacteria has been studied [14, 47, 48]. They appear to be controlled by cellular energy status indicators (ATP, NADH, NADPH) and/or intermediates of the TCA cycle. The 3-ketothiolase has also been investigated [10–14, 49, 50]. This enzyme is, above all, inhibited by CoASH [10, 14, 49]. This important feature will be further considered below.

The next step on the path to poly(3HB) is catalyzed by an NADPH- or NADH-linked reductase. These enzymes are under the control of acetoacetyl-CoA (Table 1) and are activated by ATP and isocitrate, according to *in vitro* experiments [15, 19, 20, 50–53]. These properties conform to the outline developed above for metabolic regulation of poly(3HB) synthesis (cf. [4]). However, this rationale does not consider the strong inhibition of the 3-ketothiolase by CoASH already demonstrated by Senior and Dawes [10] and Jackson and Dawes [54]. Under balanced growth conditions, the concentration of free CoA has been found to be high, so the activity of the biosynthetic 3-ketothiolase must be low. Thus, the enzyme must be activated during poly(3HB) accumulation. In support of these conclusions, it has been shown that, under growth limiting conditions in the presence of an excess carbon source, the level of free CoA is reduced. For example, it falls from 0.6 nmol per mg dry mass in *M. rhodesianum* during growth, to the detection limits of current equipment when poly(3HB) accumulates [14, 55]. To understand poly(3HB) synthesis, it is necessary to explain this decrease. For this purpose the concept of the central precursor [56–58] is helpful. According to this concept, the synthesis of all indispensable and essential macromolecules and polymers of an organism is started from a C_3 compound, e.g., 3-phosphoglycerate (Sect. 4.1.1). Depending on the energy made available from the source(s) of carbon and energy (E^a), especially from multicarbon substrates (see Sect. 4.1.1, Eq. 1), the central precursor must be converted into acetyl-CoA and finally oxidized to CO_2 to generate energy (Eq. 3). With glucose, about 40% of the substrate-carbon are dissimilated. In the absence of nitrogen, oxygen, or phosphate the C_3 compound can no longer be utilized for anabolic purposes, such as the synthesis of proteins and nucleic acids. Instead (provided its synthesis is not stopped) the central precursor is available to be converted into acetyl-CoA, and free CoA is consumed. The inhibition of the 3-ketothiolase is then lifted.

In general, unbalanced supply of nutrients for growth or deficits, of nitrogen or oxygen for instance, reduce the complexity of metabolism and channel the flow of carbon skeletons into more unidirectional (one-way) paths. Poly(3HB) is an example. For the formation of poly(3HB) it is not sufficient to diminish the flux of acetyl-CoA through the TCA cycle. The citrate synthase does not necessarily have to be inhibited (cf. acetic acid as a substrate). Overproduction of poly(3HB) needs, above all, an increased supply of acetyl-CoA. As long as appropriate substrates are available, substrate uptake is not controlled and catabolism is possible, sufficient reducing power being generated. "Sufficiency" does not concern the possible yield of poly(3HB) here. The yield of poly(3HB) depends not only on the substrate and the organism used but also on the balance between the ratio of NADH:NADPH generated, and that required for the synthesis of 3-hydroxybutyryl-CoA. Thus, the maximum yield of poly(3HB) may be limited by the NADPH availability [59]. Free CoA, more than 2/H/, appears to be the key effector in the regulatory network. It controls the 3-ketothiolase activity, and it is possible that the poly(3HB) synthase is inhibited by cellular concentration of CoASH, albeit at higher levels than 3-ketothiolase [37]. CoASH is reaction product and substrate; it is not really consumed, but only used, and insofar it functions like a catalyst.

From this perspective it would be interesting to discover if there is a relationship between the substrate used and the concentration of free CoA under conditions of unlimited growth. If there is, depending on the source of carbon and energy used and the K_i value of the 3-ketothiolase for CoASH, cell multiplication and poly(3HB) accumulation can occur simultaneously [60, 61]. If this enzyme is not involved, poly(3HB) [29] and other polyesters [28, 29] can also be synthesized during growth.

2.2 Intracellular Breakdown of Poly(3HB)

2.2.1 *Metabolic Routes*

Like other polymers synthesized by living systems, poly(3HB) can be degraded at a later stage by the organism producing it. Hence, it is very likely that some of the accumulated poly(3HB) is being degraded all the time, the equilibrium of this turnover being determined by external factors. Degradation is not simply a reversal of the synthesis (Fig. 1). Although small amounts of poly(3HB) are formed – in some bacteria – under balanced growth conditions, the polymer is generally overproduced when cells have adequate supplies of carbon but are limited by another nutrient. When the supply of the limiting nutrient is restored cells start to break down the polymer. The first step of the reversion is catalyzed by a soluble depolymerase (for reviews see [62, 63] and Chap. 10 of this book). In *Ralstonia eutropha*, D-(–)-3-hydroxybutyrate was found to be the sole hydrolysis product [64]. In some other bacteria, e.g., *Bacillus megaterium*, *Rhodospirillum rubrum*, and *Zoogloea ramigera*, a mixture of dimeric and monomeric products has been isolated (cf. [4, 62]). The monomers are then oxidized into acetoacetate by the NAD-dependent D-(–)-3-hydroxybutyrate dehydrogenase (EC 1.1.1.30) [65, 66]. Acetoacetate can be converted into acetoacetyl-CoA by several systems. The activation is an energy-consuming process. The transfer of CoASH can proceed either directly or with the help of succinyl-CoA. In the first case the activation is catalyzed by acetoacetyl-CoA synthetase (EC 6.2.1.16), as seen in *Z. ramigera* [67], and in the second case by the succinyl-CoA transferase (EC 2.8.3.5; succinyl-CoA:3-oxoacid coenzyme A transferase), as observed in *Azotobacter beijerinckii* [10] and *Methylobacterium extorquens* [47, 52, 53]. In both cases, 2 ATP are consumed for the formation of 1 acetoacetyl-CoA. Provided NADH is converted into ATP via electron transport phosphorylation with a P/O quotient of 2, the sequence from poly(3HB) back to the common intermediate of synthesis and degradation (Fig. 1) is bioenergetically neutral. In the following reaction, acetoacetyl-CoA can be hydrolyzed into two acetyl-CoA, which can then enter the TCA cycle. In this manner the polymer is recycled and the monomers can be utilized as sources of energy and carbon.

2.2.2 *"Fine" Regulation of Poly(3HB) Breakdown*

Surprisingly, the "fine" regulation of the enzymes involved in poly(3HB) degradation has been hardly investigated at all (cf. Chap. 10 of this book). Most of the research on intracellular mobilization of poly(3HB) and oxidation of the monomers was done more than 25 years ago.

Nothing seems to be known as yet of the metabolic regulation of enzymes of poly(3HB) producing bacteria that catalyze the last step of the poly(3HB) cycle, i.e., the activation of acetoacetate (Fig. 1, steps 9, 10). Since both enzymes are involved in a catabolic sequence, it may be speculated that they are inhibited, for instance, by ATP. If they are really controlled energetically, their response to energy charge [68] or adenine nucleotides [69] should follow R (ATP regenerating) curves. The same should apply to the D-(–)-3-hydroxybutyrate dehydrogenase. This enzyme is inhibited by NADH (competing with NAD) (Table 1) as shown in *A. beijerinckii* [10, 50, 70]. Thus, as long as NADH can be produced and the concentration is high, i.e., a source of energy is available, the D-(–)-3-hydroxybutyrate cannot be and is not oxidized.

The genetic or "coarse" regulation of the depolymerase has been well studied (cf. [62, 63] and Chap. 10 of this book), but relatively little is known of the "fine" control of the first step (Fig. 1, step 7). Since there is some evidence that the synthesis and degradation occur simultaneously, a certain level of depolymerase activity should be present permanently, and the specific activity of the enzyme(s) and final breakdown of poly(3HB) should be controlled by effectors of some kind. It is difficult to imagine that monomers or intermediates of carbon metabolism are effectors, but easy to accept that the activity is modulated by a phosphorylating: dephosphorylating mechanism. The depolymerase has been suggested to be inactive when phosphorylated and vice versa [71]. If so, the activity is indirectly energetically regulated. This hypothesis appears to be quite plausible and deserves to be studied in more detail.

Since it may be assumed that part of poly(3HB) synthesized is degraded during accumulation, that the equilibrium determines the content of poly(3HB), the molecular mass, and molecular mass distribution, a detailed analysis of the regulation of the poly(3HB) cycle will be useful for a better understanding as well as optimization of industrial production of poly(3HB).

3
Poly(3HB) as a Strategic Survival Polymer

If microbes or, strictly speaking, aerobic bacteria are provided with macro- and micronutrients in such a way that nothing is lacking or available in excess, and the chemo-physical factors of the environment, e.g., pH and temperature, conform to genotypically determined optima, the organisms will grow and multiply and produce only carbon dioxide, water, and heat. The cells or the biomass will have a strain-specific composition, and intermediates will appear only in catalytic amounts. Such situations should be of very short duration, and occur very seldom in natural ecosystems, if at all. Hence, microorganisms have to cope with external shortcomings, and environmental fluctuations, of widely varying frequency and intensity. To survive in such conditions they have evolved various strategies, one of which is overflow metabolism [5]. Since biological stability is generated by steady renewal (turnover), i.e., synthesis (and resynthesis) of structure-forming and functional cellular macromolecules and polymers, the organisms need, above all, energy. Poly(3HB) is generally overproduced and intracellularly accumulated when more reducing power is being generated than can be productively consumed due to limitations for other syntheses. This can be seen as an investment for the future insofar as poly(3HB) can be "mobilized" under conditions of external energy starvation. It can serve as a source of NADH (and ultimately of ATP) even though no energy is consumed for the synthesis of poly(3HB) starting from acetyl-CoA, and only the bioenergetically worthless NADPH is stored (Fig. 1) (from NADPH via poly(3HB) to NADH; formation and degradation of poly(3HB) as a quasi transhydrogenase reaction or H-transmission cycle). After depolymerization, the resulting D-(–)-3-hydroxybutyrate is oxidized by NAD-specific D-(–)-3-hydroxybutyrate dehydrogenase. This reaction is the first step of energy generation during the degradation of poly(3HB). Provided acetoacetate can be activated, the acetoacetyl-CoA obtained can be converted into acetyl-CoA by the reversibly operating 3-ketothiolase (Fig. 1). Under aerobic conditions acetyl-CoA can finally be dissimilated via the TCA cycle. The net gain of the oxidation of one D-(–)-3-hydroxybutyrate amounts to 2 NADPH + 5 NADH + 2 $FADH_2$. Under anaerobic conditions the energy-rich bond of acetyl-CoA is not necessarily bioenergetically useless; it can be used to phosphorylate ADP to form 1 ATP (plus 1 acetate).

For the assimilation of multicarbon substrates, aerobic chemo-organoheterotrophs require a complete TCA cycle for the generation of reducing power and, ultimately, energy. This catabolic function of the TCA cycle may be dispensable, provided the cells can synthesize poly(3HB), and the synthesis is energy generating [5, 72]. In this respect poly(3HB) synthesis resembles alcohol formation by fermentative utilization of glucose. Like the alcohol formation, which is induced by oxygen deficit or glucose excess (the Crabtree effect), the energy generated can be utilized for anabolic processes and even for growth and multiplication of the producer [5, 58, 73]. This is clearly true in *Methylobacterium rhodesianum* [74]. With methanol as sole source of carbon and energy the energy budget of methylotrophs is probably covered by oxidation of the methanol via the so-called linear dissimilatory sequence (cf. [75]). In facultatively methylotrophic bacteria, e.g., the pink pigmented *M. rhodesianum*, some enzymes of the TCA cycle are repressed and poly(3HB) is formed – as usual – in response to external limitations, for instance of ammonia. On fructose, however, *M. rhodesianum* starts accumulating poly(3HB) during exponential growth. The low bioenergetic capacity of the TCA cycle is seen as the reason for this growth-associated poly(3HB) synthesis. This hypothesis is supported by the fact that the simultaneous poly(3HB) formation can be suppressed by an additional supply of NADH in the form of formate, which is oxidized via a formate dehydrogenase.

The energy gain from the conversion of a certain substrate into acetyl-CoA as the precursor for the poly(3HB) synthesis depends upon its metabolism and the efficiency of the electron transport phosphorylation. It is in all cases considerably smaller and not economical in terms of maximum growth yield, but the cells are more "stable" than poly(3HB)-free, i.e., energy-poorer ones [76–78]. The synthesis of poly(3HB) is not only an alternative to the TCA cycle, or a cheap substitute: it allows the cells to use substrates as sources for growth, which properly need a complete TCA cycle, and it improves the competitiveness. It is in contrast to the above-mentioned ("altruistic") alcohol (formation) an "egoistic" principle that helps survive and enables internal bottlenecks or deficits to be overcome. The latter may not be bioenergetically determined, they may also be decided by anabolic or anaplerotic sequences, as has been shown for *Ralstonia eutropha* [79]. This organism is able to grow on phenol as sole source of carbon and energy. The assimilation relies on the chromosome-encoded *meta* cleavage of the aromatic ring [80]. The resulting intermediates are pyruvate and acetaldehyde, which are further metabolized. Pyruvate can be used for gluconeogenesis and for the synthesis of all other cell constituents or directed into the oxidative TCA cycle. Acetaldehyde can be oxidized to acetate. But the acetate-carbon cannot be utilized for gluconeogenesis and contribute to biomass formation since the glyoxylate shunt is not switched on during growth on phenol. Instead acetate is activated and dissimilated for the generation of biologically useful energy, and bioenergetically superfluous acetate is deposited in poly(3HB). (The synthesis itself acts as a hydrogen sink and avoids an excess of energy.) An alternative would be accumulation of free acetate (which is known to repress the phenol hydroxylase activity), acidification of the medium, and poisoning of the cells (by the "traumatic effect" [81]).

The growth-associated synthesis of poly(3HB) is clearly of selective advantage. The synthesis of poly(3HB) is not only – as stated above – an investment in the future, but also an opportunity to cope with and widen actually existing intrinsic bottlenecks.

If poly(3HB) formation is an effective strategy for survival the "poly(3HB) machinery" should be ready for use at all times. Its utilization (and the extent to which it is expressed) is determined by external conditions, the metabolic and regulatory disposition of a given strain, and the substrate offered. From this perspective, growth-associated synthesis of poly(3HB) (or, more precisely, poly(3HB) formation-associated growth) ought to occur more frequently than has yet been observed and reported.

4 Poly(3HB) as a Commercially Attractive Polyester

Poly(3HB) has several interesting properties. It is thermoplastically malleable, biocompatible, and biodegradable, and it has potential for a number of possible applications [82–85]. It can be synthesized from reduced C_1 compounds, e.g., methane and methanol, from *Knallgas* (detonating gas), and from multicarbon substrates, e.g., carbohydrates and hydrocarbons, from pure substrates and from waste materials, even from potentially toxic compounds such as phenol. We are already capable of tailoring its properties either by modifying its biosynthesis (e.g., using substrate mixtures for production of copolymers) [27–29, 86, 87] or afterwards through physico-chemical treatments (post-biosynthetic modification) [88–90]. Nevertheless, there is no real demand for either poly(3HB), other members of this polyester family, copolyesters, or blends. The main reason for this is the price of the products, which is heavily dependent on the costs of the raw materials, i.e., substrates.

Irrespective of whether poly(3HB) is produced discontinuously or continuously, the overall process comprises several stages:

1. Synthesis of the producer (cell factory) by growth and multiplication
2. Synthesis and accumulation of the desired product
3. Isolation of the product from the biomass (cells) and purification

The economic viability of the whole process is consequently determined, in addition to various market forces, by the efficiencies of these three stages and the velocities of the syntheses. Since the synthesis of poly(3HB) requires the cells both as a catalyst and as a reactor (cell factory), the overall yield is also influenced by the cellular content of the polymer.

4.1 Theoretical and Experimental Yield Coefficients

If we want to improve the economic viability and to reduce the price of the product, we have to maximize the yield coefficients for growth as well as for formation of the product.

Knowledge of the maximal possible yield coefficients as target quantities on the one hand, and the reasons for discrepancies between these values and experimentally obtained values on the other, can help reduce the experimental expenditure needed for optimizing yield, and recommend worthwhile approaches and measures to test.

4.1.1 *Growth and Multiplication*

The growth yield coefficients can be calculated [56, 58] by sub-dividing metabolism responsible for the whole process of growth and multiplication into:

1. Precursor synthesis, i.e., conversion of the carbon substrate ($C_aH_bO_c$) into a central metabolite, such as 3-phosphoglycerate (PGA), from which all cell constituents are thought to be synthesized:

$$C_aH_bO_c \rightarrow PGA + CO_2^a + E^a \quad (1)$$

2. Cell molecule synthesis (anabolism), i.e., formally, reduction of the precursor to the level of a hypothetical "average" cell molecule $\{C_\alpha H_\beta O_\gamma N_\delta\}$ and assimilation of nitrogen (e.g., NH_3), representing synthesis of all cell constituents:

$$PGA + NH_3 + E \rightarrow \{C_\alpha H_\beta O_\gamma N_\delta\} + CO_2^a \quad (2)$$

   The energy and reducing power necessary (E) for the synthesis of cell substance, i.e., growth and multiplication, can be made available simultaneously with the synthesis of the central precursor. If this "assimilatory" energy (E^a) is not sufficient ($E^a < E$) to convert the precursor-carbon completely into biomass, further substrate must be dissimilated.

3. Dissimilation, i.e., oxidation (or mineralization) of substrate merely for generation of reducing power and biologically useful energy (E^d "dissimilatory" energy):

$$C_aH_bO_c \rightarrow CO_2^d + E^d \quad (3)$$

   The upper limits calculated for several substrates are listed in Table 2.

The calculated values are higher than the experimentally obtained coefficients, because the substrates are deficient in energy. The extent of the deficit depends

Table 2. Comparison of energetically independent and experimentally obtained yield coefficients

Substrate	$Y_{85\%}$[a]	$Y_{75\%}$[a]	Y_{exp}	CCE_{exp}
Methane	1.35		0.9	56.7
Methanol	0.68		0.33[b]	62.5
Ethanol		0.83	0.8	72.3
Acetic acid		0.64	0.4	46.8
Glucose	0.72		0.5	59

Y = g dry mass per g substrate; exp: experimentally obtained; $Y_{85\%,75\%} = Y_{CMD}$. Cell molecule $\{C_4H_8O_2N_1\}$.
[a] Energetically independent carbon conversion efficiency (CCE); in %.
[b] Serine pathway.

upon the energy content of the substrate, the metabolic pathway, and the efficiency of the energy conservation via electron transport phosphorylation. Ethanol seems to be a fairly carbon:energy-balanced substrate, because the calculated, i.e., the energetically independent carbon conversion efficiency, can be attained experimentally [91].

4.1.2
Poly(3HB) Synthesis

The yield coefficient of the poly(3HB) synthesis determined solely from carbon metabolism can easily be calculated if the metabolic sequence from the carbon substrate to poly(3HB) $\{C_4H_6O_2\}$ is known, for instance:

$$2\ \text{Acetate} \rightarrow \{C_4H_6O_2\} - 2\ \text{ATP} - 1\ \text{NAD(P)H} \tag{4}$$

$$1\ \text{Glucose} \rightarrow \{C_4H_6O_2\} + 2\ \text{ATP} + 3\ \text{NAD(P)H} + 2\ CO_2 \tag{5}$$

Table 3 shows calculated values pertaining to several important substrates.

Table 3. Comparison of energetically independent and experimentally obtained efficiencies of the conversion of various substrates into poly(3HB)

Substrate	CCE_{CMD}	Y_{CMD}	Y_{ec}	Y_{exp}
Methane	100%	1.34	0.73	0.45
Methanol	100%	0.67	0.38–0.6[a]	0.2
Ethanol	100%	0.93	0.9	0.5
Acetic acid	100%	0.72	0.57	0.33
Glucose	66.7%	0.48	0.27	0.33

Y = g poly(3HB) per g substrate; ec: energetically corrected.
CCE = carbon conversion efficiency in %.
CMD = carbon metabolism determined, i.e., energetically independent.
[a] This range results from differences in possible energy gains derived from the oxidation of formaldehyde.

The calculated values are not reached experimentally. There are several reasons for this. First, during the synthesis of poly(3HB), metabolic pathways not involved in the synthesis are not switched off completely, so that certain amounts of by-products are formed. Second, the synthesis of poly(3HB) is seldom neutral energetically [72, 73]. Depending upon the substrate and the metabolic pathway, the synthesis either follows Eq. (4), or liberates reducing power and biologically usable energy (Eq. 5). The various forms of energy (ATP, NADH) made available (Eq. 5) must be dissipated. This may be done at the expense of extra substrate, thus consuming more substrate than needed stoichiometrically, i.e., according to the metabolic pathway. If nitrogen is still present, even growth and multiplication can occur simultaneously; thus energy-yielding poly(3HB) syntheses are potentially growth-associated. Growth-coupling can be quite desirable from the biotechnologist's point of view, insofar as it offers a chance of making the process continuous and stabilizing the producer by renewal. In syntheses which require ATP and/or NAD(P)H (Eq. 4), extra substrate is consumed merely for the purpose of energy generation. The amount of additional substrate may be estimated if the balance equation for dissimilation (Eqs. 3, 6) and the P/O-quotient are known. It depends on the energy that can be made available through the "biological combustion". In the case of acetate – provided only 1 ATP is necessary for the activation – the balance equation reads as follows:

$$1\ \text{Acetate} \rightarrow 2CO_2 + 3\ \text{NAD(P)H} + 1\ \text{FADH}_2 \qquad (6)$$

Assuming the P/O-quotient of NADH is 2 and NADPH can be used bioenergetically, about 0.5 acetate must be oxidized to "neutralize" the synthesis. This expenditure of substrate diminishes the product yield coefficient from 0.72 g poly(3HB) per g acetic acid (Table 3) to about 0.57 g per g. Since the experimentally obtained yield coefficient is lower (being, on average, about 0.33 g per g, Table 3), we may draw three conclusions. Firstly, the P/O-quotient is lower than 2. Secondly, the fate of acetate is not strictly determined, i.e., its utilization is not a one-way path and does not terminate in a dead end. Third, there is no doubt that some energy generated from acetate is necessary for homeostasis and turnover processes (maintenance) under conditions of poly(3HB) synthesis and accumulation (with acetic acid as an uncoupler).

4.2 Approaches to Increase Yield

4.2.1 *Yield Coefficient*

Poly(3HB) syntheses coupled to consumption of energy (Eq. 4) resemble the general balance equation for the synthesis of biomass:

$$C_aH_bO_c + E \rightarrow \{C_\alpha H_\beta O_\gamma N_\delta\} + CO_2 \qquad \text{(cf. 1 + 2)}$$

During growth the substrate, which must be dissimilated (Eqs. 3, 6) in order to generate the energy necessary to assimilate the substrate-carbon and, ultimately, to synthesize biomass, can be replaced by a so-called victim substrate, e.g., formate, which functions as a donor of energy and/or reducing power [58]. By adding such extra energy the growth yield coefficient of, for instance, *Acinetobacter calcoaceticus* on acetate increases from about 0.4 g dry mass per g acetic acid to the upper limit determined by carbon metabolism of about 0.64 g per g [92] (Table 2).

In poly(3HB) synthesis the maximum carbon conversion should theoretically be reached not only with the aid of a victim substrate but also by mixing energetically excess and deficient substrates, e.g., glucose + acetate [5]. If NADH and NADPH are considered to be equal energetically, and as "substrates" for reductive syntheses, provided that the P/O-quotient is 2 and that an "ideal" channeling of glucose and acetate occurs, both substrates should contribute to poly(3HB) synthesis, and compensate for each other energetically. With a molar mixing ratio of about 1 glucose:4 acetate, the poly(3HB) synthesis is neutralized energetically, and a carbon conversion efficiency of 85.7% should be reached [5, 93].

As sources of carbon and energy for growth and multiplication, glucose and acetate are energetically deficient (Table 2). Therefore, simultaneous utilization of both substrates will not result in an increase in the growth efficiency. At first glance, with a two-stage (discontinuous) process, mixing substrates seems to be a promising approach in the production phase. The following efficiencies are possible theoretically (Table 4).

As the calculated values show, mixing substrates does not increase dramatically the overall efficiency (in this example). This seems surprising at first, but

Table 4. Theoretically possible overall yield coefficients for the production of poly(3HB)

Substrate for formation of biomass/poly(3HB)	CCE %	poly(3HB) content (%)
Acetate	29.2	45.7
	42.8	62.7
	80	100
Acetate/mixture[a]	30.0	45.7
	44.5	62.7
Glucose/mixture[a]	35.0	45.7
	49.9	62.7
Glucose	33.3	48.9
Glucose[b]	39.2	47.8
Acetate + formate	42.9[c]	45.7
Acetate/(acetate + formate)	31.5[c]	45.7
(Acetate + formate)/mixture	40.0[d]	45.7

[a] Mixture: 1 glucose: 4 acetate.
[b] In a continuous process; the energy generated during the poly(3HB) formation is thought to be consumed in growth and multiplication.
[c] g poly(3HB)-carbon per g acetate-carbon.
[d] g poly(3HB)-carbon per g (acetate+glucose)-carbon.

it is quite plausible because glucose helps increase the conversion efficiency of acetate into poly(3HB) by only 5.7%, i.e., from about 80% to 85.7%. Using a victim substrate, e.g., formate, the conversion efficiency of acetate-carbon can be increased considerably, but the formate-carbon is lost. If this approach is biochemically possible and technologically feasible, project managers have to decide whether or not it is economically viable. Another possibility is presented by energy-yielding substrates, provided the energy generated during the poly(3HB) synthesis can be utilized for growth, i.e., for production of the "catalyst". In such cases the carbon conversion efficiency of turning glucose into poly(3HB) should be higher than in a batch regime (Table 4).

4.2.2
Poly(3HB) Content and Production Rate

The overall yield coefficient (Y_{ov}) must be lower than that of the single factors. Since poly(3HB) is the target product it is reasonable to assume that the poly(3HB) content has a strong influence on the final yield. This is true, as the figures calculated show (Table 4). The overall yield varies not just with the poly(3HB) content p (in g poly(3HB) per g biomass), but also with the yield coefficients of the poly(3HB)-free biomass ($Y_{X'}$), and poly(3HB) synthesis (Y_P). It can be calculated according to the following equation:

$$Y_{OV} = \frac{Y_{X'} Y_P}{Y_{X'} + Y_P \left(\frac{1}{p} - 1 \right)} \tag{7}$$

Some results are presented in Fig. 2.

As anticipated, the overall yield increases proportionally with the poly(3HB) content and, of course, at low poly(3HB) contents the slope of the overall yield is steeper with increasing growth yield coefficients. For a given pair of $Y_{X'}$ and Y_P the slopes are equal with increasing $Y_{X'}$ or Y_P at a poly(3HB) content, as it is calculated according to the equation poly(3HB)% = $Y_P \times 100\%/(Y_P+Y_{X'})$. Thus, if it is intended to reach a higher poly(3HB) content, then first of all it is advisable to optimize Y_P rather than $Y_{X'}$.

In a batch process the final poly(3HB) content is determined by the specific poly(3HB) formation rate q_p (in g poly(3HB) per g poly(3HB)-free dry mass per h) and the duration of the accumulation process (t). It can be pre-calculated according to the following equation:

$$PHB(\%) = \frac{q_P t}{q_P t + 1} 100\% \tag{8}$$

assuming q_p remains constant during the whole period of poly(3HB) formation.

However, in practice, q_p often declines some time after reaching the maximum value. It has been repeatedly reported that under conditions of ammonium-limited poly(3HB) accumulation, a small amount of ammonium feeding can prolong periods of constant q_p values, or even increase maximum values

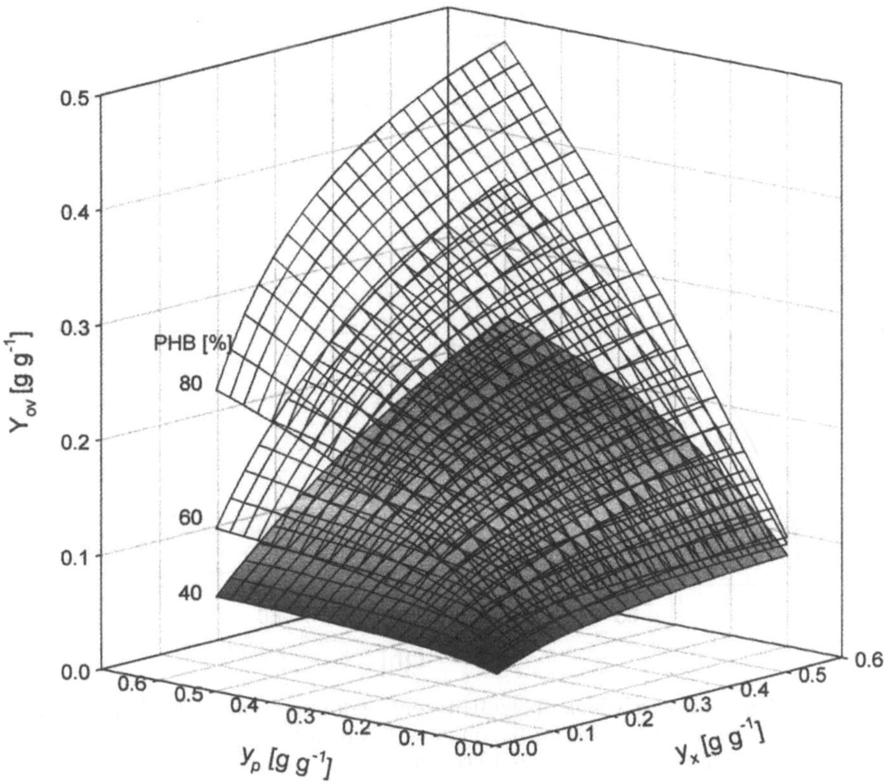

Fig. 2. Overall yield coefficients Y_{OV} as a function of $Y_{X'}$ and Y_P for poly(3HB) contents of 40%, 60%, and 80%

[86, 94–97]. The first effect can be easily explained. The feeding of ammonium allows the resynthesis of structural units, compensating for cellular components lost in ongoing turnover. Some authors [94, 95], who used glucose as substrate, have also reported increases in carbon conversion efficiency to poly(3HB) when low concentrations of ammonium were applied, compared to cultivations without ammonium feeding. These observations seem to confirm our prediction that energy made available by the conversion of energy excess substrates into poly(3HB) may be used for reproduction.

Figure 3A shows the kinetics of poly(3HB) accumulation for several specific formation rates, depending on the duration of accumulation. As can be seen, the poly(3HB) content approaches a value of 100% asymptotically. However – and this might be considered trivial – how closely the upper limit is reached is ruled by the fact that an increasing content requires increasing synthesis while the capacity is decreasing (Fig. 3B). Economically acceptable contents are reached after very different periods of accumulation; this means that the specific formation rate is a very important parameter which depends on the organism involved and, with a given genotype, on the substrate used. Hence we should fo-

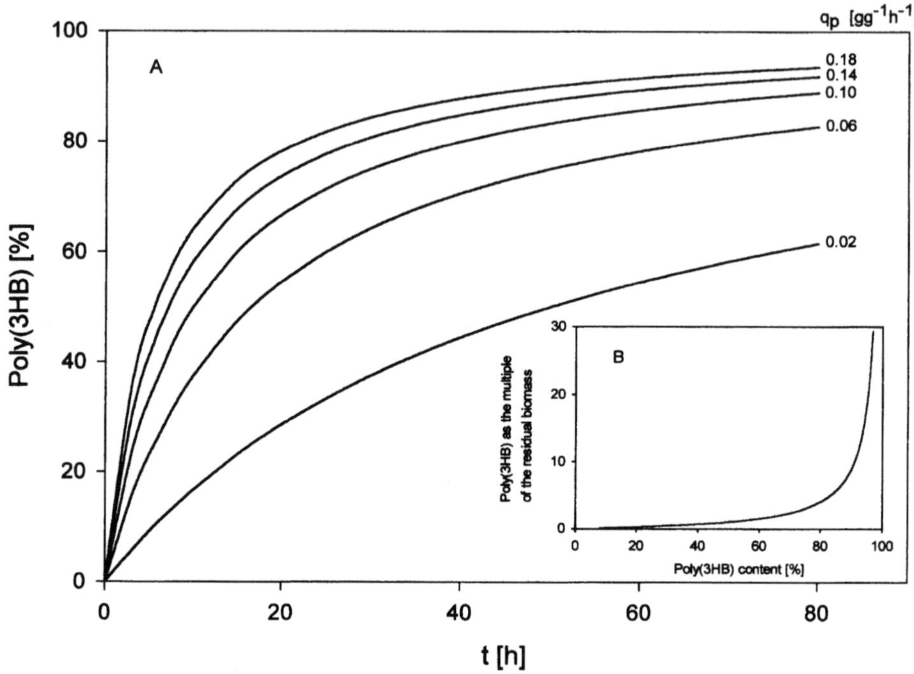

Fig. 3 A, B. **A** Time courses of cellular poly(3HB) content during the accumulation phase plotted for various specific poly(3HB) formation rates. **B** Amount of poly(3HB) as a multiple of the residual biomass at different poly(3HB) contents

cus our attention on the selection of appropriate organisms, and it seems that the yield and economic problems of poly(3HB) production can be solved by maximizing specific formation rates, and product concentration, respectively.

The specific growth rate (μ) becomes of interest for the poly(3HB) content if the poly(3HB) synthesis is growth-coupled. The possible content is then independent of time and is determined by the relation of q_p to μ [74]:

$$PHB(\%) = \frac{q_p}{q_p t + \mu} \, 100\% \tag{9}$$

Such a situation occurs in continuous processes, and can be realized in special cases with cells growing unlimited in discontinuous processes, e.g., as reported in several studies of *Alcaligenes latus* [60], *Azotobacter vinelandii* [98], or *Methylobacterium rhodesianum* [74].

4.2.3
Productivity

Depending on the scale of production, an important parameter determining the economic viability of poly(3HB) production (besides overall yield and poly(3HB) content) is productivity. Productivities of poly(3HB) production attained are in the range of about 2–5 g l^{-1} h^{-1} [61, 99, 100]. These values are two to three orders of magnitude lower than analogous chemical processes of polymer synthesis.

The productivity of the overall process depends on both the biological potential and performance of the technical equipment. Therefore, it is necessary to study the influence of the apparatus on the biological systems. The resulting productivity depends on at least five parameters, i.e., X_0, μ, q_p, t_1, and t_2 (where t_1 = duration of growth and t_2 = duration of poly(3HB) accumulation). To evaluate different values of productivity the initial biomass concentration, X_0, should always be indicated, e.g., as $pr_{0.1}$ for productivity with a X_0 value of 0.1 g l^{-1}.

It is useful to increase the biomass concentrations as much as possible at the end of the growth phase to get high absolute poly(3HB) accumulation rates during the production phase. Additionally, higher poly(3HB)-free biomass concentrations lead to higher productivity of the biomass production process. But biomass concentrations are limited by properties of the technical system, especially by the oxygen transfer rate. The biological potential, represented by the specific rates of growth and production, is the basis of the productivity of each of the two phases. To demonstrate their influence on the (overall) productivity, the other parameters must be taken into account. Figure 4 shows theoretically possible productivities, assuming constant μ and q_p during the growth and production phase, respectively, in a three-parameter plot with predefined values for X_0, X′, and the poly(3HB) content.

The mesh plots show different slopes of productivity, depending on μ and q_p, and the poly(3HB) content. How an increase in μ or q_p enhances productivity is presented in Fig. 5, starting from two different pairs of actual μ and q_p values. Each of the initial values was increased by 20% or 40%. If both specific velocities are increased, then the productivity increases in the same way. Starting from $\mu = 0.35\ h^{-1}$ and $q_p = 0.15\ g\ g^{-1}\ h^{-1}$ (A) optimization of the growth rate has a stronger effect over a broader range of final poly(3HB) contents than starting from $\mu = 0.2\ h^{-1}$ and $q_p = 0.05\ g\ g^{-1}\ h^{-1}$ (B).

Of course, the influence of each parameter depends on the proportional duration of the two phases. The following equation gives the poly(3HB) content, following an equal duration for both phases and, thus, an equal contribution from both specific rates:

$$PHB(\%) = \frac{\text{In}\,X' - \text{In}\,X_0}{\text{In}\,\dfrac{X'}{X_0} + \dfrac{\mu}{q_p}} \tag{10}$$

A common concern is to define the optimum duration of the phases. Whereas the growth phase (t_1) should clearly be as long as the possible biomass concentration is reached this is not so obvious for the production phase. Figure 6 shows the influence of t_1 and t_2 on productivity, with predefined values for X_0, μ, and q_p.

The productivity increases most quickly during the first hours of poly(3HB) accumulation, and then increases more and more slowly. The q_p value will eventually drop with increasing time of accumulation, following which the produc-

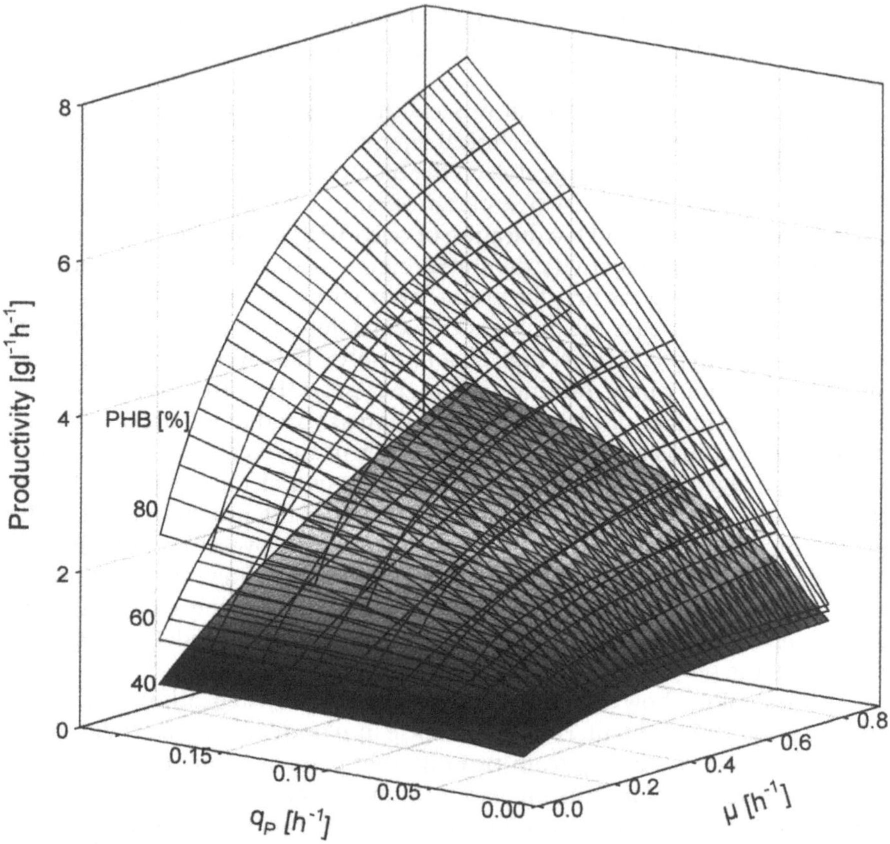

Fig. 4. Productivity of the overall process of poly(3HB) synthesis as a function of specific rates of growth and poly(3HB) accumulation for final poly(3HB) contents of 40%, 60%, and 80%, corresponding to poly(3HB) concentrations of 33.3 g l^{-1}, 75 g l^{-1}, and 200 g l^{-1}, respectively. The initial and final poly(3HB)-free biomass concentrations are 0.1 g l^{-1} and 50 g l^{-1}, respectively

tivity will drop too. This means there is an optimum duration for the poly(3HB) accumulation phase, since the productivity will eventually decline, even though the poly(3HB) concentration can be further increased. With prolongation beyond this optimum time, the poly(3HB) content increases only insignificantly (Fig. 3 A). High poly(3HB) content is of particular interest anyway as poly(3HB) must be recovered and purified. Hence, in order to develop an industrial process, it will be necessary, depending on the actual conditions, to evaluate all factors [101, 102].

Fig. 5 A, B. Enhancement of productivity achieved by increasing μ or q_p, starting from initial values of: **A** 0.35 h^{-1} and 0.15 h^{-1}; **B** 0.20 h^{-1} and 0.05 h^{-1}, respectively. Increase in μ and constant q_p = ●; increase in q_p and constant μ = ○

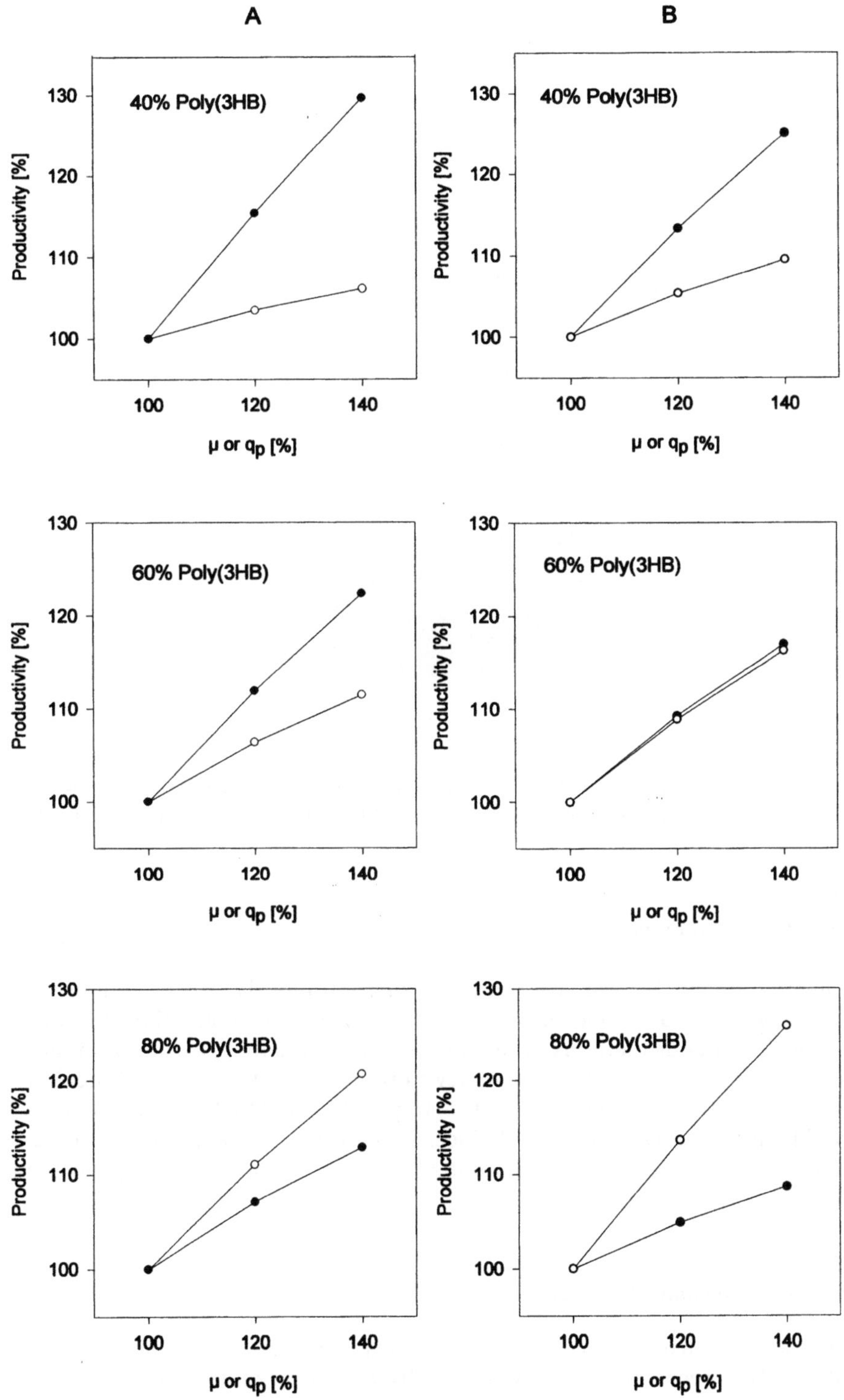
A
B
40% Poly(3HB)
60% Poly(3HB)
80% Poly(3HB)
Productivity [%]
μ or q_p [%]
100
110
120
130
140

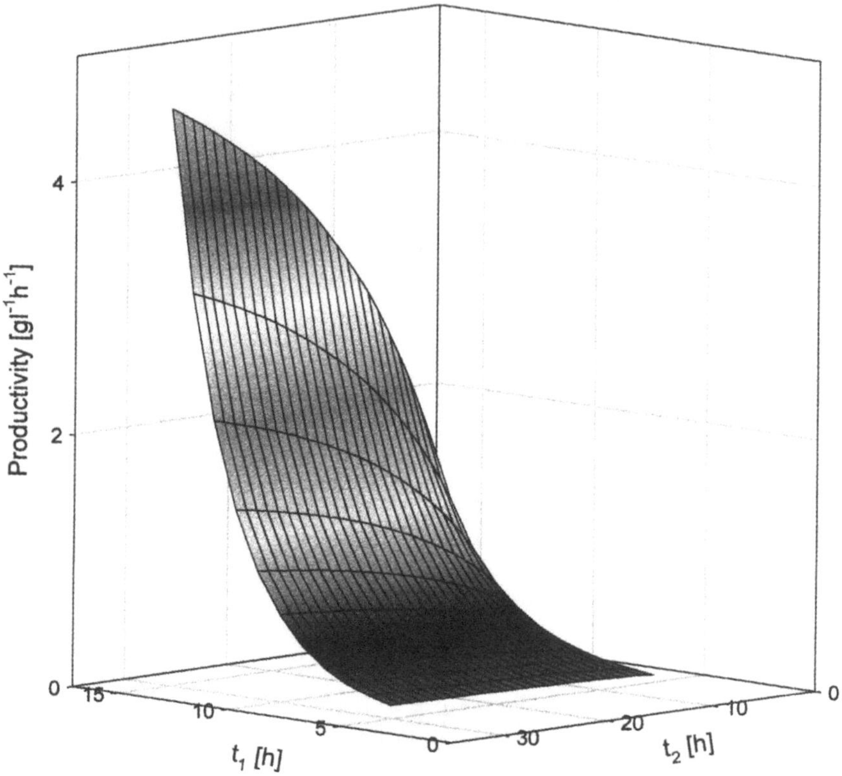

Fig. 6. Productivity of the poly(3HB) production as a function of the duration of growth (t_1) and accumulation (t_2) phases calculated for: $X_0 = 0.1\ g\ l^{-1}$, $\mu = 0.4\ h^{-1}$, and $q_p = 0.17\ g\ g^{-1}\ h^{-1}$

4.3
Technological Procedures

Generally, poly(3HB) can be produced discontinuously or continuously. To reach high biomass concentrations fed-batch processes are the method of choice. Continuous methods have only been occasionally used, but unfounded, as shown below. Comprehensive reviews on the current state of technical procedures have been given by Lee and Chang [103], Lee [99], Braunegg et al. [37], and Madison and Huisman [104]. Special conditions and approaches to maximize the exploitation of bacterial potentials with the two types of process regime are discussed below.

4.3.1
Batch (Discontinuous) Processes

Batch and fed-batch processes are the usual regimes for the production of poly(3HB). It seems they meet well the physiological requirements for synthesizing this product. Limitations required for the overproduction of poly(3HB) can easily be imposed in batch processes. The whole process comprises two phases – a growth phase and a production phase.

Although numerous bacteria synthesize and accumulate poly(3HB), using a wide range of substrates, few species and carbon substrates have attracted interest for its commercial production. The most important organisms are strains of *Ralstonia eutropha* [105], *Alcaligenes latus* [106], *Azotobacter vinelandii* [98, 107], *Methylobacterium organophilum* [108], *M. rhodesianum* [109], *Protomonas extorquens* [96, 97], *Methylocystis* [110], and recombinant strains of *Escherichia coli* [111–113]. The most commonly used substrates are glucose, sucrose, or waste products like molasses and the C_1-substrates, methanol and methane. Productivities reached are in the range of about 1–3 g l^{-1} h^{-1} [99, 100]. Using methanol as a cheap substrate such high productivities have only been reached with very high biomass concentrations, of about 100 g l^{-1}, at the end of the growth phase, since the methylotrophic strains show relatively low rates of growth and product formation. The highest productivity so far reported (4.94 g l^{-1} h^{-1}) was obtained with *A. latus* DSM1123 [61]. This strain showed a two- to three-times higher specific poly(3HB) formation rate than the best of the other producers, and synthesized poly(3HB) simultaneously with growth [60, 61].

Acetic acid is a further interesting substrate, insofar as it is cheap, available as a bulk chemical, and promises, theoretically, more efficient carbon conversion into poly(3HB) than the carbohydrates (Table 3). However, it has not played an important role in this respect, probably because of its strong inhibitory effect on bacterial growth, by acting as an uncoupler. It has been shown that with *Paracoccus denitrificans*, for instance, the acetic acid concentration in the culture medium must not exceed 0.4 g l^{-1} to obtain specific rates of growth and poly(3HB) formation of about 0.5 h^{-1} and 0.16 g g^{-1} h^{-1}, respectively [114]. These rates are similar to the specific velocities of *R. eutropha* using glucose as substrate. In order to keep the substrate concentrations low, the rate at which it is fed must be tightly correlated to the actual demand. Therefore, different feeding methods were employed in the *Paracoccus* studies during the exponential growth phase, the transitional phase with limited growth, and the production phase. Carrying out cultivation in a 100-l fermenter, a biomass concentration of 66 g l^{-1}, with a poly(3HB) content of 65% and a productivity for the process as a whole of $pr_{2.8}$ = 0.74 g l^{-1} h^{-1} was obtained [114]. This example shows that, contrary to first impressions, fed-batch processes often require sophisticated feeding methods to maintain optimum conditions throughout the process.

4.3.2
Continuous Processes

Continuous cultivation, which needs permanent multiplication of cells, seems not to be appropriate for poly(3HB) synthesis because the product is mainly formed during the stationary growth phase. But there are three opportunities, discussed below, that make continuous poly(3HB) synthesis possible.

First, not only after the complete exhaustion, but also when the growth rate decreases due to the limitation of a nutrient like ammonium poly(3HB) formation starts. Such conditions, often called growth-associated poly(3HB) synthesis, have been used for continuous, chemostatic cultivation, in conjunction with the limitation of a nutrient other than the carbon source [4, 115–121]. Strictly speaking, this poly(3HB) synthesis should not be referred to as being associated with growth, but rather with the limitation of growth. It is related to the shortfall in growth velocity due to the limitation. During chemostatic cultivation the poly(3HB) content should decrease with increasing dilution rate, as has been observed by Wilkinson and Munro [116], Senior et al. [115], Dawes and Senior [4], and Siegel and Ollis [118]. As a rule, however, only insufficiently low poly(3HB) contents are obtainable using these techniques, or very low dilution rates have to be used.

Second, in some cases cells accumulate poly(3HB) even during unlimited exponential growth. Such a strong coupling of growth and poly(3HB) synthesis is characterized by a proportionality between q_p and μ or the dilution rate [122]. As a further difference to the growth-associated type of accumulation mentioned above, poly(3HB)-containing cells can be obtained by carbon substrate limited chemostatic cultivation. This phenomenon was first observed by Wilkinson and Munro [116], cultivating *Bacillus megaterium* on glucose. We propose to define this type of poly(3HB) synthesis as "unlimited growth-related". Following the Wilkinson and Munro study [116] it has been reported only for a few strains, e.g., *A. latus* [60] and mutant strains of *A. vinelandii* [98] and of *R. eutropha* [123]. However, the occurrence of unlimited growth-related poly(3HB) synthesis not only depends on the bacterial strain, but also on the substrate used. *M. rhodesianum*, and other strains of the pink pigmented facultative methylotrophs, show this coupling of growth and poly(3HB) synthesis on fructose and other multicarbon substrates, but not on methanol [74]. A common prerequisite of growth-related poly(3HB) synthesis seems to be an internal bottleneck, e.g., impaired respiratory oxidation of NADH in the case of *A. vinelandii* [98, 107] and a low TCA capacity in the *R. eutropha* mutant [123] and the methylotrophs [74]. The conversion of glucose or fructose into poly(3HB) supports the energy budget of the cells during growth. (If formation of poly(3HB) is the sole source of energy for growth the poly(3HB) content resulting from simultaneous growth would amount to about 48–60% (Table 4) [74].) These cases verify the prediction made above that energy excess substrates are predestined for unlimited growth-related poly(3HB) synthesis.

Another example of poly(3HB) formation during unlimited growth was reported by Doi et al. [29]. These authors demonstrated that the synthesis of poly(3HB) from butyrate in *R. eutropha* occurs under balanced growth conditions in the presence of ammonium because the building blocks of poly(3HB) are available without 3-ketothiolase catalyzed conversion. This holds true basically for substrates which are assimilated according to the principle of prefabricated construction [58], e.g., pentanol/poly(3-hydroxyvalerate) [124].

The unlimited growth-related type of poly(3HB) synthesis is most suited to one-stage continuous cultivation. This can be performed in a chemostat with carbon substrate limitation. The poly(3HB) content of the cells can be further

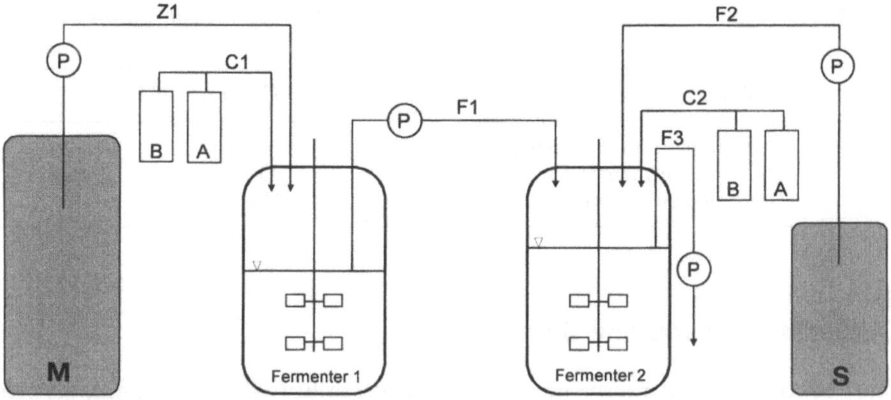

Fig. 7. Scheme for a two-stage continuous cultivation technique. *A* = acid, *B* = base, *M* = medium, *P* = pump, *S* = substrate, *C1* and *C2* = base or acid flow, *F1* = culture flow, *F2* = substrate flow, *F3* = outflow, *Z1* = medium flow

increased by operating with a limitation of another growth substrate. Thus, unlimited growth-related poly(3HB) synthesis can be augmented with limitation-induced poly(3HB) synthesis.

Third, as an alternative to the one-stage continuous cultivation technique, especially for the limitation-caused poly(3HB) synthesis discussed above, a two-stage method can be applied. The principle is shown in Fig. 7.

The cells can be grown at high dilution rates in the first fermenter, usually under the limitation of a growth substrate which triggers poly(3HB) synthesis, to obtain cells at the beginning of the poly(3HB) accumulation phase when entering the second fermenter. Product formation in the second vessel can then take place at high specific velocity.

A high level of poly(3HB) accumulation is also obtained if the cells are grown under carbon substrate limitation, and the cultivation in the second fermenter is also carried out under carbon limitation. In this case, a substrate flow rate (F2) below that corresponding to the maximum specific poly(3HB) formation rate should be chosen [114]. This cultivation strategy is especially convenient when using toxic substrates like acetic acid. Low substrate concentrations are more conveniently maintained in continuous cultivation than in fed-batch cultivation. The only additional equipment needed is a system to ensure constant working volumes and flow rates.

Using this two-stage method and acetic acid as substrate, with culture volumes of 4 l and 12.8 l, respectively, 857 ml of culture broth was harvested per hour containing a biomass concentration of 39 g l^{-1} with a poly(3HB) content of 62%. The resulting productivity was 1.23 g l^{-1} h^{-1} (Table 5).

The continuous regime of poly(3HB) production has some important advantages. High and constant productivities are obtainable because high rates of growth and product formation can be maintained over long periods. Continuous cultivation methods are also convenient for the use of toxic substrates, and constant product quality should be easily obtained.

Table 5. Flow rates in a two-stage continuous cultivation of *Paracoccus denitrificans* with acetic acid as substrate

Flow	Medium/Acetic acid concentration [g l^{-1}]	Flow rate [ml h^{-1}]
Z1	Growth medium (fermenter 1)/49 g l^{-1}	702
F2	Substrate (fermenter 2)/560 g l^{-1}	131
F3	Harvested culture broth/< 0.1 g l^{-1}	857

5 Concluding Remarks and Outlook

The commercial attractiveness and marketability of the homopolymer poly(3HB) are essentially determined by the properties and the price of the product. The properties depend on the molecular mass and molecular mass distribution – which in turn appear to depend on the organism used, the carbon source offered, the conditions set, and the duration of accumulation. However, these interrelations have not yet been carefully studied, and they are insufficiently understood. To improve and tailor its properties during synthesis and accumulation, and to maximize the intracellular poly(3HB) content and the yield, more must be found out about the physiological function of poly(3HB) and the regulation of the poly(3HB) cycle. Although much is known about the regulation of synthesis, the regulation of intracellular breakdown, the external signals, and signal processing remain to be studied properly.

Poly(3HB) is usually produced in a batch or fed-batch regime. These types of process control are derived from the general observation that overproduction of poly(3HB) occurs when cell multiplication is limited by an essential nutrient and the carbon substrate is available in excess. Batchwise production has an advantage in that a high poly(3HB) content can be reached. One disadvantage is the quality of the product, which can vary from one batch to the next. This can be overcome by a continuous process. Continuous production is basically possible if:

1. 3-Ketothiolase is not inhibited (by free CoA) and not involved in synthesis
2. Poly(3HB) synthesis is coupled with the generation of biologically useful energy, which can be consumed for growth and multiplication

Whether synthesis is energy-generating or energy-consuming and 3-ketothiolase is involved in the synthesis depends on the source of carbon and energy as well as the metabolic pathway, and can be predicted. Although whether a correlation also exists between the actual 3-ketothiolase activity and the carbon substrate is unclear, it is certainly conceivable. This should be elucidated by systematic studies and considered if continuous processes with particular substrates, e.g., potentially toxic and waste products, are to be evolved. The economic viability of poly(3HB) production from, for instance, glucose, sugar-containing products (i.e., molasses and whey), and methanol are well-known, the upper limits of the economically relevant parameters have been calculated, and the auxiliary substrate concept appears able to approach these limits. Nevertheless, the final price of the product will still be high, and so cheap raw materials must be found. Attention should be increasingly focused on waste products. Biotechnologists must develop processes to convert worthless materials, e.g., environmentally harmful materials, into harmless valuables, such as biodegradable poly(3HB).

A continuous operation seems to be the method of choice, as the catalyst (cells) will have to be stabilized by steady reproduction.

References

1. Brandl H, Gross RA, Lenz RW, Fuller RC (1990) Adv Biochem Engin/Biotechnol 41:77
2. Anderson AJ, Dawes EA (1990) Microbiol Rev 54:450
3. Steinbüchel A (1991) In: Byrom D (ed) Biomaterials. Stockton Press, New York, p 124
4. Dawes EA, Senior PJ (1973) Adv Microbial Physiol 10:135
5. Babel W (1992) FEMS Microbiol Rev 103:141
6. Macrae RM, Wilkinson JF (1958) J Gen Microbiol 19:210
7. Neijssel OM, Tempest DW (1979) Symp Soc Gen Microbiol 29:53
8. Merrick JM, Doudoroff M (1961) Nature 189:890
9. Steinbüchel A, Aerts K, Babel W, Föllner C, Liebergesell M, Madkour MH, Mayer F, Pieper-Fürst U, Pries A, Valentin HE, Wieczorek R (1995) Can J Microbiol 41(Suppl 1):94
10. Senior PJ, Dawes EA (1973) Biochem J 134:225
11. Haywood GW, Anderson AJ, Chu L, Dawes EA (1988) FEMS Microbiol Lett 52:91
12. Nishimura T, Saito T, Tomita K (1978) Arch Microbiol 116:21
13. Williams DR, Anderson AJ, Dawes EA (1993) In: Schlegel HG, Steinbüchel A (eds) Proc Internat Symp Biol Polyhydroxyalkanoates '92 (ISBP'92) Göttingen. Golze-Druck, Göttingen, p 387
14. Mothes G, Skinfill Rivera I, Babel W (1997) Arch Microbiol 166:405
15. Haywood GW, Anderson AJ, Chu L, Dawes EA (1988) FEMS Microbiol Lett 52:259
16. Moskowitz GJ, Merrick JM (1969) Biochemistry 8:2748
17. Mothes G, Babel W (1995) Can J Microbiol 41 (Suppl 1):124
18. Doi Y, Kitamura S, Abe H (1995) Macromolecules 28:4822
19. Saito T, Fukui T, Ikeda F, Tanaka Y, Tomita K (1977) Arch Microbiol 114:211
20. Mothes G, Babel W (1994) Arch Microbiol 161:68
21. Bloomfield G, Sandhu G, Carr NG (1969) FEBS Lett 5:246
22. Ritchie GAF, Senior PJ, Dawes EA (1971) Biochem J 121:309
23. Amos DA, McInerney MJ (1993) Arch Microbiol 159:16
24. Liebergesell M, Steinbüchel A (1992) Eur J Biochem 209:135
25. Rehm BHA, Steinbüchel A (1999) Int J Biol Macromol 25:3
26. Haywood GW, Anderson AJ, Dawes EA (1989) Biotechnol Lett 11:471
27. Doi Y, Kunioka M, Nakamura Y, Soga K (1987) Macromolecules 20:2988
28. Doi Y, Tamaki A, Kunioka M, Soga K (1987) J Chem Soc Chem Commun 1635
29. Doi Y, Tamaki A, Kunioka M, Soga K (1988) Appl Microbiol Biotechnol 28:330
30. Haywood GW, Anderson AJ, Dawes EA (1989) FEMS Microbiol Lett 57:1
31. Valentin HE, Schönebaum A, Steinbüchel A (1992) Appl Microbiol Biotechnol 36:507
32. Huisman GW, de Leeuw O, Eggingk G, Witholt B (1989) Appl Environ Microbiol 55:1949
33. Gerngross TU, Martin DP (1995) Proc Natl Acad Sci USA 92:6279
34. Su L, Lenz RW, Martin DP (2000) Macromolecules (in press) (refer to [99])
35. Sim SJ, Snell KD, Hogan SA, Stubbe J, Rha C, Sinskey AJ (1997) Nat Biotechnol 15:63
36. Kraak MN, Smits THM, Kessler B, Witholt B (1997) J Bacteriol 179:4985
37. Braunegg G, Lefebvre G, Genser KF (1998) J Biotechnol 65:127
38. Kunioka M, Kawagushi Y, Doi Y (1989) Appl Microbiol Biotechnol 30:569
39. Valentin HE, Zwingmann G, Schönebaum A, Steinbüchel A (1995) Eur J Biochem 227:43

40. Williams DR, Anderson AJ, Dawes EA, Ewing DF (1994) Appl Microbiol Biotechnol 40:717
41. Haywood GW, Anderson AJ, Ewing DF, Dawes EA (1990) Appl Environ Microbiol 56: 3354
42. Huijberts GN, Eggink G, de Waard P, Huisman GW, Witholt B (1992) Appl Environ Microbiol 58:536
43. DeSmet MJ, Eggink GM, Witholt B, Kingma J, Wynberg H (1983) J Bacteriol 154:870
44. Lageveen RG, Huisman GW, Preusting H, Ketelae P, Eggingk G, Witholt B (1988) Appl Environ Microbiol 54:2924
45. Weitzman PDJ (1981) Adv Microbial Physiol 22:185
46. Müller-Kraft G, Babel W (1986) Biol Rundsch 24:165
47. Belova LL, Sokolov AP, Morgunov IG, Trotsenko YA (1997) Biochemistry 62:71
48. Henderson RA, Jones CW (1997) Arch Microbiol 168:486
49. Oeding V, Schlegel HG (1973) Biochem J 134:239
50. Tomita K, Saito T, Fukui T (1983) In: Lennon DLF, Stratman FW, Zahlten RN (eds) Biochemistry of metabolic processes. Elsevier Science Publishing, p 353
51. Fukui T, Ito M, Saito T, Tomita K (1987) Biochim Biophys Acta 917:365
52. Belova LL, Trotsenko YA, Sokolov AP, Sidonov IA (1997) FEMS Microbiol Lett 156:275
53. Belova LL, Sokolov AP, Trotsenko YA (1997) Appl Biochem Microbiol 33:70
54. Jackson FA, Dawes EA (1976) J Gen Microbiol 97:303
55. Mothes G, Ackermann J-U, Babel W (1998) Arch Microbiol 144:62
56. Stouthamer AH (1973) Antonie van Leeuwenhoek 39:545
57. Van Dijken JP, Harder W (1975) Biotech Bioeng 17:15
58. Babel W, Brinkmann U, Müller RH (1993) Acta Biotechnol 13:211
59. Shi H, Shimizu K, Shiraishi M (1997) J Ferment Bioeng 84:579
60. Braunegg G, Bogensbergen B (1985) Acta Biotechnol 5:339
61. Wang F, Lee SY (1997) Appl Environ Microbiol 63:3703
62. Jendrossek D, Schirmer A, Schlegel HG (1996) Appl Microbiol Biotechnol 46:451
63. Jendrossek D (1998) Polym Degrad Stabil 59:317
64. Hippe H, Schlegel HG (1976) Arch Mikrobiol 56:278
65. Daniel M, Choi JH, Kim JH, Lebeault JM (1992) Appl Microbiol Biotechnol 37:702
66. Kovar J, Matyskova I, Matyska L (1986) Biochim Biophys Acta 871:302
67. Fukui T, Ito M, Tomita K (1982) Eur J Biochem 127:423
68. Atkinson DE (1966) Ann Rev Biochem 35:85
69. Knowles JC (1977) Symp Soc Gen Microbiol 27:241
70. Ritchie GAF (1968) PhD thesis, University of Hull
71. Pries A, Priefert H, Krüger N, Steinbüchel A (1991) J Bacteriol 173:5843
72. Babel W (1986) Acta Biotechnol 6:215
73. Babel W (1990) Biotech Adv 8:261
74. Ackermann J-U, Babel W (1997) Appl Microbiol Biotechnol 47:144
75. Anthony C (1982) The biochemistry of methylotrophs. Academic Press
76. Tal S, Okon Y (1985) Can J Microbiol 31:608
77. Macrae RM, Wilkinson JF (1958) Proc R Phys Soc Edin 27:73
78. Schlegel H-G, Gottschalk G, von Bartha R (1961) Nature 191:463
79. Leonard D, Lindley ND (1998) Microbiology 144:241
80. Hughes EJ, Bayly RC (1983) J Bacteriol 154:1363
81. Hueting S, Tempest DW (1977) Arch Microbiol 155:73
82. Byrom D (1987) Tibtech 5:246
83. Babel W, Riis V, Hainich E (1990) Plaste und Kautschuk 37:109
84. Knowles JC (1993) J Med Engin Technol 17:129
85. Babel W (1997) BioWorld 4:16
86. Hilger U, Sattler K, Littkowski U (1991) Zentralbl Mikrobiol 146:83
87. Lee SY (1996) Biotechnol Bioeng 49:1
88. De Koning GJM, Lemstra PJ (1993) Polymer 34:4089
89. Lengweiler UD, Fritz MG, Seebach D (1996) Helv Chim Acta 79:670

90. Seebach D, Fritz MG (1999) Int J Biol Macromol 25:217
91. Müller RH, Babel W (1988) Acta Biotechnol 8:249
92. Müller RH, Babel W (1986) Acta Biotechnol 144:62
93. Ackermann J-U, Babel W (1998) Polym Degrad Stabil 59:183
94. Bitar A, Underhill S (1990) Biotechnol Lett 12:563
95. Aragao GMF, Lindley ND, Uribelarrea JL, Pareilleux A (1996) Biotechnol Lett 18:937
96. Suzuki T, Yamane T, Shimizu S (1986) Appl Microbiol Biotechnol 24:366
97. Suzuki T, Yamane T, Shimizu S (1986) Appl Microbiol Biotechnol 24:370
98. Page WJ, Knosp O (1989) Appl Environ Microbiol 55:1334
99. Lee SY (1996) Tibtech 14:431
100. Ryu HW, Hahn SK, Chang YK, Chang HN (1997) Biotechnol Bioeng 55:28
101. Lee SY, Choi J (1998) Polym Degrad Stabil 59:387
102. Choi J, Lee SY (1999) Appl Microbiol Biotechnol 51:13
103. Lee SY, Chang HN (1995) Can J Microbiol 41(Suppl 1):207
104. Madison LA, Huisman GW (1999) Microbiol Mol Biol Rev 63:21
105. Byrom D (1992) FEMS Microbiol Rev 103:247
106. Hrabak O (1992) FEMS Microbiol Rev 103:251
107. Manchak J, Page WJ (1994) Microbiol 140:953
108. Kim SW, Kim P, Lee HS, Kim JH (1996) Biotechnol Lett 18:25
109. Patent: DE 97-19704045
110. Wendlandt K-D, Jeschorek M, Helm J, Stottmeister U (1998) Polym Degrad Stabil 59:191
111. Kim BS, Chang HN, Lee SY (1992) Biotechnol Lett 14:811
112. Choi JI, Lee SY, Han K (1998) Appl Environ Microbiol 64:4897
113. Wang F, Lee SY (1998) Biotechnol Bioeng 58:325
114. Ackermann J-U, Mothes G, Babel W (1999) ISEB '99 Meeting Biopolymers Leipzig (in press)
115. Senior PJ, Beech GA, Ritchie GA, Dawes EA (1972) Biochem J 128:1193
116. Wilkinson JF, Munro AS (1967) In: Powell EO, Evans CGT, Strange RE, Tempest DW (eds) Microbial physiology and continuous culture. HMSO, London, p 173
117. Morinaga Y, Yamanaka S, Ishizaki A, Hirose Y (1978) Agric Biol Chem 42:439
118. Siegel RS, Ollis DF (1984) Biotechnol Bioeng 26:764
119. Duchars MG, Attwood MM (1989) J Gen Microbiol 135:787
120. Ramsay BA, Lomaliza K, Chavarie C, Dube B, Bataille P, Ramsay JA (1990) Appl Environ Microbiol 56:2093
121. Egli T (1991) Antonie van Leeuwenhoek 60:225
122. de Hollander JA (1993) Antonie van Leeuwenhoek 63:375
123. Park J-S, Lee YH (1996) J Ferment Bioeng 81:197
124. Yamane T, Fukunaga M, Lee YW (1996) Biotechnol Bioeng 50:197

Received: January 2000

Production of Microbial Polyesters: Fermentation and Downstream Processes

B. Kessler[1], R. Weusthuis[2], B. Witholt[1], G. Eggink[2]

[1] Institute of Biotechnology, ETH Zürich, Hönggerberg HPT, 8093 Zürich, Switzerland
E-mail: bw@biotech.biol.ethz.ch

[2] Agrotechnological Research Institute (ATO-DLO), Bornsesteeg 59, 6700 AA Wageningen, The Netherlands

Poly(3-hydroxyalkanoates) (PHAs) constitute a large and versatile family of polyesters produced by various bacteria. PHAs are receiving considerable attention because of their potential as renewable and biodegradable plastics, and as a source of chiral synthons since the monomers are chiral. Industrial PHA production processes have been developed for poly(3-hydroxybutyrate) (poly(3HB)) and poly(3-hydroxybutyrate-*co*-3-valerate) (poly(3HB-*co*-3HV). More than 100 other poly(3HA_{MCL})s, characterized by monomers of medium chain length, have been identified in the past two decades. These monomers typically contain 6–14 carbon atoms, are usually linked via 3-hydroxy ester linkages, but can occasionally also exhibit 2-, 4-, 5-, or 6-hydroxy ester linkages. Such polyesters are collectively referred to as medium chain length PHAs poly(3HA_{MCL})s.

The vast majority of these interesting biopolyesters have been studied and produced only on the laboratory scale. However, there have been several attempts to develop pilot scale processes, and these provide some insight into the production economics of poly(3HA_{MCL})s other than poly(3HB) and poly(3HB-*co*-3HV). These processes utilize diverse fermentation strategies to control the monomer composition of the polymer, enabling the tailoring of polymer material properties to some extent. The best studied of these is poly(3-hydroxyoctanoate) (poly(3HO)), which contains about 90% 3-hydroxyoctanoate. This biopolyester has been produced on the pilot scale and is now being used in several experimental applications.

Keywords. PHA, Polyester, Fermentation, Downstream process, Economics

Advances in Biochemical Engineering/
Biotechnology, Vol. 71
Managing Editor: Th. Scheper

1 Introduction

The first microbial polyhydroxyalkanoate (PHA) to be discovered was poly(3-hydroxybutyrate) (poly(3HB)), described in 1926 [1]. Since then poly(3HB) accumulation was found to occur in various microorganisms, representatives of Gram-negative and Gram-positive species (i.e., autotrophs, heterotrophs, phototrophs, aerobes, anaerobes), and archaebacteria (as reviewed elsewhere [2–4]). Bacteria synthesize and accumulate PHAs as carbon and energy storage materials under conditions of limiting nutrients in the presence of excess carbon source. When the supply of the limiting nutrient is restored, the PHA can be degraded by intracellular depolymerases and subsequently metabolized as carbon and energy source [5]. The 3-hydroxyalkanoic acid monomer units in these microbial polyesters, where determined, are all in *R*-configuration, indicating high stereospecificity of the biosynthetic enzymes. The molecular weights of the polymers range from 2×10^5 to 3×10^6, depending on the specific polymer, the microorganism, and growth conditions.

The discovery of a polyester consisting mainly of hydroxyoctanoate monomers by de Smet et al. [6] was the first example of a new group, poly $(3HA_{MCL})$s, that contains a wide variety of different medium-chain-length monomers. To date, more than 100 different monomers have been found in poly$(3HA_{MCL})$s [7]. These include 3-hydroxy acids of 6–14 carbon atoms with a large variety of saturated, unsaturated, straight, or branched aliphatic chains, in some cases containing aromatic side groups. Furthermore, monomers with various different functional groups in the side chain such as halogen atoms, hydroxy-, epoxy-, cyano-, carboxyl-, phenoxy-, cyanophenoxy-, nitrophenoxy-, and esterified carboxyl groups have been introduced into poly$(3HA_{MCL})$s (for reviews see [3, 7, 8]). The poly$(3HA_{MCL})$s are of potential interest for specific uses where chirality and elastomeric properties of the polymers are important. In addition, the monomers of PHAs that contain different functional groups in their side chain are a potentially valuable source of chiral synthons [9, 10]. In this chapter we will focus on microbial production of these polyesters by fermentation, recovery of the material, and present economic considerations. A general overview of the different production steps is presented in Fig. 1.

2 Fermentation Processes

2.1 Production of Poly(3HB) and Poly(3HB-*co*-3HV)

Many bacteria have been screened to produce poly(3HB) or poly(3HB-*co*-3HV) (a copolymer consisting of 3-hydroxybutyrate and 3-hydroxyvalerate). How-

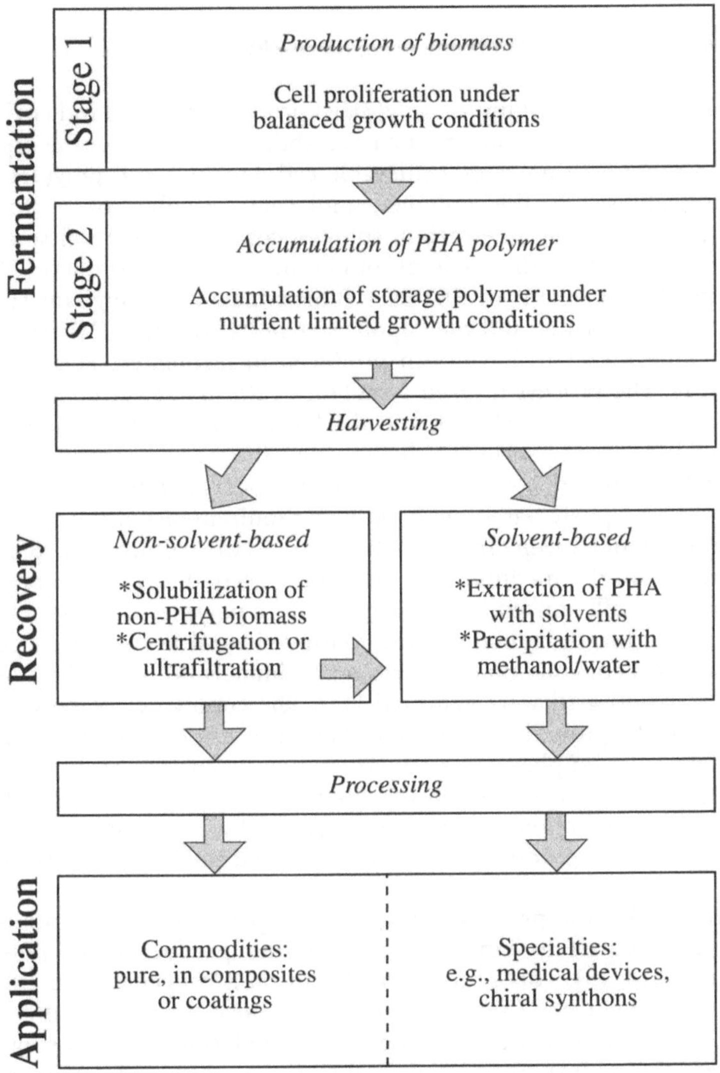

Fig. 1. General overview of a PHA production process. All PHA production processes consist of a fermentation and a recovery step, followed by polymer processing for specific applications. In many cases the fermentation is divided into two stages, a biomass production and a PHA accumulation stage (for further detail see text). Several methods for recovery of the material have been described, of which solvent-based and non-solvent-based recovery protocols are illustrated

ever, there are only a few which have been used for biotechnological production of these polyesters on a large scale. We describe below the two examples which have found application in an industrial process and also present several alternative processes and bacterial strains that have been considered for poly(3HB) and poly(3HB-*co*-3HV) production.

2.1.1
Industrial Production of Poly(3HB) and Poly(3HB-co-3HV)

The suitability of a bacterium for poly(3HB) production on an industrial scale depends on many different factors such as stability and safety of the organism, growth and accumulation rates, achievable cell densities and poly(3HB) contents, extractability of the polymer, molecular weights of accumulated poly(3HB), range of utilizable carbon sources, costs of the carbon source and the other components of the medium, and occurrence of byproducts [11].

In the production plant of ZENECA BioProducts at Billingham (UK) the poly(3HB) homopolymer and the copolyester poly(3HB-*co*-3HV) (sold under the trade name BIOPOL) have been produced in fermenters at scales up to 200,000 l [11]. The process utilized a mutant of *Ralstonia eutropha* (formerly called *Alcaligenes eutrophus*) [12], and the fermentation was carried out in a two-step fed batch process. During the first step, the cells were grown in a mineral salt medium with glucose as sole carbon and energy source, and a calculated amount of phosphate based on the known requirements of the organism to allow production of a given amount of biomass. As the culture grew, phosphate was depleted from the medium, and during the second step, when phosphate was limiting, the cells started to produce and store polymer. Glucose was fed to the culture, and the fermentation was continued until the required polymer content was reached. Each phase lasted approximately 48 h and final biomass dry weights of 100 g l^{-1} were achieved [13]. The copolymer of poly(3HB-*co*-3HV) was made by providing a mixed feed of glucose and propionic acid in the polymer accumulation phase. The 3-hydroxyvalerate content of the copolymer was regulated by adjusting the ratio of the two substrates in the feed [13]. In 1996 the entire production process was acquired by Monsanto (USA), which however stopped the production of poly(3HB) and poly(3HB-*co*-3HV) at the end of 1998.

A different industrial process has been developed at the Biotechnologische Forschungsgesellschaft in Linz (Austria) for the production of poly(3HB) homopolymer [14, 15]. This process was based on *Alcaligenes latus* DSM1124 which can accumulate poly(3HB) to up to 80% of the cellular dry matter during balanced cell growth in a one-step fed batch fermentation process using a mineral salts medium with sucrose as sole carbon source. Although a biomass density of 60 g l^{-1} was achieved [15], the company stopped the production of poly(3HB) in 1993.

2.1.2
Alternative Processes and Bacterial Strains for Poly(3HB) and Poly(3HB-co-3HV) Production

Various other bacterial strains and processes have been studied by academic groups for the production of poly(3HB) or poly(3HB-*co*-3HV), several of which are presented here. Some methylotrophic and methanotrophic bacteria are interesting for poly(3HB) production purposes. Methanol is an inexpensive substrate and there is considerable experience in methanol fermentation technology because methylotrophic bacteria have been considered for the large-scale industrial production of single cell protein. It has been found that only methylotrophic bacteria using the Calvin-Benson-Bassham pathway or the serine pathway for the assimilation of C1 compounds are able to accumulate poly(3HB) [16]. *Pseudomonas* sp. K could be cultivated in a fed batch process with nitrogen deficiency to cell densities as high as 233 g l^{-1} cell dry weight and with a polymer content of 64% (w/w) for a product yield of 0.2 g poly(3HB) per g methanol [17]. *Protomonas extorquens* could also be grown to cell densities of 190 g l^{-1} cell dry weight with a polymer content of 60% [18]. The methylotrophic bacteria *Paracoccus denitrificans* and *Methylobacterium extorquens* synthesize a poly(3HB-*co*-3HV) copolyester when methanol and *n*-amyl alcohol are fed simultaneously in nitrogen-limited medium. The 3-hydroxyvalerate content of the formed poly(3HB-*co*-3HV) reached up to 91.5 mol% in *P. denitrificans,* dependent on the substrate ratio, whereas in *M. extorquens* the maximum 3-hydroxyvalerate content did not exceed 38.2 mol% [19].

P. denitrificans is able to produce a poly(3HV) homopolymer when grown on *n*-pentanol, the polymer reaching 24% of cell dry weight [20]. *Paracoccus* sp. 12-A is even able to accumulate poly(3HB-*co*-3HV) when grown on formate [21]. A microcomputer-aided automatic fed batch culture system under potassium-limited conditions was set up for poly(3HB) production with *Methylobacterium organophilum* using methanol as substrate [22]. Interestingly, the capability of methanotrophs to grow on chlorinated hydrocarbons, a characteristic which has been used for bioremediation purposes, can be exploited for poly(3HB) production as well. It could be shown that *Methylosinus trichosporium* OB3b is capable of degrading trichloroethylene (TCE), a common groundwater contaminant, and produce poly(3HB) therefrom [23].

Azotobacter vinelandii UWD is another strain worth mentioning. Transformation of strain *A. vinelandii* UW with DNA from *A. vinelandii* 113 resulted in *A. vinelandii* UWD [24] which turned out to be an interesting organism for the production of poly(3HB) from inexpensive carbon sources. It synthesizes poly(3HB) during exponential growth on unrefined carbon sources such as corn syrup, cane molasses, beet molasses, or malt extract with high yield [25]. Even ultra-high molecular weight (up to 4 million Da) poly(3HB) could be produced when the strain was grown on a medium containing beet molasses [26]. Addition of valerate as a co-substrate allowed production of a poly(3HB-*co*-3HV) copolymer containing up to 23 mol% 3-hydroxyvalerate [27]. When grown on fish peptone, which was used as a nitrogen source by this nitrogen-fixing bacterium, poly(3HB) accumulation but not cell growth was enhanced [28]. Furthermore, addition of fish peptone generated large, pleomorphic, osmotically sensitive cells and caused partial cell lysis. The fragility of these cells was exploited in a simple procedure for the extraction of poly(3HB) [28].

In order to produce poly(3HB) from inexpensive substrates, a two-stage fed batch method employing two different microorganisms has also been explored [29]. In the first stage, the pentose xylose was converted by strain *Lactococcus lactis* to a mixture of lactic and acetic acids. After removal of the cells, *Ralstonia eutropha* was inoculated in the supernatant in the same fermentor. Cost calculations must be performed to determine whether this might be a viable strategy to produce poly(3HB) from inexpensive carbon sources.

Formation of poly(3HB) is also a common feature of anoxygenic phototrophic bacteria. Studied examples include *Chromatium, Thiocystis, Thiocapsa* and *Rhodococcos, Rhodobacter, Rhodospirillum* species of the purple sulfur and purple non-sulfur groups, respectively. On acetate most purple sulfur bacteria produce only a poly(3HB) homopolymer. Interestingly, *Chromobacterium violaceum,* which is a purple pigment containing heterotrophic bacterium, is capable of producing a homopolyester consisting of 3-hydroxyvalerate when cultivated on valeric acid as sole carbon source [30]. A fed batch process has been developed that allows formation of 41 g l^{-1} cellular biomass with approximately 65% of polymer [31]. In contrast, some purple non-sulfur bacteria incorporate small amounts of 3-hydroxyvalerate in poly(3HB) even when grown on acetate as sole carbon source [32]. The effect of the culture medium pH on intracellular accumulation of poly(3HB) has been studied in pH-stat cultures of *Rhodobacter sphaeroides* RV. A sub-optimal pH for growth (pH 8.0–8.5) resulted in a higher content of poly(3HB) which suggested a pH-controlled poly (3HB) accumulation process [33].

R. eutropha is actually an autotrophic hydrogen-oxidizing bacterium which can also produce poly(3HB) from CO_2, H_2, and O_2 [34]. The critical factor in such autotrophic cultivation processes is to avoid possible gas explosions. Therefore, a recycled gas, closed circuit culture system equipped with several safety features was developed and the oxygen concentration in the substrate gas phase was kept below the lower limit for gas explosions. A bacterial biomass of 91.3 g l^{-1} has been achieved and the poly(3HB) content reached up to 67% per cell dry weight under these oxygen-limited conditions [35].

Cyanobacteria which are capable of producing poly(3HB) using CO_2 as carbon source are also worth mentioning. For example, *Synechococcus* sp. MA19, a thermophilic strain, produces 20% poly(3HB) (w/w) when cultivated in a nitrogen-free inorganic medium aerated with CO_2 [36]. Enrichment of CO_2 and increasing light intensity had a stimulating effect on poly(3HB) accumulation. In addition, darkness increased the poly(3HB) content to 27% because cell dry weight was reduced due to glycogen decomposition [36].

Finally, a novel process has been described for efficient photoconversion of low-grade organic materials such as waste biomass into polyesters. In this process, dry biomass has been thermally gasified which resulted in gas mixtures composed primarily of CO and H_2. Photosynthetic bacteria photoassimilated components of the synthesized gas into new cell mass. Under unbalanced culture conditions, when growth was limited by several nutrients, up to 28% of the new biomass was found to be poly(3HB) [37].

In summary, many alternative poly(3HB) and poly(3HB-*co*-3HV) producing strategies have been demonstrated in the past which might be considered for economic evaluation and future production processes. However, it should not be forgotten that all of these bacterial processes may some day have to compete with future alternative processes based on the production of poly(3HB) and poly(3HB-*co*-3HV) in transgenic plants [38–40].

2.2
Production of Poly(3HAMCL)s other than Poly(3HB) and Poly(3HB-*co*-3HV)

Fluorescent pseudomonads are capable of synthesizing poly($3HA_{MCL}$)s from a large number of substrates. Work on the biotechnological production of poly($3HA_{MCL}$) has focused mainly on two model systems – *Pseudomonas oleovorans* and *P. putida*. *P. oleovorans* is able to use alkanes and alkenes as substrate due to the presence of the OCT-plasmid while *P. putida*, which does not have this plasmid, cannot. In contrast to *P. oleovorans*, however, *P. putida* can use carbohydrates such as glucose and fructose for the production of poly($3HA_{MCL}$).

2.2.1
Production of Saturated and Unsaturated Poly(3HAMCL)s by Pseudomonads

Lageveen et al. [41] showed that the monomer composition of aliphatic saturated poly($3HA_{MCL}$) produced by *P. oleovorans* is depended on the type of *n*-alkane used. It appeared that the *n*-alkanes were degraded by the subsequent removal of C2-units and it was therefore proposed that the β-oxidation pathway was involved in poly($3HA_{MCL}$) biosynthesis. Preusting et al. [42] confirmed these results but also showed that with hexane as substrate some 3-hydroxyoctanoate and 3-hydroxydecanoate were produced, indicating that additional pathways were involved in poly($3HA_{MCL}$) biosynthesis (Table 1).

Comparable results were found with poly($3HA_{MCL}$) production by *P. putida* KT2442 using fatty acids as substrates [43]. Studies with ^{13}C-labeled decanoic acid and inhibitors of β-oxidation and fatty acid synthesis showed that this substrate was converted into poly($3HA_{MCL}$) by the β-oxidation pathway exclusively. Experiments with ^{13}C-labeled hexanoic acid as substrate showed that three pathways are involved in its conversion into poly($3HA_{MCL}$). Hexanoic acid can be incorporated directly into poly($3HA_{MCL}$) as 3-hydroxyhexanoic acid after half a cycle of the β-oxidation. In addition, it appeared that some of the hexanoic acid is partly or fully degraded by the β-oxidation cycle and that the generated acetyl-CoA is used for de novo fatty acid synthesis to produce C6 to C14 monomers. Finally, the presence of unsaturated monomers suggests that de novo fatty acid synthesis is also involved. There was also evidence that hexanoic acid was elongated to 3-hydroxyoctanoic acid [43].

The affinity of the PHA polymerase that performs the polymerization reaction differs clearly for the various 3-hydroxy fatty acids. Based on the composition data of PHA synthesized from alkanes or fatty acids it is clear that the 3-hydroxy fatty acids with chain lengths of 8, 9, or 10 carbon atoms are preferred. The affinity of the polymerase decreases for 3-hydroxy fatty acids with longer or shorter chains.

As mentioned in the introduction, 3-hydroxy fatty acids with functional groups can also be incorporated in poly($3HA_{MCL}$). Table 2 illustrates this with many examples of alkenes, 3-hydroxyalkenoic acids, and substituted 3-hydroxyalkanoic acids that are readily integrated in poly($3HA_{MCL}$). Long chain fatty acids have also been used successfully as substrates for poly($3HA_{MCL}$) production. De Waard et al. [44] used oleic acid and linoleic acid to produce

Table 1. Monomer composition[a] of PHAs produced by *Pseudomonas oleovorans* grown on *n*-alkanes as sole carbon and energy source[b]

Carbon source	C4	C5	C6	C7	C8	C9	C10
n-Hexane			83.1 ± 0.5	< 1.0	12.0 ± 0.2	< 1.0	4.9 ± 0.4
n-Heptane		2.5 ± 0.1		97.5 ± 0.1	< 1.0	< 1.0	
n-Octane	< 1.0		12.2 ± 0.2		87.8 ± 0.2		< 1.0
n-Nonane		2.3 ± 0.1		40.6 ± 0.4	1.4 ± 0.2	55.7 ± 0.4	
n-Decane	< 1.0		11.1 ± 0.4	1.1 ± 0.1	65.8 ± 0.2	1.2 ± 0.1	20.8 ± 0.7

[a] Given in mol % of total PHA.
[b] Taken from [41, 42].
C4 = 3-hydroxybutyric acid, C5 = 3-hydroxyvaleric acid, C6 = 3-hydroxyhexanoic acid, C7 = 3-hydroxyheptanoic acid, C8 = 3-hydroxyoctanoic acid, C9 = 3-hydroxynonanoic acid; C10 = 3-hydroxydecanoic acid.

Table 2. Examples of used substrates[a] and resulting types of polymer

Functional group	Copolymer			Homopolymer	Blend
	Single substrate	Two good[b] substrates	Good[b] + poor[b] substrates		

Saturated alkyl	Hexanoate or hexane;
	Octanoate or octane;
	Nonanoate or nonane;
	Decanoate or decane;
	Undecanoate or undecane;
	Dodecanoate or dodecane
		Octanoate/nonanoate
		
		poly(3HB-*co*-3HV): glucose/propionate (or many others, see Sect. 2.1)
				Heptanoate or heptane[d];
				
				poly(3HB): glucose (or many others, see section 2.1);
				poly(3HV): pentanol or valerate;
				poly(4HB): 1-4-butanediol or 4-butyrate
					poly(3HB) + poly ($3HA_{MCL}$): glucose[e]

Table 2 (continued)

Unsaturated alkyl	Octenoate or octene; Decenoate or decene; Oleate	Octanoate/undecenoate; Nonanoate/octenoate; Octane/octene			Nonanoate+10-undecenoate[f]
	poly(3HA$_{MCL}$): glucose or gluconate				
Methyl-branched	7-Methyl-octanoate	Octanoate/7-methyl-octanoate	Octanoate/5-methyl-octanoate or 6-methyl-octanoate		
Alkylester	Heptanoate-7-methylester; Octanoate-8-methylester; Nonanoate-9-methylester; Decanoate-10-methylester; Heptanoate-7-ethylester; Octanoate-8-ethylester; Decanoate-10-ethylester; Heptanoate-7-propylester	Octanoate/octanoate-8-methylester or nonanoate-9-methylester or decanoate-10-methylester or octanoate-8-ethylester or decanoate-10-ethylester			
Phenyl	6-Phenylhexanoate; 7-Phenylheptanoate; 9-Phenylnonanoate; 11-Phenyldecanoate; 9-Tolylnonanoate	5-Phenylvalerate/ 5-(4-tolyl)valerate	Nonanoate/5-(4-bi-phenyl) valerate	5-Phenylvalerate	Nonanoate + 5-phenyl pentanoate or 6-phenylhexanoate or 7-phenylheptanoate or 9-phenylnonanoate or 11-phenyldecanoate
Tolyl		5-Phenylvalerate/ 5-(4-Tolyl)valerate		5-(4-Tolyl)valerate	Nonanoate/9-tolyl-nonanoate

Table 2 (continued)

Functional group	Copolymer			Homopolymer	Blend
	Single substrate	Two good[b] substrates	Good[b] + poor[b] substrates		
Nitrophenyl			Nonanoate/5-(2,4-dinitrophenyl) valerate		
Phenoxy	6-Phenoxyhexanoate; 8-Phenoxyoctanoate; 11-Phenoxyundecanoate	Octanoate/11-phenoxyundecanoate			
Cyanophenoxy			Octanoate/6-(4-cyanophenoxy) hexanoate[c] or 5-(4-cyanophen-oxyvalerate or 4-(4-cyanophen-oxybutyrate		
Nitrophenoxy			Octanoate/6-(4-nitrophenoxy-hexanoate		
Cyano			Octanoate/11-cya-noundecanoate; Nonanoate/11-cya-noundecanoate		

Table 2 (continued)

Halogenated	Nonanoate/6-bromohexanoate or 8-bromooctanoate or 11-bromoundecanoate; Nonane/1-fluorononane; Nonanoate/trifluorohexanoate or nonafluorononanoate or tridecafluorodecanoate; Octaoate/1-chlorooctane

[a] Frequently used substrates, list not complete.

[b] Good, substrates which support cell growth and PHA formation; poor, substrates which support PHA formation only in the presence of cell growth supporting substrates.

[c] Block polymer with enriched segments of 3-hydroxycyanophenoxyalkanoates.

[d] Polymer consists of >98% 3-hydroxyheptanoate, small traces of other components.

[e] Strain *Pseudomonas* sp. 61–3.

[f] Sequential feeding of both substrates.

poly($3HA_{MCL}$) with *P. putida* KT2442. It was found that oleic acid was degraded via the enoyl-CoA isomerase-dependent route to form mono-unsaturated monomers which were incorporated in poly($3HA_{MCL}$).

Linolenic acid was degraded via the dienoyl-CoA reductase dependent route, as indicated by the fact that unsaturated monomers containing two double bonds were detected. Casini [45] used hydrolyzed linseed oil as substrate for *P. putida* KT2442. The presence of three double bonds in linolenic acid led to the incorporation of C14:3 and C16:3 3-hydroxy fatty acids in poly($3HA_{MCL}$). This was the first time that C16 3-hydroxy fatty acids were found to be incorporated in poly($3HA_{MCL}$). The fact that only unsaturated (3-hydroxy) fatty acid containing three double bonds is found suggests that the β-oxidation system or the poly($3HA_{MCL}$) polymerase have a higher affinity for (multiple) unsaturated 3-hydroxy fatty acids than for the corresponding di- or mono-unsaturated fatty acids.

Poly($3HA_{MCL}$)s have also been produced from free fatty acid mixtures derived from industrial by-products which are potentially interesting low-cost renewable resources. Isolation and analysis of the polymer allowed the identification of 16 different saturated, mono-unsaturated and di-unsaturated monomers [46]. Except for the presence of diene-containing monomers and a large number of minor components, the composition of the fatty acid mixture derived PHA did not differ significantly from oleic acid derived PHAs.

When cells are grown on non-aliphatic substrates, such as glucose, fructose, acetate, or glycerol, these are converted to appropriate precursors that can be incorporated into poly($3HA_{MCL}$)s via fatty acid synthesis. The resulting PHAs have a monomer composition that is similar to that seen after growth on alkanes, often with 3-hydroxydecanoic acid as the major monomer. (β-Oxidation does not seem to play a role in the conversion of these substrates into poly($3HA_{MCL}$) since the addition of a β-oxidation inhibitor did not affect the monomer composition [47].

When a mixture of fatty acids or hydrocarbons is used as substrate, all compounds are simultaneously used for growth and poly($3HA_{MCL}$) production. This makes it possible to control the monomer composition (monomer carbon chain length; number, nature and position of unsaturations; other functionalities) of poly($3HA_{MCL}$) to some extent, potentially enabling the tailoring of material properties to meet the demands of specific applications. A recent example of such tailoring employs a LabVIEW-based feedback control system which uses online gas chromatography data to maintain continuously fed substrates at desired levels. This has allowed co-feeding of 10-undecenoate and octanoate such that both were incorporated into a poly($3HA_{MCL}$) copolymer at desired composition ratios [48].

The development of fermentation processes for the production of poly($3HA_{MCL}$) started with the experiments of Preusting et al. [49]. *P. oleovorans* was grown in two liquid phase fed-batch cultures. The two phases consisted of a watery phase containing mineral nutrients and an organic phase of octane. Using an organic phase is convenient because this results, without extra addition during the process, in a constant availability of the carbon source for the microorganisms in the watery phase. The feed rate of the growth limiting substrate was constant. After an initial growth period nitrogen became limiting. A biomass concentration of 37.1 g l^{-1} was reached in 48 h, containing 33% of poly($3HA_{MCL}$), resulting in a productivity of 0.25 g l^{-1} h^{-1}. Continuous cultivations were performed with a comparable set-up [50]. The optimal growth rate was 0.05 h^{-1}. The maximum poly($3HA_{MCL}$) productivity was 0.58 g l^{-1} h^{-1}, at a maximum biomass concentration of 11.6 g l^{-1}. However, the poly($3HA_{MCL}$) content decreased to 20% (w/w) compared with the fed-batch experiments.

The medium composition used in the fed-batch process was optimized, resulting in cell densities near 100 g l^{-1}. By using an exponential feed rate resulting in a growth rate of 0.05 h^{-1}, a maximum biomass concentration of 112 g l^{-1} was attained, with a biomass productivity of 1.8 g l^{-1} h^{-1}. The poly($3HA_{MCL}$) productivity however was low, 0.34 g l^{-1} h^{-1}, caused by a steady decrease of the poly($3HA_{MCL}$) content during the last part of the fermentation [51]. When this optimized medium composition was used in the continuous culture system described above, a maximum biomass concentration of 18 g l^{-1} was reached. The PHA content however remained low at approximately 10% [51]. It is still unclear what causes these low PHA contents.

In order to develop a more efficient PHA production process a two-stage continuous culture system was set up. Biomass was produced in the first phase, while in the second stage poly($3HA_{MCL}$) was synthesized in the absence of a nitrogen source. A maximum polymer content of 63% was reached, at a productivity of 1.06 g l^{-1} h^{-1}. This polymer content is the highest reported for poly($3HA_{MCL}$) to date [51, 52].

Fed-batch fermentations with *P. oleovorans* have been carried out using octanol and octanoate as substrate [53]. To ensure high oxygen transfer rates pure oxygen was used. With octanoate as substrate 41.8 g l^{-1} biomass with a cellular PHA content of 37% and a productivity of 0.34 g l^{-1} were reached. Higher biomass concentrations could not be achieved due to accumulation of the toxic octanoate.

poly($3HA_{MCL}$) production processes with *P. putida* have been developed in parallel. In contrast to *P. oleovorans*, *P. putida* does not have to be grown under nutrient-limited conditions in the presence of a carbon excess to produce poly($3HA_{MCL}$). In addition, as stated above, *P. putida* uses fatty acids rather than alkanes or alkenes as substrate for poly($3HA_{MCL}$) formation. These fatty acids however cannot be used as a second phase during fermentation because the resulting high fatty acid concentrations are toxic to *P. putida*. In high cell density continuous cultures *P. putida* has been grown to 30 g l^{-1} and contained 23% poly($3HA_{MCL}$) with oleic acid as substrate, corresponding to a productivity of 0.67 g l^{-1} h^{-1} [54].

To perform fed-batch experiments with *P. putida* a method had to be developed to prevent carbon limitation and to prevent a buildup of the concentration of the fatty acids to inhibitory levels. HPLC methods to measure the concentration of aliphatic substrates and octanoic acid have been reported, but these are not suitable for the detection of long chain fatty acids in a watery phase due to their low solubility. Instead Huijberts et al. [55, 56] developed a method in which discrete pulses of fatty acids were added to fed-batch cultures. Substrate exhaustion was detected by a sudden increase in dissolved oxygen tension and this signal was used to trigger the injection of another fatty acid pulse into the fermentor. This minimized the time during which the culture was carbon limited, while the maximum concentration of fatty acids could be kept below toxic levels. With oleic acid as substrate, a maximum biomass concentration of 92 g l^{-1} containing 45% poly(3HA_{MCL}) was reached after 26 h, for an average productivity of 1.6 g poly(3HA_{MCL}) l^{-1} h^{-1}. The same experiment has also been performed with fatty acids derived from linseed oil, coconut oil, tallow oil, and mixtures of these with comparable results. This allows the production of poly(3HA_{MCL}) with various monomer compositions. These results show that for the production of poly(3HA_{MCL}) with *P. putida*, fed-batch cultivation is the method of choice. The low PHA content of the biomass grown in chemostat cultures renders this cultivation method unsuitable for large-scale production.

Most publications on the biotechnological production of poly(3HA_{MCL}) stress the importance of adequate oxygen transfer. Substrates such as alkanes and long chain fatty acids are highly reduced and require more oxygen than does glucose to reach the oxidation state of poly(3HA_{MCL})s and biomass. The high oxygen transfer rates which can be reached in laboratory fermentors are however not easily reached on production scales. There are two promising possibilities to reduce the oxygen consumption rate, so that successful laboratory PHA production processes can be scaled up. First, maximizing the cellular PHA content, which is desirable to optimize the PHA productivity, also reduces the oxygen consumption per unit total cell mass, because more oxygen is utilized in the production of non-PHA biomass than for the production of PHA; reduced substrates are oxidized further in the production of proteins, carbohydrates, and nucleic acids than during incorporation into PHA. Second, the oxygen consumption can be decreased by using more highly oxidized co-substrates. Durner [57] showed that citrate and octanoic acid can be used simultaneously in batch cultures of *P. oleovorans* and Weusthuis et al. (unpublished results) demonstrated the same for the combination of glucose and oleic acid in high cell density fed-batch processes with *P. putida*. These findings indicate that pseudomonads are able to use different unrelated substrates simultaneously, even under carbon excess conditions.

2.2.2
Production of "Functionalized" Poly(3HAMCL)s by Pseudomonads

It has been shown that pseudomonads can incorporate more than 60 different monomers into PHA [7]. PHAs containing a functional group in the monomer side chain are often referred to as "functional" PHAs. A widespread technique for the production of "functionalized" PHAs is to use two substrates, one of which mostly supports bacterial cell growth while the other is utilized as a precursor of specific monomers which are incorporated into PHA [58–65]. The substrates can be fed simultaneously or sequentially in two-stage cultivation processes. Bacterial cell mass is produced in the first stage and in a second stage PHA forming substrates are added to the culture, as has been reported for the production of PHAs containing multi-fluorinated, cyano, or nitrophenoxy side chain substituents [61, 64]. Table 2 shows a number of functional PHAs produced from single or two substrates. In this section we will discuss a few of the more significant and recent examples which use different cultivation strategies for incorporation of specific monomers.

In principle, three types of substrates can be differentiated [8]:

1. Substrates which support cell growth and PHA production
2. Substrates which support growth but not PHA production
3. Substrates which support PHA production but not growth

Depending on the type of substrate combination to be used, different cultivation processes and feeding strategies are required.

As mentioned above, when carbon source mixtures such as citrate/octanoate [57] or glucose/oleic acid (Weusthuis et al., unpublished results) were utilized simultaneously in batch cultures, fatty acids were used mostly for PHA synthesis and carbohydrates were dissimilated for non-PHA cell growth and to supply maintenance energy. Such co-feeding strategies have been used not only to produce specific random copolymers, but block polymers or polymer blends have also been obtained. Growth of *P. oleovorans* or *P. putida* on a mixture of 5-phenylvaleric acid (or other arylalkyl acids) and nonanoic acid resulted in a poly(3-hydroxy-5-phenylvalerate) homopolymer and a random copolymer containing 3-hydroxynonanoate and 3-hydroxyheptanoate [66–68]. It has been shown that both types of polyester occur in the same granule [67]. Interestingly, it has even been proposed that by sequential feeding of nonanoic acid and 10-undecenoic acid, a physical mixture of two different polymers is produced, with small amounts of an additional PHA containing repeating units from both substrates [69]. In contrast, co-feeding of octanoate and cyanophenoxyalkanoates resulted in PHA block polymers containing chain segments enriched in 3-hydroxycyanophenoxyalkanoate monomers [62].

Production of PHAs from toxic organic compounds requires other cultivation strategies. A cultivation method was developed to improve growth of *P. oleovorans* on toxic organic solvents, such as 1-hexene. This method includes dilution of 1-hexene with a non-metabolizable second organic phase to lower the toxic effect of the apolar carbon source and a long-term chemostat enrichment culture to increase the solvent tolerance and the specific growth rate [70]. Although very little 1-hexene was ultimately incorporated into PHA, this was probably due to the fact that 1-hexene is also converted to epoxyhexane by *P. oleovorans*, this compound being even more toxic to the cells than 1-hexene. When the same approach was used with several other toxic substrates which are not converted to even more toxic reaction products, significant amounts of functionalized poly($3HA_{MCL}$)s were formed. Furthermore, application of dual-carbon/nitrogen-limited conditions for cell growth and PHA production on volatile and toxic substrates resulted in decreased cell lysis, side product formation, and biosurfactant production, and therefore higher cell and PHA yields [71].

2.2.3
Production of Poly(3HAMCL)s with Varied Polymer Backbones

As described above, most PHAs are 3-hydroxy esters. However, certain variations in the polymer backbone are possible (Fig. 2). The most common varia-

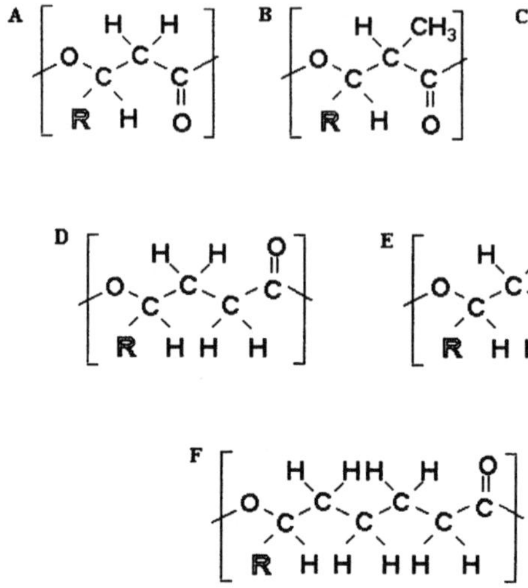

Fig. 2 A–F. Hydroxy acid monomers of PHAs with varied polymer backbones: **A** standard 3-hydroxyalkanoic acid monomer; **B** 2-methyl-3-hydroxyalkanoic acid monomer; **C** 2,2-dimethyl-3-hydroxyalkanoic acid monomer; **D** 4-hydroxyalkanoic acid monomer; **E** 5-hydroxyalkanoic acid monomer; **F** 6-hydroxyalkanoic acid monomer. (R indicates alkyl residue or H)

tion is the occurrence of 4-hydroxy rather than 3-hydroxy ester linkages. Among many different strains, *Ralstonia eutropha*, *Alcaligenes latus*, and *Comamonas acidovorans* are capable of synthesizing random copolymers consisting of 3-hydroxybutyrate (3HB) and 4-hydroxybutyrate (4HB) when grown in media containing 4-hydroxybutyric acid, 1,4-butanediol, butyrolactone, or 4-chlorobutyric acid [72–76]. It could be shown that a random sequence distribution of 3HB and 4HB units is produced by *R. eutropha* from butyrolactone and fructose mixtures, whereas a mixture of random copolyesters with two different fractions is produced when butyrolactone and butyric acid are used as carbon sources [73]. Homopolymers of 4HB can also be produced if 1,4-butanediol or 4-butyric acid are used as sole carbon source [75, 77]. A more sophisticated three-stage cultivation process was developed for the production of a 4HB homopolymer by *Hydrogenophaga pseudoflava* [78]. This process consists of a cell mass formation step, a nitrogen addition step in order to remove completely the accumulated 3HB containing polymer, and a polymer formation step on butyrolactone. Furthermore, cultivation conditions can also influence the composition of the polymer. It has been shown that addition of 2% polyethylene glycol resulted in increased 4-hydroxybutyrate incorporation and formation of a product with predominantly 3HB-3HB and 4HB-4HB diads instead of random copolyesters [79]. In addition to 4HB containing PHAs, polymers containing 4-hydroxyvalerate monomers have been synthesized by bacteria belonging to the genus *Ralstonia* and by the strain *Pseudomonas oxalaticus* [80]. The accumulated terpolymer of 3-hydroxybutyrate, 3-hydroxyvalerate, and 4-hydroxyvalerate was synthesized from 4-valerolactone or 4-hydroxyvalerate as sole carbon source [80].

Polyesters containing 2-hydroxy, 5-hydroxy, or 6-hydroxy monomers have also been described. A PHA copolymer containing 2-hydroxybutyric acid has recently been prepared by Metabolix [81]. Terpolymers of 3-hydroxybutyrate, 3-hydroxyvalerate, and 5-hydroxyvalerate can be produced by *R. eutropha* fed with either 5-chlorovalerate alone or 5-hydroxyvalerate in combination with valeric acid [82]. NMR analysis of PHAs isolated from *Pseudomonas aeruginosa* 44T1 cultivated on euphorbia oil and castor oil strongly suggested the presence of 3,6-dihydroxydodecanoate, 6-hydroxy-3c-dodecenoate, and 4-hydroxydecanoate [83]. Furthermore, PHAs containing other monomers such as 5-hydroxyhexanoic acid, 4-hydroxyheptanoic acid, and 4-hydroxyoctanoic acid [84] can be produced by certain recombinant bacteria. However, the considerable range of recombinant PHA producers will not be discussed in this chapter.

Another polymer backbone variant contains methyl groups at carbon atom 2. *R. eutropha* and *Burkholderia cepacia*, for example, accumulate a terpolyester consisting of 3-hydroxybutyric acid, 3-hydroxyvaleric acid, and 2-methyl-3-hydroxybutyric acid when the cells are cultivated in a mineral salts medium containing tiglic acid as sole carbon source or in combination with gluconic acid [85]. Furthermore, when grown in a mineral salts medium containing glucose and 3-hydroxypivalic acid, a hydroxyalkanoic acid with two methyl group substituents at the alpha-carbon atom, as carbon sources, *Rhodococcus ruber* and related Gram-positive bacteria accumulate novel copolyesters that contain 3-hydroxypivalic acid as a constituent [86]. PHA containing 2-methyl-3-hydroxyvalerate monomers have also been found to be formed by anaerobic-aerobic activated sludge. In general, however, PHA-related metabolic processes observed in activated sludge systems have thus far not been seen in pure cultures [87].

3 Downstream Processes

After the fermentation step, the PHA-containing cells must be separated from the broth by conventional procedures such as centrifugation, filtration, or flocculation-centrifugation, and cells are then disrupted to recover the polymer. Several methods have been developed for the recovery of PHAs. The most often used method involves extraction of the polymer from the biomass with solvents (e.g., chloroform, methylene chloride, propylene carbonate, dichloroethane) [88–91]. Large amounts of solvent are required in this process. Several other methods have been developed, such as sodium hypochlorite treatments for the differential digestion of non-PHA cellular materials [92]. Although this method is effective, it causes severe degradation of PHA, resulting in a reduction in the molecular weight [93].

ZENECA has developed a non-solvent based recovery process as an alternative to solvent extraction for the commercial production of poly(3HB) and poly(3HB-*co*-3V) by *A. eutrophus* [94, 95]. In this process the cells were first exposed to a temperature of 80 °C and subsequently treated with a cocktail of various hydrolytic enzymes consisting of lysozyme, phospholipase, lecithinase, the proteinase alcalase, and others. Most of the cellular components were hydrolyzed by these enzymes, whereas the polymer remained intact. After washing, flocculation, and drying, the polymer was recovered as a white powder which was melted, extruded, and converted into chips – the form in which the polymer was supplied to the fabricators.

In contrast, the Biotechnologische Forschungsgesellschaft has used a solvent-based process for the recovery of poly(3HB) from *A. latus*. The cells were harvested by centrifugation and the poly(3HB) was subsequently extracted from the suspended cells with methylene chloride and precipitated from the solvent by the addition of water. After drying a polyester with 99% purity was obtained [15]. The process also included the recovery of the solvent.

Recovery procedures for poly(3HA_{MCL}) resemble those originally developed for the production of poly(3HB). A number of solvent extraction processes have been assessed to separate poly(3HA_{MCL})s from biomass. These usually involve the use of a chlorinated solvent such as chloroform [41] or methylene chloride. Recently, it has been reported that poly(3HA_{MCL})s can be extracted with hexane or acetone instead of chlorinated solvents [81], and subsequently precipitated by the addition of a non-solvent for the PHA, such as methanol. Using this method, the resulting polymer can be obtained in high purity. An alternative, non-solvent-based extraction process was developed by de Koning et al. [96, 97] and further optimized to make the overall production process more attractive [46]. The biomass is separated from the medium by centrifugation, and treated with a protease cocktail and a detergent to solubilize all cell components. Removal of the solubilized cell material and concentration of the resulting PHA suspension is achieved by crossflow microfiltration. The submicron poly(3HA_{MCL}) granules display a density close to that of water [98], as a result in which a poly(3HA_{MCL}) suspension does not settle [99], and in fact forms a highly stable polymer latex. The overall purity of such latexes is of the order of 95%. Furthermore, supercritical CO_2 is highly effective at extracting lipids and other hydrophobic contaminants from PHA-containing cells, and 100% purity can be reached in a single step [81].

Since the liberation of chromosomal DNA during lysis causes a dramatic increase in viscosity, a nuclease-encoding gene from *Staphylococcus aureus* was integrated into the genomes of several PHA producers. The nuclease is directed to the periplasm, and occasionally to the culture medium, and significantly reduces the viscosity of the lysate during the downstream process without affecting PHA production or strain stability [100].

4 Economics

In contrast to poly(3HB), poly(3HA_{MCL}) has not yet been produced on a commercial scale. Consequently, process development for the production and purification of poly(3HB) has received much more attention than processes for the production of poly(3HA_{MCL}). It is interesting therefore to compare parameters for the production of poly(3HA_{MCL}) with those for the production of poly(3HB). Table 3 summarizes the key parameters for the best poly(3HB) [101] and poly(3HA_{MCL}) [51, 55, 56] production processes.

Table 3. Process parameters of poly(3HB) and poly($3HA_{MCL}$) production

Organism Fermentation type	poly(3HB) *A. latus* Fed batch	poly($3HA_{MCL}$) *P. putida* Fed batch	 *P. oleovorans* Two-stage continuous
Substrate	Sucrose	Oleic acid	Octane
Culture time (h)	20	25	–
Cell concentration ($g\,l^{-1}$)	111.7	93	17
PHA content (%)	88	43	66
Productivity ($g\,l^{-1}\,h^{-1}$)	4.94	1.6	1.06
Yield ($g\,g^{-1}$)	0.42	0.3–0.4	nd
Reference	[101]	[55, 56]	[51]

nd = not determined.

When comparing fed-batch operated cultures it can be seen that the main difference between poly(3HB) and poly($3HA_{MCL}$) production that affect the process parameters is the lower cellular poly($3HA_{MCL}$) content, compared to the cellular poly(3HB) content. Not surprisingly, it has been reported that a low poly($3HA_{MCL}$) content decreases the productivity and yield and increases the costs for downstream processing and waste disposal [102].

It should be noted in such comparisons that the PHA content is always expressed as a weight ratio between PHA and total biomass. The density of poly(3HB) however is 1.24 $g\,ml^{-1}$ [103] whereas the density of poly($3HA_{MCL}$), depending on the monomer composition, is close to 1.00 $g\,ml^{-1}$. On a volume basis, a PHA content of 66% (Table 3) in *P. oleovorans* corresponds to a poly(3HB) content of 82%! This is relevant because in the development of new applications, the volume of the material is generally more important than its weight. Thus, if poly(3HB) and poly($3HA_{MCL}$) could be used for the same application, 24% more poly(3HB) than poly($3HA_{MCL}$) would be necessary on a weight basis. Another complicating factor is that in the method used for the determination of the PHA content the biomass is first centrifuged. Since cells with a high poly($3HA_{MCL}$) content can have densities lower than the density of the fermentation broth and therefore may in some cases not be fully pelleted, the determined poly($3HA_{MCL}$) content can sometimes be underestimated.

The material properties of poly(3HB) and poly(3HB-*co*-3HV) make it a polymer that has to compete with conventional commodity plastics such as polyethylene and polypropylene, with biodegradability as the main distinguishing feature. The bulk prices of polyethylene and polypropylene are 0.39–0.49 $ lb^{-1} and 0.29–0.47 $ lb^{-1} respectively (Chemical Market Reporter, August 1999). Poly(3HB) and poly(3HB-*co*-3HV) have been produced on a commercial scale (BIOPOL, Monsanto) using bacterial fermentations, with production in plants as a longer term goal. However, it has been estimated that the production costs of poly(3HB) and poly(3HB-*co*-3HV) in plants would in the long run still be 25–50% higher than polymers such as polyethylene and polypropylene. Based on this estimate, Monsanto decided in late 1998 to stop the BIOPOL program (M. Cannon, personal communication).

Poly($3HA_{MCL}$) and functionalized PHAs are a family of many different polymers, which can all be produced using the same or similar fermentation processes by simply changing the type of substrate(s) used. This enables the production of tailor made poly($3HA_{MCL}$) variants for specific applications. It is likely that this flexibility to incorporate various monomers will decrease when genetically modified plants are used for poly($3HA_{MCL}$) production. Even then, however, the potential of (bio)chemical modifications of the polymers will make it possible to adapt the material characteristics to meet the demands of a multitude of applications, sidestepping direct competition with low-priced commodity polymers and enabling improved competitiveness on the specialties market.

5 Conclusion and Outlook

PHAs have now been around since 1926, and serious efforts to commercialize poly(3HB) and poly(3HB-*co*-3HV) were started more than two decades ago. The early impetus for the production of bacterial polyesters came from a combination of the availability of large-scale fermentation equipment at ICI, the instability of the oil market and looming oil shortages since the early 1970s, and growing interest in biodegradability and sustainability in the 1980s and early 1990s. The cost of poly(3HB) remained high, however, decreasing from around US$ 30 kg^{-1} in the late 1980s to about US$ 15 kg^{-1} in the mid-1990s (informal communications from ICI and Monsanto). With the perspectives of plant-based PHA production, cost estimates were lowered by one order of magnitude, and in 1993 Monsanto embarked on a significant research program to develop PHA-producing recombinant crops. This program was abandoned in late 1998. It now appears that even if PHA can be produced in plants, it will remain a major challenge to achieve sufficiently high yields per hectare and sufficiently cost-effective crop processing to permit prices below US$ 1 kg^{-1} biopolymer.

Two other factors contribute to the reduced urgency of developing alternatives to petrochemically derived plastics. First, although biodegradability and sustainability have been major societal priorities from the mid-1980s to the late 1990s, consumers have not generally been willing to pay a premium for biodegradable or sustainable products. Thus, although they are political priorities, biodegradability and sustainability have not been major industrial priorities. Second, alternative energy sources, and especially solar energy via photovoltaic conversion, appear much more promising at the end of 1999 than only a few years earlier. Thus, non-fossil fuel based energy may well be viable and competitive within two or three decades, in which case the pressure on available petroleum and coal reserves will diminish significantly. Given that petrochemistry accounts for less than 10% of our total oil consumption, that the consumption of oil for new plastic production is likely to stabilize or even decline as petroplastic recycling develops further, and that petroplastics do not enter the atmosphere, petroplastics may well be considered much more environmentally benign in the coming decade than they appear now. Consuming only a few percent of our present oil production, the production of petroplastics can be carried on for several centuries if and when oil is no longer used for energy production at the rates of the past few decades.

Thus, even if PHAs can be produced in plants, they are not likely to succeed in the market based only on sustainability or environmental considerations. Instead, if PHAs are to be viable they will have to compete based on unique properties not to be found in existing petroplastics. Specialty applications of PHAs include hydrophobic coatings, specialty elastomers, medical implants, functionalized polymers for chromatography, microgranules to be used as binders in paints or in blends that incorporate latexes, and as sources of chiral monomers. Such materials will probably continue to be produced in bioreactors rather than in plants. The agroproduction of PHAs will become interesting only after significant markets have developed for specific PHAs, at which point the effort necessary to modify the intermediate metabolic pathways to produce the required PHA precursors and the challenges of producing the resulting polyesters as large-scale agrocrops can be justified. The latter will not be trivial. There has recently been a shift in public perceptions concerning transgenic crops, which is likely to slow the introduction not only of genetically altered food crops but may well also affect the introduction of new crops for the production of agrochemicals and agropolymers.

In conclusion, where only a few years ago the way seemed clear for the development of PHAs as a useful and major source of sustainable polymers, recent developments now suggest that we are back where we were in the late 1980s or early 1990s. PHAs are interesting biopolyesters, which can be produced by fermentation. They will not compete with petroplastics directly. The development of higher value applications depends on the production of kg amounts of as large a range of PHAs as possible. Successful applications will then drive the production of kiloton amounts of material, which is likely to be produced by fermentation until production levels in the range of 20–100 kilotons per year are reached. Once markets are established, PHA production in plants will be considered again. This is likely to take at least another decade.

Acknowledgement. Part of the above text was taken from the review published previously by B. Kessler and B. Witholt in: Flickinger MC, Drew SW (eds) Encyclopedia of bioprocess technology: fermentation, biocatalysis and bioseparation, vol 5, pp 2024–2040. We thank O. Peoples (Metabolix Inc.) for supplying information.

References

1. Lemoigne M (1926) Bull Soc Chem Biol 8:770
2. Lee SY (1996) Biotechnol Bioeng 49:1
3. Sasikala C, Ramana CV (1996) Biodegradable polyesters. In: Neidleman SL, Laskin AI (eds) Advances in applied microbiology. Academic Press, San Diego, p 97
4. Steinbüchel A (1991) Polyhydroxyalkanoic acid. In: Byrom D (ed) Biomaterials. Novel materials from biological sources. Macmillan, Basingstoke, p 123
5. Merrick JM, Doudoroff M (1964) J Bacteriol 88:60
6. de Smet MJ, Eggink G, Witholt B, Kingma J, Wynberg H (1983) J Bacteriol 154:870
7. Steinbüchel A, Valentin HE (1995) FEMS Microbiol Lett 128:219
8. Lenz RW, Kim YB, Fuller RC (1992) FEMS Microbiol Rev 103:207

9. Ohashi T, Hasegawa J (1992) D(–)-β-Hydroxycarboxylic acids as raw materials for captopril and β-lactams. In: Collins AN, Sheldrake GN, Crosby J (eds) Chirality in industry. ZENECA Specialties, Manchester, UK, p 269
10. Lageveen R, Witholt, B (1986) European Patent, US Patent; Witholt B, Eggink G, Huisman GW (1992) US Patent
11. Byrom D (1992) FEMS Microbiol Rev 103:247
12. Schlegel HG, Gottschalk G (1965) Biochem Zeitung 342:249
13. Byrom D (1990) Industrial production of copolymer from *Alcaligenes eutrophus.* In: Dawes EA (ed) Novel biodegradable microbial polymers. Kluwer, Dordrecht, p 113
14. Hänggi UJ (1990) Pilot scale production of poly(3HB) with *Alcaligenes latus.* In: Dawes EA (ed) Novel biodegradable microbial polymers. Kluwer, Dordrecht, p 65
15. Hrabak O (1992) FEMS Microbiol Rev 103:251
16. Babel W (1992) FEMS Microbiol Rev 103:141
17. Suzuki T, Yamane T, Shimizu S (1986) Appl Microbiol Biotechnol 23:322
18. Suzuki T, Yamane T, Shimizu S (1986) Appl Microbiol Biotechnol 24:366
19. Ueda S, Matsumoto S, Takagi A, Yamane T (1992) Appl Environ Microbiol 58:3574
20. Yamane T, Chen XF, Ueda S (1996) Appl Environ Microbiol 62:380
21. Mineki S, Fukutome N, Oinuma N, Nagashima H, Iida M (1997) Macromolecules 30: 319
22. Kim SW, Kim P, Lee HS, Kim JH (1996) Biotechnol Lett 18:25
23. Shah NN, Hanna ML, Taylor RT (1996) Biotechnol Bioeng 49:161
24. Page WJ, Knosp O (1989) Appl Environ Microbiol 55:1334
25. Page WJ (1989) Appl Microbiol Biotechnol 31:329
26. Chen GQ, Page WJ (1994) Biotechnol Lett 16:155
27. Page WJ, Manchak J, Rudi B (1992) Appl Environ Microbiol 58:2866
28. Page WJ, Cornish A (1993) Appl Environ Microbiol 59:4236
29. Tanaka K, Katamune K, Ishizaki A (1993) Biotechnol Lett 15:1217
30. Steinbüchel A, Debzi EM, Marchessault RH, Timm A (1993) Appl Microbiol Biotechnol 39:443
31. Steinbüchel A, Schmack G (1995) J Environ Polymer Degradation 3:243
32. Brandl H, Knee EJ, Fuller RC, Gross RA, Lenz RW (1989) Int J Biol Macromol 11:49
33. Suzuki T, Tsygankov AA, Miyake J, Tokiwa Y, Asada Y (1995) Biotechnol Lett 17:395
34. Schlegel HG, Gottschalk G, von Bartha R (1961) Nature 191:463
35. Tanaka K, Ishizaki A, Kanamaru T, Kawano T (1995) Biotechnol Bioeng 45:268
36. Miyake M, Erata M, Asada Y (1996) J Ferment Bioeng 82:512
37. Maness PC, Weaver PF (1994) Appl Biochem Biotechnol 45–6:395
38. Poirier Y (1999) Curr Opin Biotechnol 10:181
39. van der Leij FR, Witholt B (1995) Can J Microbiol 41:222
40. Riesmeier J, Kossmann J, Trethewey R, Heyer A, Landschutze V, Willmitzer L (1998) Polym Degrad Stabil 59:383
41. Lageveen RG, Huisman GW, Preusting H, Ketelaar P, Eggink G, Witholt B (1988) Appl Environ Microbiol 54:2924
42. Preusting H, Nijenhuis A, Witholt B (1990) Macromolecules 23:4220
43. Huijberts GNM, de Rijk TC, de Waard P, Eggink G (1995) J Bacteriol 176:1661
44. de Waard P, van der Wal H, Huijberts GNM, Eggink G (1993) J Biol Chem 268:157
45. Casini E (1997) J Environ Polymer Degrad 5:153
46. Kellerhals MB (1999) PhD thesis, ETH Zürich, Switzerland
47. Huijberts GNM, Eggink G, de Waard P, Huisman GW, Witholt B (1992) Appl Environ Microbiol 58:536
48. Kellerhals MB, Kessler B, Witholt B (1999) Biotechnol Bioeng 62:306–315
49. Preusting H, van Houten R, Hoefs A, Kool van Langenberghe E, Favre-Bulle O, Witholt B (1993) Biotechnol Bioeng 41:550
50. Preusting H, Hazenberg W, Witholt B (1993) Enzyme Microb Technol 15:311
51. Hazenberg WM (1997) PhD thesis, ETH Zürich, Switzerland
52. Jung K, Hazenberg W, Prieto MA, Witholt B Biotechnol Bioeng (in press)

53. Lee SY, Chang HN (1995) Adv Biochem Eng Biotechnol 52:27
54. Huijberts GNM, Eggink G (1996) Appl Microbiol Biotechnol 46:233
55. Huijberts GNM (1996) PhD thesis, Rijksuniversiteit Groningen, The Netherlands
56. Weusthuis RA, Huijberts GNM, Eggink G (1997) Production of mcl-poly(hydroxyalkanoates) (review). In: Eggink G, Steinbüchel A, Poirier Y, Witholt B (eds) 1996 International Symposium on Bacterial Polyhydroxyalkanoates. NRC Research Press, Ottawa
57. Durner RA (1998) PhD thesis, ETH Zürich, Switzerland
58. Scholz C, Fuller RC, Lenz RW (1994) Macromol Chem Physics 195:1405
59. de Koning GJM, van Bilsen HHM, Lemstra PJ, Hazenberg W, Witholt B, Preusting H, van der Galiën JG, Schirmer A, Jendrossek D (1994) Polymer 35:2090
60. Hori K, Soga K, Doi Y (1994) Biotechnol Lett 16:501
61. Kim O, Gross RA, Hammar WJ, Newmark RA (1996) Macromolecules 29:4572
62. Gross RA, Kim O, Rutherford DR, Newmark RA (1996) Polym Int 39:205
63. Song JJ, Yoon SC (1996) Appl Environ Microbiol 62:536
64. Kim OY, Gross RA, Rutherford DR (1995) Can J Microbiol 41:32
65. Curley JM, Hazer B, Lenz RW, Fuller RC (1996) Macromolecules 29:1762
66. Kim YB, Lenz RW, Fuller RC (1991) Macromolecules 24:5256
67. Curley JM, Lenz RW, Fuller RC (1996) Int J Biol Macromol 19:29
68. Hazer B, Lenz RW, Fuller RC (1996) Polymer 37:5951
69. Kim YB, Rhee YH, Lenz RW (1997) Polym J 29:894
70. Jung K (1999) PhD thesis, ETH Zürich, Switzerland
71. Jung K, Sierro N, Egli T, Kessler B, Witholt B (unpublished)
72. Renner G, Haage G, Braunegg G (1996) Appl Microbiol Biotechnol 46:268
73. Doi Y, Segawa A, Kunioka M (1990) Int J Biol Macromol 12:106
74. Kang CK, Kusaka S, Doi Y (1995) Biotechnol Lett 17:583
75. Saito Y, Doi Y (1994) Int Biol Macromol 16:99
76. Saito Y, Nakamura S, Hiramitsu M, Doi Y (1996) Polym Int 39:169
77. Nakamura S, Doi Y, Scandola M (1992) Macromolecules 25:4237
78. Choi MH, Yoon SC, Lenz RW (1999) Appl Environ Microbiol 65:1570
79. Shi FY, Gross RA, Rutherford DR (1996) Macromolecules 29:10
80. Valentin HE, Schönebaum A, Steinbüchel A (1992) Appl Microbiol Biotechnol 36:507
81. Williams SF, Martin DP, Horowitz DM, Peoples OP (1999) Int J Biol Macromol 25:111
82. Doi Y, Tamaki A, Kunioka M, Soga K (1987) Macromol Chem Rapid Commun 8:631
83. Eggink G, de Waard P, Huijberts GNM (1995) Can J Microbiol 41:14
84. Valentin HE, Schönebaum A, Steinbüchel A (1996) Appl Microbiol Biotechnol 46:261
85. Füchtenbusch B, Fabritius D, Steinbüchel A (1996) FEMS Microbiol Lett 138:153
86. Füchtenbusch B, Fabritius D, Waltermann M, Steinbüchel A (1998) FEMS Microbiol Lett 159:85
87. Satoh H, Mino T, Matsuo T (1999) Int J Biol Macromol 25:105
88. Barham PJ, Selwood A (1982) European Patent Application
89. Holmes PA, Jones E (1982) European Patent Application
90. van Lautem N, Gilian J (1986) European Patent Application
91. Walker J, Whitton JR, Alderson B (1982) European Patent Application
92. Williamson DR, Wilkinson JF (1958) J Gen Microbiol 19:198
93. Berger E, Ramsay BA, Ramsay JA, Chaverie C, Braunegg G (1989) Biotechnol Tech 3: 227
94. Byrom D (1987) Trends Biotech 5:246
95. Holmes PA, Lim GB (1985) European Patent Application
96. de Koning GJM, Kellerhals M, van Meurs C, Witholt B (1997) Bioprocess Engineering 17:15
97. de Koning GJM, Witholt B (1997) Bioprocess Engineering 17:7
98. Preusting H, Kingma J, Huisman G, Steinbüchel A, Witholt B (1993) J Environ Polym Degrad 1:11
99. Marchessault RH, Morin FG, Wong S, Saracovan I (1995) Can J Microbiol 41:138

100. Boynton ZL, Koon JJ, Brennan EM, Clouart JD, Horowitz DM, Gerngross TU, Huisman GW (1999) Appl Environ Microbiol 65:1524
101. Wang FL, Lee SY (1997) Appl Environ Microbiol 63:3703
102. Choi J, Lee SY (1999) Appl Microbiol Biotechnol 51:13
103. Marchessault RH, Monasterios CJ, Morin FG, Sundararajan PR (1990) Int J Biol Macromol 12:158

Received: March 2000

Production of Microbial Polyester by Fermentation of Recombinant Microorganisms

Sang Yup Lee, Jong-il Choi

Department of Chemical Engineering and BioProcess Engineering Research Center, Korea Advanced Institute of Science and Technology, 373-1 Kusong-dong, Yusong-gu, Taejon 305-701, Korea,
E-mail: leesy@mail.kaist.ac.kr

Polyhydroxyalkanoates (PHAs) can be produced from renewable sources and are biodegradable with similar material properties and processibility to conventional plastic materials. With recent advances in our understanding of the biochemistry and genetics of PHA biosynthesis and cloning of the PHA biosynthesis genes from a number of different bacteria, many different recombinant bacteria have been developed to improve PHA production for commercial applications. For enhancing PHA synthetic capacity, homologous or heterologous expression of the PHA biosynthetic enzymes has been attempted. Several genes that allow utilization of various substrates were transformed into PHA producers, or non-PHA producers utilizing inexpensive carbon substrate were transformed with the PHA biosynthesis genes. Novel PHAs have been synthesized by introducing a new PHA biosynthesis pathway or a new PHA synthase gene. In this article, recent advances in the production of PHA by recombinant bacteria are described.

Keywords. Polyhydroxyalkanoates, PHA biosynthesis genes, Recombinant bacteria, Fermentation

Advances in Biochemical Engineering/
Biotechnology, Vol. 71
Managing Editor: Th. Scheper

1 Introduction

Polyhydroxyalkanoates (PHAs) are the polymers of hydroxyalkanoates, which are accumulated as a carbon and energy storage material in various microorganisms [1–6]. PHAs are synthesized when one of nutritional elements such as N, P, S, O, or Mg is limiting in the presence of excess carbon source. After the discovery of poly(3-hydroxybutyrate) [P(3HB)] in 1926 [7], a large variety of different PHAs consisted of different number of main chain carbon atoms and different types of alkyl-pendent groups have been reported [8]. The molecular weight of PHAs varies depending on the PHA producer, but is generally in the range 50,000–1,000,000 Da, which is sufficiently high to give polymer characteristics similar to conventional plastics and elastomers [4, 9, 10]. PHAs can be produced from renewable sources and are completely biodegradable. PHAs can be processed using the equipment commonly used for the processing of polyolefins and other synthetic plastics, indicating that PHAs may be suitable for the application in several areas as a partial substitute for nondegradable synthetic plastics [4, 5, 9, 11]. However, the applications of PHAs are not restricted to those areas, and the use of PHAs as osteosynthetic materials, bone plates, surgical sutures, and other materials in medicine has also been proposed [11, 12]. In addition, PHAs can be used as intermediates for synthesis of enantiomerically pure (R)-(–)-chemicals [13].

Even though there are more than 300 different microorganisms that are known to synthesize PHAs in nature, only a few bacteria including *Ralstonia eutropha*, *Alcaligenes latus*, *Azotobacter vinelandii*, *Chromobacterium violaceum*, methylotrophs and pseudomonads, can produce PHA to an extent that meets commercial interest [5, 14–16]. However, from the economical evaluation of the processes for the production of PHA by these bacteria, it was shown that the production cost of PHA was much higher compared with that of the conventional plastic materials [17–19]. Another problem is that P(3HB), the best characterized member of PHAs, does not possess desirable material properties for its use in a wide range of applications [4, 9]. Even though several bacteria could produce PHAs having better polymer properties than P(3HB) homopolymer, the PHA concentration and the PHA content obtained were too low, except for poly(3-hydroxybutyrate-*co*-3-hydroxyvalerate) [P(3HB-*co*-3HV)] copolymer, for the development of various applications [5, 14, 15]. With the recent advances in our understanding of the biochemistry and genetics of PHA biosynthesis and the cloning of the PHA biosynthesis genes from a number of different bacteria [15, 20–25], a variety of recombinant bacteria have been developed to enhance PHA production from inexpensive carbon sources and to produce novel PHA having better material properties. In this article, recent advances in the development of recombinant bacteria for the production of PHA are reviewed. Particularly, advances made after the developments described in several recent review articles [22, 24, 25] are emphasized.

2
Strategies for the Development of Recombinant Bacteria

After the first cloning of PHA biosynthesis genes from *R. eutropha* [26–28], the PHA synthase genes, encoding the most important enzyme in PHA biosynthesis, were cloned from more than 40 different bacteria [21, 25]. Various recombinant microorganisms were developed by transforming natural PHA producers, PHA negative mutant and non-PHA producers with the cloned PHA synthase genes. Using these recombinant strains harboring homologous or heterologous PHA biosynthesis genes, the biochemistry of PHA biosynthesis and the characteristics of PHA synthase have been studied in detail. Initially, these recombinant strains were rarely employed for the commercial production of PHAs, but mostly used for studying biochemical and biological aspects of PHA biosynthesis.

By adapting the concept of metabolic engineering, a multidisciplinary effort towards the development of super-organisms possessing desired metabolisms, many different recombinant bacteria were developed for enhancing PHA synthetic capacity, for broadening the utilizable substrate ranges, and for producing novel PHAs. With the aim of producing PHAs more efficiently, homologous or heterologous overexpression of the PHA biosynthetic enzymes in various organisms has been attempted. Also, PHA biosynthetic pathways were introduced into non-PHA producers having more robust central metabolic pathways for more efficient production of PHA. To produce PHA from inexpensive carbon sources, PHA producers were transformed with several genes, allowing the utilization of various substrates, or non-PHA producers utilizing inexpensive carbon sources were transformed with the PHA biosynthesis genes. Novel PHAs have been synthesized by recombinant PHA producers harboring a new PHA biosynthesis pathway or a new PHA synthase gene. These metabolic engineering strategies for the production of PHAs have recently been reviewed in detail [25]. Therefore, this review focuses on the development of recombinant strains for fermentation studies.

3
Economic Consideration for the Production of PHA

Several factors affect the overall economics of PHA production. These include PHA productivity, PHA content, yield of PHA on carbon source, carbon substrate cost, and recovery method employed. Figure 1 shows the production costs of P(3HB) by various P(3HB) contents and P(3HB) productivities [29]. The effect of P(3HB) productivity on the production cost is only related to the cost of the fermentation equipment [18]. However, the P(3HB) content has multiple effects on the volume of the fermentation equipment and the recovery process [17, 18]. The increase of P(3HB) yield on carbon source and the use of less expensive carbon substrates reduce the cost of carbon substrate [17, 29]. Development of an efficient recovery method, which will be different for each bacterium employed, is also important to overall economics of PHA production. When the actual fermentation processes employing many different re-

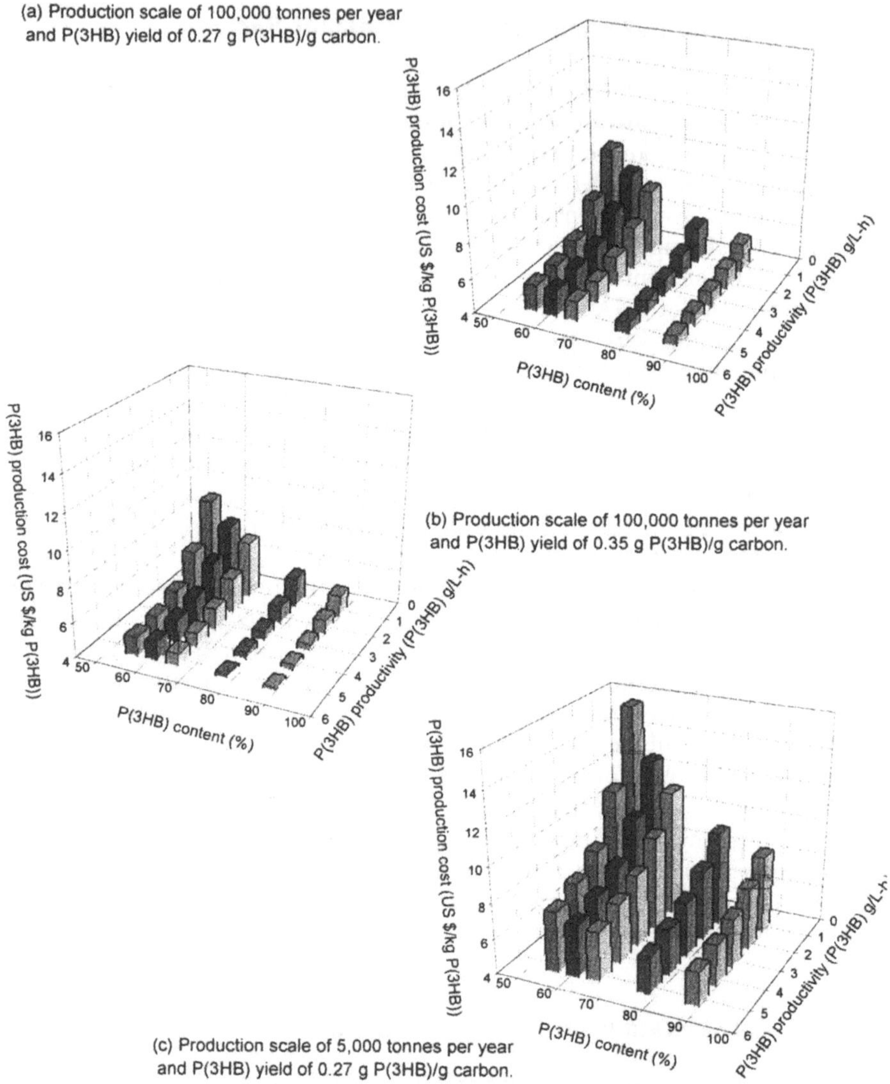

Fig. 1 a–c. Sensitivity analysis on the effect of P(3HB) content and productivity on the production cost of P(3HB) at different yield and production scale. Surfactant-hypochlorite digestion method was used for the recovery of P(3HB) (reproduced from [29] with permission)

combinant bacteria are developed, the factors described above need to be carefully considered.

4 Recombinant *Escherichia coli*

Escherichia coli has been a workhorse for the production of various proteins because it has been studied most extensively [30, 31]. During the studies on the cloning of the *R. eutropha* PHA biosynthesis genes, it was found that these genes were constitutively expressed in *E. coli* and a large amount of P(3HB) could be accumulated in recombinant *E. coli* harboring these genes [26–28]. After these initial findings, production of PHAs by recombinant *E. coli* has been intensively investigated. Several advantages of employing recombinant *E. coli* for the production of PHAs were found [5, 14, 32, 33]:

1. Fast growth to a high cell density
2. Accumulation of a large amount of PHAs
3. Ability to utilize several inexpensive carbon sources
4. Relatively easy purification of PHAs from *E. coli* cells
5. The lack of intracellular depolymerases which degrade the accumulated PHAs

4.1 Production of P(3HB)

There appeared a series of papers that described the development of recombinant *E. coli* strains and strategies for the production of P(3HB) [34–41]. The P(3HB) concentration of higher than 80 g/l with the productivity of greater than 2 g P(3HB)/l/h could be obtained by the pH-stat fed-batch culture of recombinant *E. coli* in a complex medium containing glucose [35]. In a defined medium containing glucose, however, less than 25 g/l of P(3HB) could only be produced by the similar fed-batch culture of recombinant *E. coli* [36]. P(3HB) biosynthesis was significantly enhanced in a defined medium by the addition of small amounts of various complex nitrogen sources, amino acids, or oleic acid [36, 37, 40]. It was found that in these conditions citrate synthase was inhibited, and more acetyl-CoA and/or NADPH, a cofactor of the reductase in the PHA biosynthetic pathway, were available for PHA biosynthesis. Recombinant *E. coli* accumulated a large amount of P(3HB) when the NADPH level and NADPH/NADP ratio were high [42]. Cells of recombinant *E. coli* underwent considerable filamentation during synthesis and accumulation of P(3HB) [41, 43]. Since cell filamentation results in slow or no further growth and lower metabolic activity, it was thought that P(3HB) production might be enhanced by suppressing cell filamentation. Suppression of filamentation was possible by overexpressing an essential cell division protein FtsZ [43]. This filamentation-suppressed recombinant *E. coli* strain allowed production of P(3HB) to a high concentration of 104 g/l with P(3HB) content of 70 wt% of cell dry weight (CDW) and the productivity of 2 g P(3HB)/l/h in a defined medium [44].

Further optimization of growth medium and culture condition resulted in P(3HB) concentration, P(3HB) content, and productivity of 149 g/l, 77 wt%, and 3.4 g P(3HB)/l/h, respectively, in a defined medium [45].

During the fed-batch culture of recombinant *E. coli*, a large amount of oxygen was necessary to maintain the dissolved oxygen concentration above 20% of air saturation. Since the use of a large amount of pure oxygen is economically unfavorable and oxygen transfer is generally poor in a larger scale fermentor, P(3HB) production under insufficient oxygen supply was considered [45]. When the insufficient oxygen was supplied during the active cell growth phase by maintaining the dissolved oxygen concentration at 1–3% of air saturation, cell growth stopped and no apparent enhancement of P(3HB) accumulation was observed. Furthermore, insufficient oxygen supply during the active cell growth phase resulted in the accumulation of acetate to a concentration of 14 g/l, which was higher than the critical acetate concentration that shows growth inhibition. However, cell growth and P(3HB) production were not hampered when insufficient oxygen was supplied during the active P(3HB) synthesis phase, during which P(3HB) is actively accumulated with the concomitant increase of the P(3HB) content. Acetate concentration was less than 4 g/l under this condition. Cell concentration, P(3HB) concentration, and P(3HB) content obtained were as high as 204.3 g/l, 157.1 g/l, 77 wt%, respectively, resulting in the P(3HB) productivity of 3.2 g/l/h [45]. These results suggested that the insufficient oxygen supply during the active P(3HB) synthesis phase did not have any harmful effect on cell growth and P(3HB) production. To examine if this strategy of oxygen supply could be employed for the production of P(3HB) in a larger scale fermentor, fed-batch culture was carried out in a 50-l pilot-scale fermentor [45]. This fermentor was a typical stirred tank bioreactor equipped with the bottom-driven three Rushton turbine impellers having the maximum agitation speed of 500 rpm. Instead of using pure oxygen, oxygen-enriched air (83% O_2) was supplied when required. Although the dissolved oxygen concentration decreased to zero when cell concentration reached 54.8 g/l at the maximum gas flow rate and agitation speed, a high concentration of P(3HB) (101.3 g/l) was still obtained in 36 h, resulting in a high P(3HB) productivity of 2.8 g/l/h [45].

Control of oxygen supply in the cultivation of recombinant *E. coli* also caused the variation of oxygen usage in the cell. Because oxygen consumption was related to CO_2 production, ATP generation, and other metabolism, P(3HB) yield on carbon substrate was affected by oxygen supply during the cultivation. Under the oxygen supplying condition, where gas flowrate was held at a constant value (4 l/min) and inlet gas composition of air and oxygen was changed to maintain the dissolved oxygen concentration above 20%, high P(3HB) yield on carbon of 0.35 g P(3HB)/g glucose was obtained with P(3HB) concentration of 177 g/l, P(3HB) content of 78 wt%, and productivity of 3.39 g P(3HB)/l/h [46]. In the fed-batch cultures of recombinant *E. coli* in a chemically defined medium, the P(3HB) yield on carbon was generally in the range of 0.27–0.29 g P(3HB)/g glucose. Therefore, this result suggests that P(3HB) yield on carbon could be enhanced by optimizing the oxygen supply during the cultivation.

Even though high P(3HB) concentration could be obtained by recombinant *E. coli*, the P(3HB) productivity was lower than that of *A. latus*. Since *A. latus* was able to produce P(3HB) with the highest productivity reported to date (4.94 g P(3HB)/l/h) [47], it was assumed that this bacterium possesses PHA biosynthetic enzymes that allow more efficient polymer synthesis. It was therefore thought that recombinant *E. coli* harboring the *A. latus* PHA biosynthesis genes might allow production of P(3HB) with higher productivity. By the pH-stat fed-batch culture of recombinant *E. coli* harboring a stable high copy number plasmid containing the *A. latus* PHA biosynthesis genes, the final cell and P(3HB) concentrations of 194.1 g/l and 141.6 g/l, respectively, were obtained, resulting in the high productivity of 4.63 g P(3HB)/l/h (Fig. 2) [48].

One of the advantages of recombinant *E. coli* over other natural PHA producers is that there are many different *E. coli* strains utilizing inexpensive carbon sources. Therefore, it is possible to produce PHA from inexpensive carbon sources, such as whey, hemicellulose, and molasses, by recombinant *E. coli*. There was a report on the production of P(3HB) from hydrolyzed molasses by recombinant *E. coli* harboring the *R. eutropha* PHA biosynthesis genes. By the fed-batch culture, the final cell concentration, P(3HB) content obtained were 39.5 g/l and 80 wt %, respectively, resulting in the productivity of 1 g P(3HB)/l/h [49]. Whey is a major byproduct from the cheese manufacture industry, and it contains approximately 4.5% (w/v) of lactose, which can be a good carbon source for PHA production. Several lactose-utilizing recombinant *E. coli* strains harboring the *R. eutropha* PHA biosynthesis genes were grown to accumulate P(3HB) in a whey-based medium. One of these strains accumulated P(3HB) up

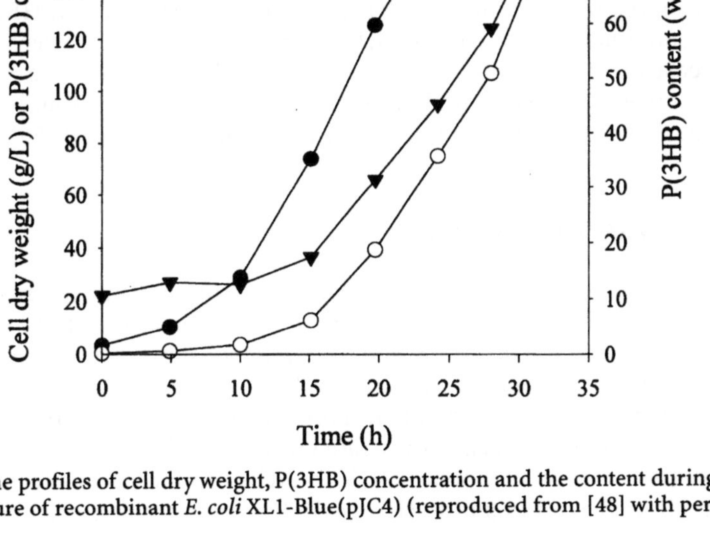

Fig. 2. Time profiles of cell dry weight, P(3HB) concentration and the content during the fed-batch culture of recombinant *E. coli* XL1-Blue(pJC4) (reproduced from [48] with permission)

to 81 wt% of CDW [50]. By fed-batch culture of this strain using concentrated whey solution as the nutrient feed, cell and P(3HB) concentrations of 87 g/l and 69 g/l, respectively, were obtained, resulting in the productivity of 1.4 g P(3HB)/l/h [51]. Recently, a recombinant *E. coli* strain harboring the *R. eutropha* PHA biosynthesis genes which can produce P(3HB) from xylose was also developed. Supplementation of a small amount of complex nitrogen source such as soybean hydrolysate improved P(3HB) production to a level of 74 wt% of CDW [52].

The molecular weight of P(3HB) could be controlled by modulating the activity of the PHA synthase in recombinant *E. coli*. The molecular weight of P(3HB) decreased with increasing PHA synthase activity [53]. An extremely high molecular weight (4×10^3 kDa) of P(3HB) could be obtained with a recombinant *E. coli* expressing the PHA synthase at low level. The polydispersity index (the ratio of weight average molecular weight to number average molecular weight) could also be controlled. In addition, there was a report on controlling the molecular weight of P(3HB) in recombinant *E. coli* by varying culture pH [54].

There has been a report on the release of PHA granules from recombinant *E. coli* cells by the introduction and expression of the bacteriophage lysis gene E [55]. It was shown that small P(3HB) granules in a semi-liquid stage were squeezed out of the cells through the E-lysis tunnel structure, which was characterized by a small opening in the envelope with borders of fused inner and outer membranes. All envelope components remained intact after E-lysis and could be removed from the mixture of released P(3HB) granules by density gradient centrifugation. In addition, a modified E-lysis procedure was described, which did enable the release of P(3HB) from cell pellet in pure water or low ionic strength buffer. There has been another report on simple P(3HB) recovery from recombinant *E. coli*. By digesting cells with NaOH, P(3HB) with purity of greater than 98.5% was obtained at an optimal condition without polymer degradation [56]. This NaOH digestion method also efficiently removed the endotoxin during the purification of P(3HB) [57]. Combination of P(3HB) production by recombinant *E. coli* with NaOH digestion considerably lowered the production cost of P(3HB). When the process of P(3HB) production by recombinant *E. coli* with recovery method of NaOH digestion was simulated (Fig. 3), the production cost of P(3HB) was as low as US$3.31/kg P(3HB) [56].

4.2
Production of PHAs other than P(3HB)

A mutant *E. coli* strain LS5218 (*fadR atoC*) was employed for the synthesis of P(3HB-*co*-3HV) copolymer since this mutant strain constitutively expresses the enzymes involved in the transport and utilization of short chain fatty acids [58, 59]. P(3HB-*co*-3HV) could be synthesized by a recombinant *E. coli* strain LS5218 harboring the *R. eutropha* PHA biosynthesis genes when propionic acid or valeric acid was added as a cosubstrate [58, 60]. The P(3HB-*co*-3HV) copolymer consisting of up to 40 mol% of 3HV could be produced. An alternative method that allowed synthesis of P(3HB-*co*-3HV) using propionic acid or valeric

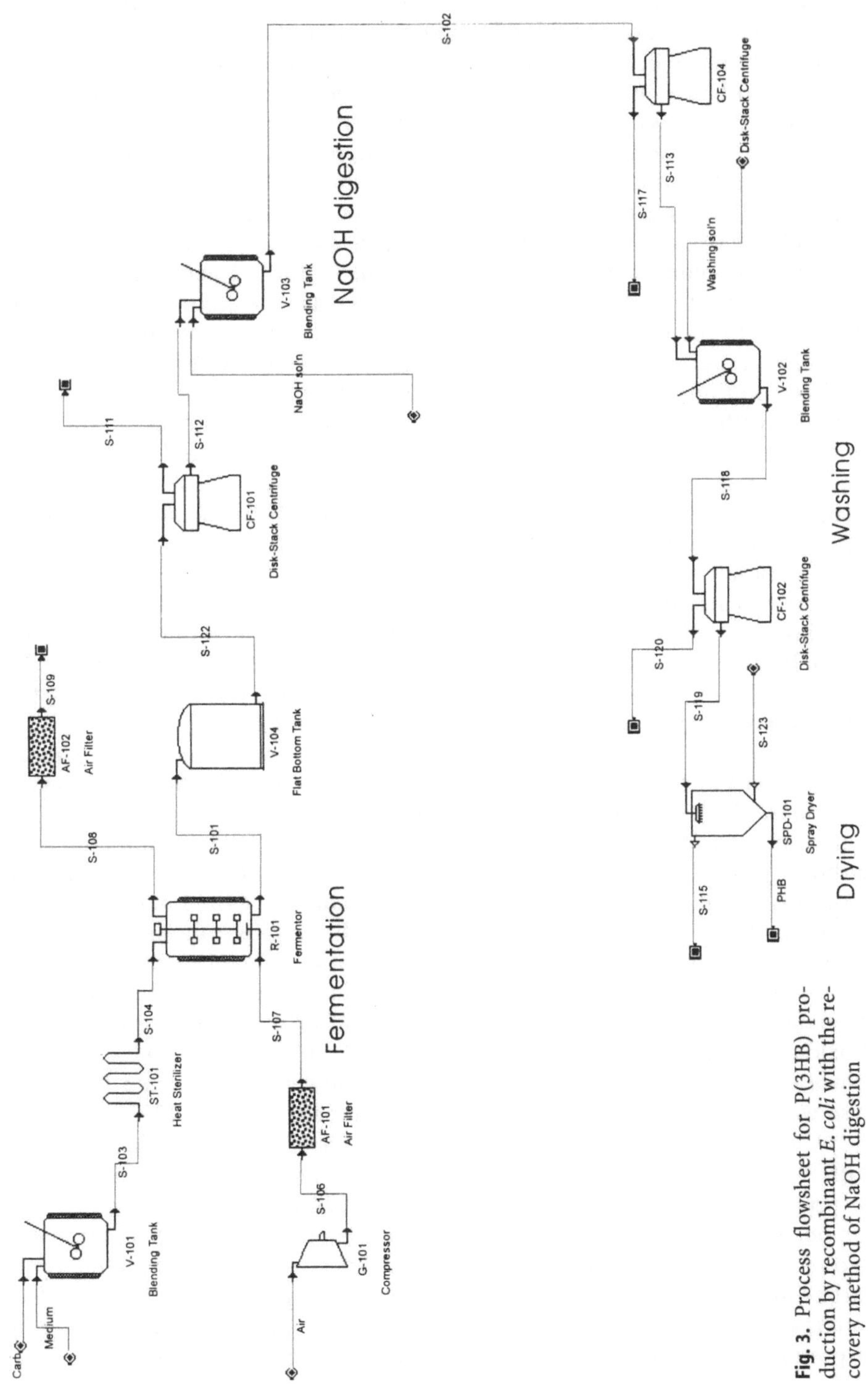

Fig. 3. Process flowsheet for P(3HB) production by recombinant *E. coli* with the recovery method of NaOH digestion

acid as a cosubstrate in non-*fadR atoC E. coli* strains was also investigated [61]. By acetic acid and/or oleic acid induction, non-*fadR atoC* strains of *E. coli* could efficiently produce P(3HB-*co*-3HV) from propionic acid [61].

In the fed-batch culture of recombinant *E. coli* harboring the *A. latus* PHA biosynthesis genes with the feeding strategy that increased glucose and propionic acid concentrations to 20 g/l and 20 mmol/l, respectively, after each feeding, cell concentration obtained was 120.3 g/l, but the PHA content was rather low (42.5 wt%) [62]. By acetic acid or oleic acid induction, the concentrations of cell and PHA, the PHA content and 3HV fraction were all increased. However, it was found that propionic acid was accumulated during the fed-batch culture, hampering cell growth and PHA production. To decrease the accumulation of propionic acid in the medium, fed-batch cultures with different feeding solutions containing lower concentration of propionic acid were carried out. When the feeding solution was added to increase propionic acid concentration to 5 mmol/l with oleic acid supplementation after acetic acid induction, the concentration of PHA and PHA content obtained were 158.8 g/l and 78.2 wt%, respectively, resulting in the PHA productivity of 2.88 g PHA/l/h (Fig. 4) [62]. Also, it was found that high PHA concentration and PHA content could be produced by feeding solution with low concentration of propionic acid, but the 3HV fraction was relatively low.

There has also been a report on the production of P(3HB-*co*-3HV) in a recombinant *E. coli* by supplementing a small amount of amino acid. When 1 mmol/l valine was added to 1% glucose medium, recombinant *E. coli* incorporated 3HV up to 2.5 mol %. When 1 mmol/l threonine was added as well, the 3HV monomer up to 4 mol% could be incorporated [63].

Production of medium chain length (MCL)-PHA consisting of six or more carbon atoms in recombinant *E. coli* has also been reported. The *Pseudomonas aeruginosa* PHA synthase gene *phaC1* under the control of the *E. coli lac* promoter was transferred into *E. coli*. When induced with IPTG, recombinant *E. coli* harboring the *P. aeruginosa phaC1* accumulated MCL-PHA from decanoate [64]. This study provided evidence that intermediates of the fatty acid β-oxidation pathway could be directed to MCL-PHA biosynthesis in *E. coli*. Recombinant *E. coli* LS1298, a fatty acid degradation mutant strain, accumulated MCL-PHA to 21% of CDW, when cultivated in LB medium containing 0.5% (w/v) decanoate. Also, the addition of acrylic acid, an inhibitor of the fatty acid β-oxidation, increased the accumulation of PHA up to about 60 wt% of CDW grown on decanoate [65]. Another PHA synthase gene *phaC2* from *P. aeruginosa* was also expressed in *E. coli* [66]. Recombinant *E. coli* LS1298 harboring the *phaC2* gene accumulated MCL-PHA up to 15% of CDW from dodecanoate. Both PHA synthases PhaC1 and PhaC2 from *P. aeruginosa* exhibited very similar properties, resulting in a similar extent of PHA accumulation, and similar composition of monomers and molecular mass when expressed in *E. coli*. It was also recently shown that recombinant *E. coli*, defective in the β-oxidation cycle harboring PHA synthase encoded by the *phaC1* or *phaC2* gene from *P. oleovorans* and the cytosolic thioesterase I-encoding *tesA* gene from *E. coli*, was able to produce MCL-PHA from gluconate [67]. The thioesterase hydrolyzes acyl-ACPs, producing enhanced intracellular levels of free fatty acids, which can then

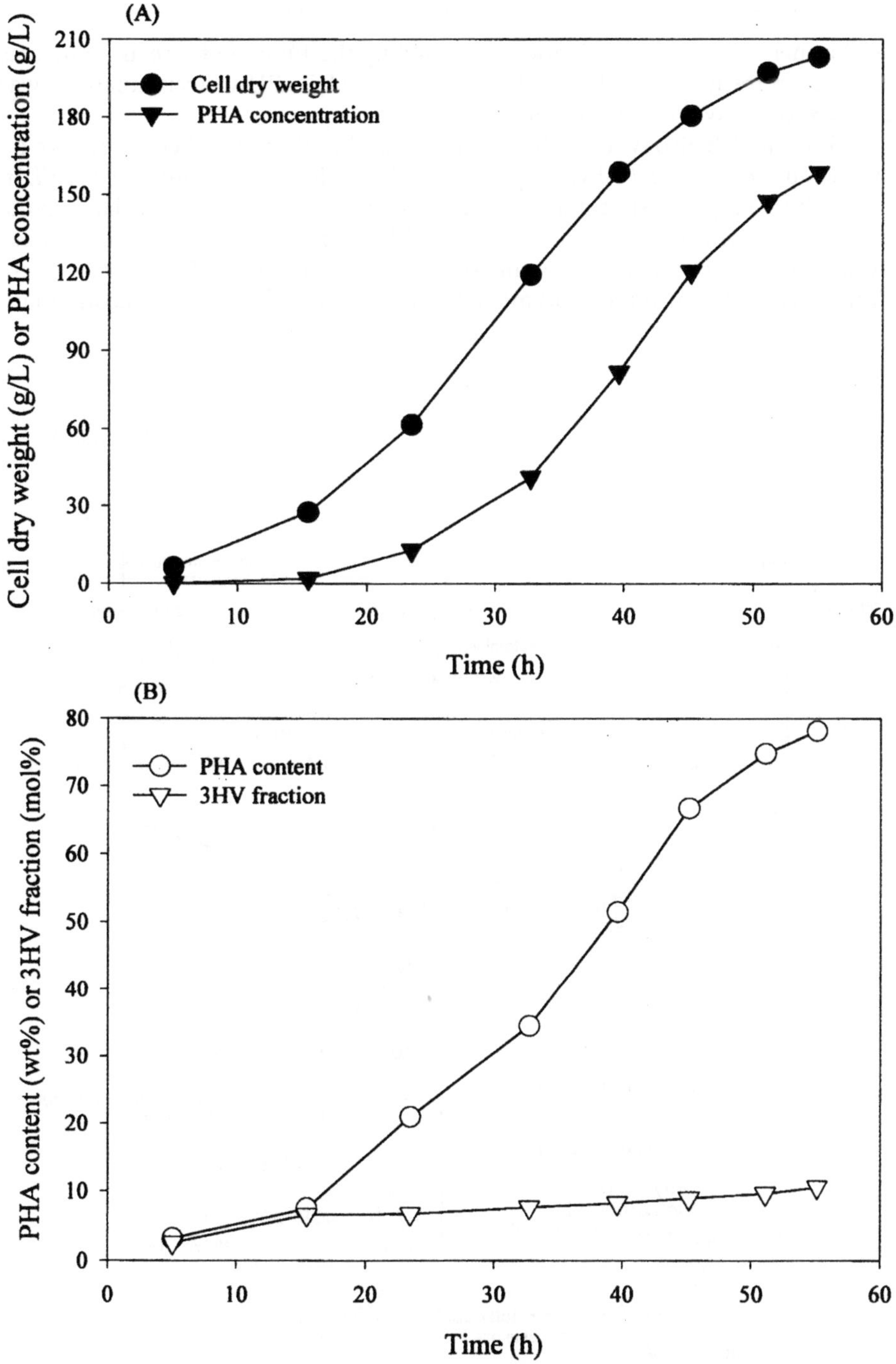

Fig. 4 A, B. Time profiles of: **A** cell dry weight and PHA concentration; **B** PHA content (wt%) and 3HV fraction in PHA (mol%) during the fed-batch culture of XL1-Blue(pJC4) with oleic acid supplementation after acetic acid induction. The feeding solution was added to increase the concentrations of glucose and propionic acid to 20 g/l and to 5 mmol/l, respectively, after each feeding (reproduced from [62] with permission)

be channeled into the β-oxidation and used by the PHA synthase as substrates for incorporation into PHA. The maximal MCL-PHA accumulation was 2.3 wt% of CDW from gluconate.

Using the 4-hydroxybutyl-CoA dehydrogenase, there has been a report on the production of poly(4-hydroxybutyrate) [P(4HB)] in recombinant *E. coli*. In *Clostridium kluyveri*, succinate is fermented to butyrate by sequential enzyme reactions. Recently, the gene encoding 4-hydroxybutyl-CoA dehydrogenase from *C. kluyveri* and the gene encoding the *R. eutropha* PHA synthase were cloned in *E. coli* to produce polymer consisting of 4HB [68]. When recombinant

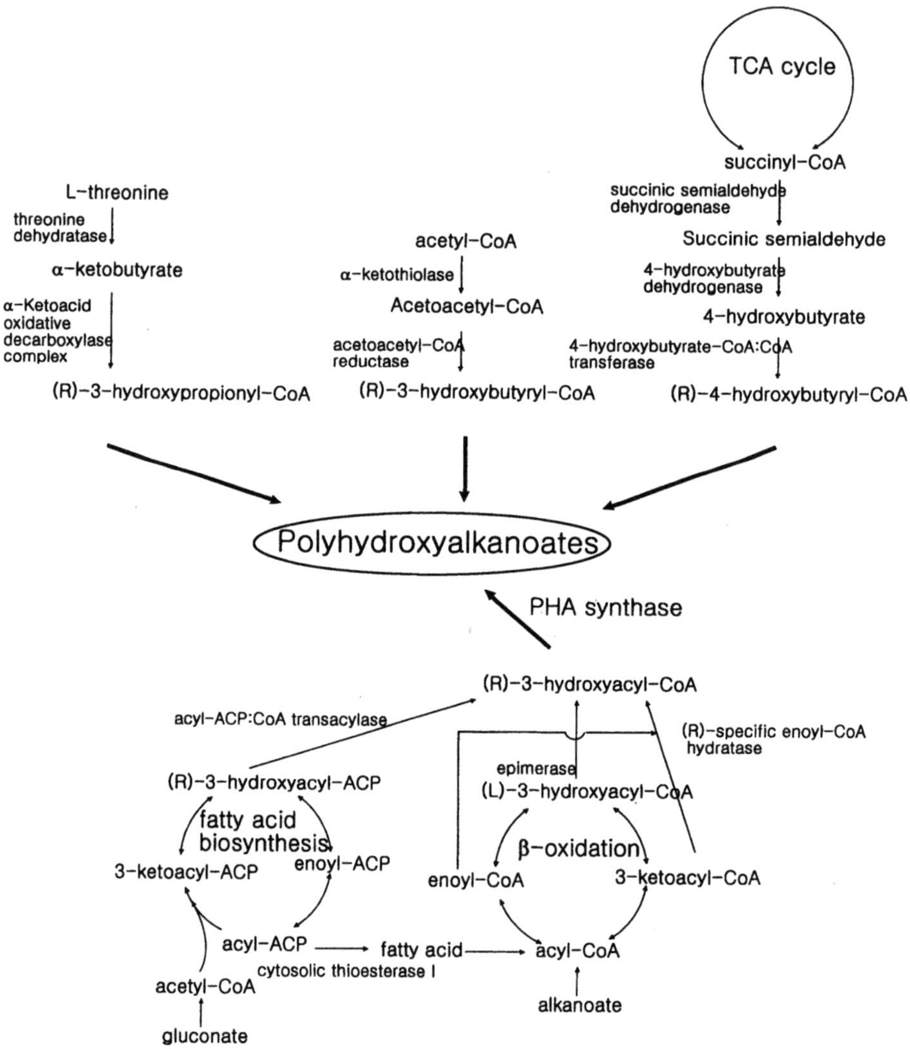

Fig. 5. PHA biosynthesis pathway in recombinant *E. coli* synthesizing novel PHA with various monomers

E. coli harboring the *C. kluyveri* 4-hydroxybutyl-CoA dehydrogenase gene and the *R. eutropha* PHA synthase gene was cultivated in a medium containing 0.4% of 4-hydroxybutyric acid (4HB) and 0.5% of glucose, homopolymer P(4HB) with the content up to 58.5 wt % was produced. Using this recombinant *E. coli* strain, fed-batch culture was carried out in M9 salt medium containing glucose and 4HB as carbon sources [69]. The final cell and P(4HB) concentrations and P(4HB) content obtained were 12.6 g/l, 4.4 g/l, and 36 wt%, respectively. It has also been reported that copolymer poly(3-hydroxybutyrate-*co*-4-hydroxybutyrate) [P(3HB-*co*-4HB)] could be produced by recombinant *E. coli* harboring the *R. eutropha* PHA biosynthetic pathway and the *C. kluyveri* succinate degradation pathway [70]. When three enzymes, 4-hydroxybutyrate dehydrogenase, succinic semialdehyde dehydrogenase, and 4-hydroxybutyryl-CoA:CoA transferase of *C. kluyveri*, were cloned and expressed in *E. coli* with the *R. eutropha* PHA biosynthesis genes, P(3HB-*co*-4HB) containing 2.8 mol% 4HB was accumulated up to 50 wt% of CDW in complex medium containing glucose.

Recombinant *E. coli* harboring the *Aeromonas caviae* PHA biosynthesis genes produced poly(3-hydroxybutyrate-*co*-3-hydroxyhexanoate) [P(3HB-*co*-3HHx)] [71]. A recombinant *E. coli* LS5218 strain harboring only the *A. caviae* *phaC* did not accumulate any PHA from dodecanoate in spite of the existence of translated PHA synthase, whereas coexpression of PHA synthase (*phaC*) and enoyl-CoA hydratase gene (*phaJ*) resulted in the accumulation of P(3HB-*co*-3HHx) to 7–11 wt% of CDW from octanoate and dodecanoate. These results indicated that both *phaC* and *phaJ* are essential for *E. coli* to establish the PHA biosynthesis pathway from MCL-alkanoic acid. The copolymer content in the strain expressing both the genes under the *lac* promoter reached 38 wt% from dodecanoate. Enzyme assays suggested that efficient monomer formation via *β*-oxidation by a high level expression of *phaJ* was important to achieve a high PHA content in the recombinant *E. coli*.

In Fig. 5, the biosynthetic pathways for the production of PHA with novel composition of hydroxyalkanoates and the enzymes involved are shown. The level of metabolic intermediates, which is determined by the cellular metabolic activities, is important for the synthesis of a desired PHA. Having the engineered metabolic pathways at hand, PHA synthase plays an important role affecting the composition of PHAs, because of substrate specificity of PHA synthase.

5 Recombinant *Ralstonia eutropha*

R. eutropha, formerly known as *Alcaligenes eutrophus*, has been used for the commercial production of P(3HB-*co*-3HV) [72]. This bacterium grows well in a relatively inexpensive minimal medium and accumulates a large amount of P(3HB) under the unbalanced growth condition. In *R. eutropha*, acetyl-CoA is converted to P(3HB) by three enzymes (genes): *β*-ketothiolase (*phaA*), acetoacetyl-CoA reductase (*phaB*), and PHA synthase (*phaC*) [6].

High cell density culture of *R. eutropha* has been studied extensively. To maintain glucose concentration within the optimal range, several feeding strategies were developed during the fed-batch cultures [73, 74]. It was important to find when to apply the nutrient limitation. By the fed-batch culture of *R. eutropha* with DO-stat feeding strategy under phosphate limitation, the final cell concentration, P(3HB) concentration and the P(3HB) content of 281 g/l, 232 g/l, and 80 wt%, respectively, could be obtained, resulting in the productivity of 3.14 g P(3HB)/l/h [74]. Since *R. eutropha* can accumulate a significant amount of PHA (up to 80 wt%), several studies were carried out to increase PHA synthesis rates or PHA yield on the carbon substrate. *R. eutropha* was transformed with a broad-host range plasmid containing its own PHA biosynthesis genes (*phaCAB*) for the homologous amplification of the enzyme activities involved in PHA biosynthesis [75]. The specific activities of three enzymes in recombinant *R. eutropha* were all increased. However, in the batch culture for the production of P(3HB), the final cell concentrations of the recombinant *R. eutropha* strains were not much different from that of the parent strain. Among these recombinant *R. eutropha* strains harboring *phaABC*, *phaAB*, or *phaC*, strains harboring *phaCAB* and *phaC* gave slightly higher P(3HB) concentration and P(3HB) content compared with the wild type [76, 77]. The P(3HB) concentration and P(3HB) content were increased by 18–22% and 8–10%, respectively. However, the reported P(3HB) contents were only 33% for the wild type and 36–40% for the recombinants, which were too low considering that the P(3HB) content typically obtainable with *R. eutropha* is 60–80%. In a similar approach to improving the P(3HB) synthesis rate by recombinant *R. eutropha*, recombinant *R. eutropha* was constructed by transforming with broad host range plasmid containing the *R. eutropha* PHA biosynthesis genes (*phaABC*). The P(3HB) synthesis rate was higher in the recombinant strain than in the wild type (Fig. 6) [78]. There has been no report on the production of P(3HB) to high concentration by these recombinant *R. eutropha* strains. Using the recombinant *R. eutropha* strains harboring *phaC*, P(3HB-*co*-3HV) and P(3HB-*co*-4HB) were also produced [79]. The cell concentration and the PHA content did not increase, but only molar fraction of 3HV and 4HB, respectively, increased in the recombinant *R. eutropha*. Recombinant *R. eutropha* harboring *phaC* produced P(3HB-*co*-3HV) having 3HV molar fraction of 50.5% from 20 g/l of propionic acid, while the wild type produced P(3HB-*co*-3HV) having 3HV molar fraction of 34.1%. From 20 g/l of 4-hydroxybutyric acid, the 4HV fraction in PHA increased from 24.1 mol% in the wild type to 57.6 mol% in recombinant strain. There have been other examples that recombinant *R. eutropha* harboring the PHA biosynthesis genes produced PHA with increased PHA content and molar fraction of co-monomer. The PHA-leaky mutant of *R. eutropha* JMP222-PHB-102 could accumulate a small amount of P(4HB) (11 wt% of CDW). When the *R. eutropha* PHA biosynthesis genes were introduced into *R. eutropha* JMP222-PHB-102, the content of P(4HB) increased to 27 wt% [80]. In the production of PHA composed of 3HB, 3HV, and 4HV from 4-hydroxyvaleric acid, *R. eutropha* mutant accumulating a terpolyester with 4HV molar fraction of up to 22.7 mol% was transformed with the hybrid plasmid harboring the *R. eutropha* PHA biosynthesis genes [81]. The molar fraction of 4HV in the polymer accumulated in the recombinant increased up to 30 mol% with a little increased PHA content.

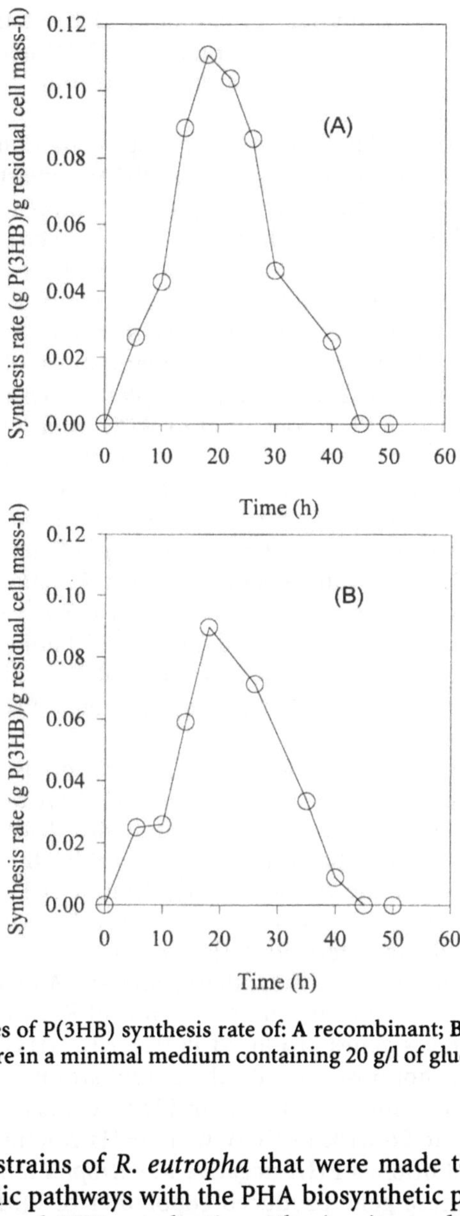

Fig. 6 A, B. Time profiles of P(3HB) synthesis rate of: **A** recombinant; **B** wild type *R. eutropha* during the batch culture in a minimal medium containing 20 g/l of glucose (reproduced from [78] with permission)

Several mutant strains of *R. eutropha* that were made to possess defective competing metabolic pathways with the PHA biosynthetic pathway were developed for the enhanced PHA production. The isocitrate dehydrogenase leaky mutant of *R. eutropha* accumulated P(3HB) more favorably at a lower carbon/nitrogen molar ratio and at a lower carbon concentration than the parent strain [82]. In batch culture, the final cell and P(3HB) concentrations, and P(3HB) yield on glucose were slightly increased. Also, in the P(3HB-*co*-3HV) biosynthesis, the molar fraction of 3HV and the 3HV yield on propionic acid increased due to the enhanced conversion of propionic acid to 3-hydroxyvaleryl-CoA rather than to acetyl-CoA and CO_2 in this mutant. Another mutant *R. eu-*

tropha strain unable to assimilate propionic acid for cell growth produced P(3HB-*co*-3HV) with increased molar fraction of 3HV and the 3HV yield on propionic acid compared with the parent strain, but the concentration of P(3HB-*co*-3HV) decreased [83].

Bacterial PHAs can be classified in two groups depending on the monomer composition [6, 14]. Many bacteria synthesize PHAs consisting of short-chain-length (SCL-) hydroxyalkanoates consisting of from three to five carbon atoms, while pseudomonads belonging to the rRNA homology group I and some other bacteria synthesize medium-chain-length (MCL-) PHAs consisting of six or more carbon atoms. The best studied PHA synthase of *R. eutropha* had previously been shown to produce only SCL-PHAs in vivo and in vitro [84, 85]. However, it was recently shown that recombinant strain of a PHA negative *R. eutropha* mutant expressing its own *phaB* and *phaC* genes was able to accumulate P(3HB-*co*-3HHx) with PHA content of 81 wt% when even chain fatty acids (chain length greater than or equal to C6) were supplied as carbon sources [86]. The molar fraction of 3HHx in copolymer was 8 mol%. Similar results were obtained with *Klebsiella aerogenes* and *Pseudomonas putida* GPp104, a negative PHA mutant, expressing the *R. eutropha phaB* and *phaC* genes. These results demonstrated a broader substrate specificity of the *R. eutropha* PHA synthase than previously thought. However, a recombinant *E. coli* expressing the *R. eutropha phaB* and *phaC* genes synthesized only P(3HB) homopolymer under the same condition. It will be interesting to see if P(3HB-*co*-3HHx) could be produced by recombinant *E. coli* expressing the *R. eutropha* PHA synthase with the enoyl-CoA hydratase or epimerase.

Several bacteria including *A. caviae* [87], *Rhodococcus ruber* [88], *Pseudomonas* sp. [89, 90], and *Rhodocyclus gelatinosus* [91] have been reported to produce copolymers consisting of both SCL- and MCL-hydroxyalkanoates. The PHA synthase genes of these bacteria will be useful for the construction of recombinant organisms for the efficient production of SCL- and MCL-PHA copolymers. Recently, the genes involved in the biosynthesis of PHAs in *A. caviae* have been cloned and characterized [92]. The cluster of the cloned DNA fragment contained the genes for PHA synthase, enoyl-CoA hydratase (*phaJ*) and a phasin (ORF1). The PHA negative mutant strains of *R. eutropha* harboring the *A. caviae* PHA synthesis genes produced P(3HB-*co*-3HHx) from hexanoate or octanoate, which was not possible with the parent strain. The PHA negative *R. eutropha* mutant expressing the *A. caviae* PHA synthase genes accumulated P(3HB-*co*-3HHx) up to 96 wt% of CDW with 3HHx fraction of 15 mol% from octanoate [92]. In addition, a PHA negative *R. eutropha* mutant expressing the *A. caviae* PHA synthase gene produced copolymer of P(3HB-*co*-3HHx) from plant oils to a high content (approximately 80 wt% of CDW) [93]. The molar fraction of 3HHx was 3–5 mol% regardless of the structure of the triglycerides fed. Furthermore, recombinant strain of a PHA negative *R. eutropha* mutant harboring the *A. caviae* PHA synthase genes synthesized poly(3-hydroxybutyrate-*co*-3-hydroxyvalerate-*co*-3-hydroxyheptanoate) terpolymer up to 78 wt% of CDW from alkanoic acid of odd carbon numbers [94].

Pseudomonas sp. 61-3 produced a blend of P(3HB) homopolymer and P(3HB-*co*-3HA) copolymer from sugars such as glucose, fructose, and mannose

[89]. Two different PHA biosynthesis gene loci were recently cloned [95]. The PHA negative mutant of *R. eutropha* harboring the *Pseudomonas* sp. PHA biosynthesis locus produced PHA consisting of 3HB, 3HHx, 3HO, and 3-hydroxydecanoate (3HD), 3-hydroxydodecanoate (3HDD) with the PHA content of up to 14 wt% of CDW.

Even though high cell density fed-batch culture of recombinant *R. eutropha* has not been reported, there is no reason for the difficulty of growing the recombinants to a high density. Development of a proper nutrient feeding strategy will allow production of various PHAs to a high concentration with high productivity.

6 Recombinant Pseudomonads

Many *Pseudomonas* strains accumulate MCL-PHAs from alkane, alkene, alkanoate, alkenoate, or alkanol [5, 6, 14, 96]. The composition of the PHAs formed by the pseudomonads of the rRNA homology group I is directly related to the structure of the carbon substrate used [6]. These results suggested that MCL-PHAs are synthesized from the intermediates of the fatty acid oxidation pathway. In almost all pseudomonads belonging to the rRNA homology group I except *Pseudomonas oleovorans*, MCL-PHA can also be synthesized from acetyl-CoA through de novo fatty acid synthetic pathway [97]. The β-oxidation pathway and de novo fatty acid synthetic pathway function independently in PHA biosynthesis.

There have been several reports on the production of MCL-PHAs by high cell density culture of *P. oleovorans* and *P. putida* [96]. Several recombinant pseudomonads have also been developed for possible improvement in PHA production. In the recombinant *P. putida* harboring either of the two PHA synthase genes (*phaC1* or *phaC2*) in a multicopy plasmid, the molecular weight of PHAs decreased, suggesting that the molecular weight of PHAs is determined by the activity of the PHA synthase [98]. It was also reported that the high level expression of PHA synthase in recombinant *E. coli* harboring the *R. eutropha* PHA biosynthesis genes decreased the molecular weight of P(3HB) [53]. There has been a report on PHA biosynthesis by recombinant *Pseudomonas* strains overexpressing the PHA synthase PhaC1 [99]. Recombinant *P. oleovorans* produced PHA to 64 wt% of CDW under non-nitrogen limiting condition while the PHA content in the wild type *P. oleovorans* was 34 wt% under this condition. However, under nitrogen limiting condition, the PHA content increased only slightly from 44 wt% in the parent strain to 47 wt% in the recombinant strain. Recombinant *Pseudomonas* strains produced PHA having higher fraction of 3HHx monomer compared with the parent strain.

A PHA negative mutant strain of *P. putida* harboring the PHA synthase gene of *P. oleovorans* produced poly(3-hydroxyhexanoate-*co*-3-hydroxyoctanoate) [P(3HHx-*co*-3HO)] having different fractions of 3HHx monomer depending on the *P. oleovorans* PHA synthase employed, suggesting that it is possible to modulate the monomer composition of PHAs by using different PHA synthase [100].

The *R. eutropha* PHA biosynthesis genes were also functionally expressed in pseudomonads strains, and P(3HB) was produced from gluconate [101]. Recombinant *P. oleovorans* harboring the *R. eutropha* PHA biosynthesis genes accumulated PHA composed of 3HB, 3HHx, and 3HO from sodium octanoate, but P(3HB) homopolymer and P(3HHx-*co*-3HO) copolymer were stored in separate granules, indicating the exclusive substrate specificities of the two different PHA synthases [102, 103]. However, when recombinant strains of the PHA-negative mutant *P. putida* GPp104 harboring the *Thiocapsa pfennigii* PHA synthase genes were cultivated on octanoate, a PHA copolymer consisting of almost equimolar amounts of 3HB and 3HHx plus small amounts of 3HO was produced [104]. When this PHA synthase gene was introduced into a PHA-negative mutant strain of *R. eutropha*, a terpolyester consisting of 3HB, 3HHx, and 4HHx was produced from 4-hydroxyhexanoic acid as a carbon source. From these results, it was concluded that the PHA synthase of *T. pfennigii* could accept CoA thioesters of both SCL and MCL-hydroxyalkanoates.

When fluorescent pseudomonads of rRNA homology group I were grown on sugars, PHA consisted primarily of C8 and C10 monomers derived from the intermediates of fatty acid biosynthesis was formed [6]. In *P. oleovorans*, the *phaC1* gene was expressed efficiently in the presence of octanoic acid while its expression is repressed when glucose or citrate is used as the carbon source. From the insertional mutagenesis studies with *P. oleovorans*, it was found that the product of *phaF*, located at the downstream of PHA biosynthesis genes, regulated the expression of *phaC1* [105]. In the *P. oleovorans* mutant having defective PhaF, higher expression level of PhaC1 was observed than in the wild-type strain in the presence of glucose or citrate.

There has been a report on the production of PHA by recombinant pseudomonads harboring genes allowing the utilization of new substrate. *Pseudomonas acidophila* is a bacterial strain producing PHA copolymers from low-molecular-weight organic compounds such as formate and acetate. When a plasmid containing the genes for chemolithoautotrophic growth of the hydrogen-oxidizing bacterium *Alcaligenes hydrogenophilus* was introduced into *P. acidophila*, this recombinant *P. acidophila* was able to grow under a gas mixture of H_2, O_2, and CO_2 in a mineral salts medium, and accumulated PHA copolymers [106]. The composition of the polymer was 52.8 mol% 3HB, 41.1 mol% 3HO, and 6.1 mol% 3HD. However, the polymer content was rather low at 7 wt% of CDW, which needs to be considerably increased to make this process more attractive.

Heterologous expression of the *A. caviae* PHA biosynthesis genes was investigated in the PHA negative mutant of *P. putida*. Recombinant *P. putida* produced P(3HB-*co*-3HHx) from gluconate, hexanoate, or octanoate up to 48 wt% of CDW with relatively high 3HHx molar fraction compared with recombinant *R. eutropha* harboring the *A. caviae* PHA biosynthesis genes [92]. In addition, the PHA negative mutant of *P. putida* harboring the *Pseudomonas* sp. PHA biosynthesis genes accumulated PHA consisting of 3HB, 3HHx, 3HO, 3HD, 3HDD, and 3-hydroxy-*cis*-5-dodecenoate up to 43 wt% of CDW [95].

Recombinant *Pseudomonas* strains harboring the nuclease gene from *Staphylococcus aureus* was developed to facilitate downstream processing [107].

The recombinant strain expressed the *Staphylococcus* nuclease directed to the periplasm without affecting PHA production. During downstream processing, the viscosity of the recombinant lysate was significantly reduced to facilitate the subsequent purification steps.

As summarized above, various approaches were taken to alter the PHA production or to facilitate downstream processing by employing recombinant pseudomonads. However, as in the case with recombinant *R. eutropha*, no study has been carried out on the high cell density culture of recombinant pseudomonads or on the scale-up of fermentation.

7
Other Recombinant Bacteria

P(3HB) synthesis by *Mycoplana rubra* B346, a pink pigmented facultatively methylotrophic bacterium, was inefficient, and the specific activity of the PHA synthase was extremely low in *M. rubra*, which was thought to be the reason for the inefficient production of PHAs. The recombinant strains of *M. rubra* harboring the PHA biosynthesis genes of *R. eutropha*, *Chromatium vinosum* D, or *Methylobacterium extorquens* could accumulate P(3HB) up to 25 wt% of CDW, which was higher than that (9 wt%) obtainable in the parent strain [108]. However, only the overexpression of PHA biosynthesis genes without engineering the metabolic pathway could not enhance the P(3HB) production further.

Some cyanobacteria have been reported to accumulate P(3HB), which suggested that it might be possible to produce P(3HB) from CO_2. However, the level of P(3HB) accumulation was rather low, and therefore attempts to increase the P(3HB) production by recombinant cyanobacteria have been made. Recombinant *Synechococcus* sp. harboring the *R. eutropha* PHA biosynthesis genes was able to accumulate P(3HB) up to 1% of CDW under nitrogen-starved condition [109]. Even though the P(3HB) content obtained was extremely low, improved P(3HB) accumulation was reported under optimized culture conditions. Under photoautotrophic and nitrogen-starved conditions, P(3HB) content of recombinant *Synechococcus* sp. was improved by CO_2-enrichment in sparging. A supplement of acetic acid under nitrogen-starvation enhanced its P(3HB) content by more than 25 wt% [110]. However, P(3HB) productivity by this recombinant *Synechococcus* strain was extremely low.

Paracoccus denitrificans, a facultatively methylotrophic bacterium, is able to grow and accumulate PHAs on methanol. Recombinant *P. denitrificans* strains with increased expression levels of all PHA synthetic enzymes were investigated for the enhanced production of PHA. The PHA content and PHA accumulation rate of recombinant *P. denitrificans* with homologous overexpression of PHA synthase were 2 and 2.7 times higher, respectively, than those of the wild strain, suggesting that the step of PHA synthase was limited in PHA biosynthesis [111].

A recombinant *Klebsiella aerogenes* strain was also developed by transforming with a plasmid containing the *R. eutropha* PHA biosynthesis genes, and was subsequently used for the production of P(3HB) from sugarcane molasses [112]. When grown on molasses, the *K. aerogenes* strain exhibited excellent growth characteristics and P(3HB) production. By fed-batch culture, P(3HB)

Table 1. Recombinant bacteria harboring the heterologous PHA biosynthesis genes and synthesized PHA

Host strain	Introduced PHA biosynthesis genes	PHA	Carbon source	Reference
PHA negative *R. eutropha*	*R. eutropha phaB* and *phaC*	P(3HB-*co*-3HHx)	Octanoate	86
P. oleovorans	*R. eutropha phaCAB*	Blend of P(3HB) and P(3HB-*co*-3-HO)	Sodium octanoate	102, 103
PHA negative *P. putida*	*T. pfenigii phaC*	P(3HB-*co*-3HHx-*co*-4HHx)	Octanoate	104
PHA negative *R. eutropha*	*T. pfenigii phaC*	P(3HB-*co*-3HHx-*co*-4HHx)	4-hydroxyhexanoic acid	104
PHA negative *R. eutropha*	*A. caviae phaC*	P(3HB-*co*-3HHx)	Octanoate	92
PHA negative *R. eutropha*	*A. caviae phaC*	P(3HB-*co*-3HC-*co*-3HHp)	Alkanoic acid of odd carbon number	94
PHA negative *R. eutropha*	*Pseudomonas* sp. 61–3	P(3HB-*co*-3HA)	Alkanoic acid of odd carbon number	95
E. coli	*R. eutropha phaABC* or *A. latus phaABC*	P(3HB-*co*-3HV)	Glucose and propionic acid	58, 61
E. coli	*A. caviae phaC* and *phaJ*	P(3HB-*co*-3HHx)	Sodium dodecanoate	71
E. coli	*C. Kluyveri* 4-hydroxybutyryl-CoA dehydrogenase and *R. eutropha phaC*	P(4HB)	4-hydroxybutyric acid and glucose	68
E. coli	*C. Kluyveri* succinate degradation pathway and *R. eutropha phaABC*	P(3HB-*co*-4HB)	glucose	70
E. coli	*P. oleovorans phaC1* (or *phaC2*) and *E. coli* 'tesA	MCL-PHA	Gluconate	67
E. coli	*P. aeruginosa phaC1* or *phaC2*	MCL-PHA	Alkanoic acid of carbon number more than C8	64, 66

Table 2. Summary of PHA production by various recombinant *E. coli* strains

Strain	PHA	Fermentation strategy	Substrate	Time (h)	Cell concentration (g/l)	PHA concentration (g/l)	PHA content (%)	Productivity (g/l/h)	Reference
Recombinant *E. coli*	P(3HB)	Fed-batch	Glucose	49	204.3	157.1	77	3.2	45
Recombinant *E. coli*	P(3HB)	Fed-batch	Glucose	30.6	194.1	141.6	73	4.63	48
Recombinant *E. coli*[a]	P(3HB)	Fed-batch	Glucose	36	153.7	101.3	65.9	2.8	45
Recombinant *E. coli*	P(3HB)	Fed-batch	Hydrolyzed molasses	31.5	39.5	31.6	80	1	49
Recombinant *E. coli*	P(3HB)	Fed-batch	Whey	49	87	69	80	1.4	51
Recombinant *E. coli*	P(3HB-*co*-3HV)	Fed-batch	Glucose and propionic acid	55.1	203.1	158.8	78.2	2.88	62
Recombinant *E. coli*	P(4HB)	Fed-batch	4HB	60	12.6	4.4	36.	0.07	69

[a] Fermentation was carried out in 50-l stirred tank fermentor.

concentration of 24 g/l with the P(3HB) content of 70% could be obtained in 32 h. A *fadR* mutant of *Klebsiella oxytoca* harboring the *R. eutropha* PHA biosynthesis genes was used for the production of P(3HB-*co*-3HV) copolymer by co-feeding propionic acid [112].

Other than plants, there have been some examples of PHA production in eucaryotes [113, 114]. However, these studies were not carried out for the enhanced or economical production of PHAs, but rather for understanding PHA synthesis and redesigning metabolic pathways in eucaryotes.

In Table 1, the recombinant bacteria harboring the heterologous PHA biosynthesis genes and synthesized PHAs with novel hydroxyalkanoate monomer have been summarized.

8
Conclusions

Recent advances in our understanding of the biochemistry of PHA biosynthesis and the cloning of the PHA biosynthesis genes from various bacteria have allowed the construction of a variety of recombinant bacteria. In the early stage, to investigate the characteristics of PHA biosynthesis genes and PHA biosynthetic pathway, the PHA biosynthesis genes were only introduced into bacteria including the PHA producers, PHA negative mutant strain, and non-PHA producers. With the advances in our knowledge on the PHA biosynthetic metabolism, several recombinant bacteria have been developed for enhanced production of PHA and production of novel PHA. *E. coli* has been the best studied host strain for the production of PHA. Large amounts of PHA could be produced by recombinant *E. coli* with high productivity. In Table 2, the PHA production by various recombinant *E. coli* strains is summarized. Recently, cloning of PHA biosynthesis genes from bacteria accumulating PHA with novel composition of hydroxyalkanoates identified several new genes involved in the biosynthesis of PHA. With these genes and knowledge of the metabolic pathway, recombinant bacteria are being developed to produce novel PHA having better polymer properties in large amounts. All these efforts and successful results suggest that recombinant bacteria will be the ultimate choice for the production of PHAs.

Acknowledgements. Our work presented in this article was supported by the Ministry of Science and Technology and by LG Chemicals, Ltd.

References

1. Anderson AJ, Dawes EA (1990) Microbiol Rev 54:450
2. Brandl H, Gross RA, Lenz RW, Fuller RC (1990) Adv Biochem Eng Biotechnol 41:77
3. Byrom D (1991) Biomaterials: novel materials from biological sources. Stockton, New York
4. Doi Y (1990) Microbial polyesters. VCH, New York
5. Lee SY (1996) Biotechnol Bioeng 49:1
6. Steinbuchel A (1991) Polyhydroxyalkanoic acids. In: Byrom D (ed) Biomaterials: novel materials from biological sources. Stockton, New York, p 124
7. Lemoigne M (1926) Bull Soc Chem Biol 8:770

8. Steinbuchel A, Valentin HE (1995) FEMS Microbiol Lett 128:219
9. Holmes PA (1988) Biologically produced PHA polymers and copolymers. In: Bassett (ed) Developments in crystalline polymers. Elsevier, London, p 1
10. Marchessault RH (1996) Trends Polym Sci 4:163
11. Hocking PJ, Marchessault RH (1994) In: Roehr M (ed) Chemistry and technology of biodegradable polymers. Blackie Academic, p 48
12. Steinbuchel A (1996) In: Roehr M (ed) Biotechnology. VCH, p 403
13. Muller HM, Seebach D (1993) Angew Chem 105:483
14. Lee SY, Chang HN (1995) Adv Biochem Eng Biotechnol 52:27
15. Lee SY (1996) Trends Biotechnol 14:431
16. Lee SY, Choi J, Wong HH (1999) Int J Biol Macromol 25:31–36
17. Choi J, Lee SY (1997) Bioprocess Eng 17:335
18. Lee SY, Choi J (1998) Polymer Degrad Stabil 59:387
19. Lee SY, Choi J, Chang HN (1997) In: Eggink G, Steinbuchel A, Poirier Y, Witholt B (eds) Proceedings of the 1996 International Symposium on Bacterial Polyhydroxyalkanoates. NRC Research Press, p 127
20. Byrom D (1994) In: Mobley DP (ed) Plastics from microbes: microbial synthesis of polymers and polymer precursors. Hanser Munich, p 5
21. Steinbuchel A, Aerts K, Babel W, Follner C, Liebergesell M, Madkour M, Mayer F, Pieper-Furst U, Pries A, Valentin HE, Wieczorek (1995) Can J Microbiol 41:94
22. Madison LL, Huisman GW (1999) Microbiol Mol Biol Rev 63:21
23. Braunegg G, Lefebvre G, Genser KF (1998) J Biotechnol 65:127
24. Steinbuchel A, Fuchtenbusch B (1998) Trends Biotechnol 16:419
25. Lee SY, Choi J (1999) In: Lee SY, Papoutsakis ET (eds) Metabolic engineering. Marcel Dekker, p 113
26. Peoples OP, Sinskey AJ (1989) J Biol Chem 264:15,298
27. Schubert P, Steinbuchel A, Schlegel HG (1988) J Bacteriol 170:5837
28. Slater SC, Voige WH, Dennis DE (1988) J Bacteriol 170:4431
29. Choi J, Lee SY (1999) Appl Microbiol Biotechnol 51:13
30. Shatzman AR (1990) Curr Opin Biotechnol 1:5
31. Yee L, Blanch HW (1992) Bio/Technol 10:1550
32. Fidler S, Dennis D (1992) FEMS Microbiol Rev 103:231
34. Lee SY (1997) Nature Biotechnol 15:17
34. Kim BS, Lee SY, Chang HN (1992) Biotechnol Lett 14:811
35. Lee SY, Yim KS, Chang HN, Chang YK (1994) J Biotechnol 32:203
36. Lee SY, Chang HN (1994) J Environ Polymer Degrad 2:169
37. Lee SY, Lee YK, Chang HN (1995) J Ferment Bioeng 79:177
38. Lee SY, Chang HN, Chang YK (1994) Ann. NY Acad Sci 721:43
39. Lee SY, Chang HN (1996) Ann NY Acad Sci 782:133
40. Lee SY, Chang HN (1995) Can J Microbiol 41:207
41. Lee SY, Lee KM, Chang HN, Steinbuchel A (1994) Biotechnol Bioeng 44:1337
42. Lee IY, Kim MK, Park YH, Lee SY (1996) Biotechnol Bioeng 52:707
43. Lee SY (1994) Biotechnol Lett 16:1247
44. Wang F, Lee SY (1998) Biotechnol Bioeng 58:325
45. Wang F, Lee SY (1997) Appl Environ Microbiol 63:4765
46. Wong HH, van Wegen RI, Choi J, Lee SY, Middelberg APJ (1999) J Microbiol Biotechnol (in press)
47. Wang F, Lee SY (1997) Appl Environ Microbiol 63:3703
48. Choi J, Lee SY, Han K (1998) Appl Environ Microbiol 64:4897
49. Liu F, Li W, Ridgway D, Gu T (1998) Biotechnol. Lett 20:345
50. Lee SY, Middelberg APJ, Lee YK (1997) Biotechnol Lett 19:1033
51. Wong HH, Lee SY (1998) Appl Microbiol Biotechnol 50:30
52. Lee SY (1998) Bioprocess Eng 18:397
53. Sim SJ, Snell KD, Hogan SA, Stubbe J, Rha C, Sinskey AJ (1997) Nature Biotechnol 15:63
54. Kusaka S, Abe H, Lee SY, Doi Y (1997) Appl Microbiol Biotechnol 47:140

55. Resch S, Gruber K, Wanner G, Slater S, Dennis D, Lubitz W (1998) J Biotechnol 65:173
56. Choi J, Lee SY (1999) Biotechnol Bioeng 62:546
57. Lee SY, Choi J, Han K, Song JY (1999) Appl Environ Microbiol 65:2762
58. Slater S, Gallaher T, Dennis D (1992) Appl Environ Microbiol 58:1089
59. Rhie HG, Dennis D (1995) Can J Microbiol 41:200
60. Yim KS, Lee SY, Chang HN (1995) Kor J Chem Eng 12:264
61. Yim KS, Lee SY, Chang HN (1996) Biotechnol Bioeng 49:495
62. Choi J, Lee SY (1999) Appl Environ Microbiol 65:4363
63. Eschenlauer AC, Stoup SK, Srienc F, Somers DA (1996) Int J Biol Macromol 19:121
64. Langenbach S, Rehm BHA, Steinbuchel A (1997) FEMS Microbiol Lett 150:303
65. Qi Q, Steinbuchel A, Rehm BHA (1998) FEMS Microbiol Lett 1677:89
66. Qi Q, Rehm BHA, Steinbuchel A (1997) FEMS Microbiol Lett 157:155
67. Klinke S, Ren Q, Witholt B, Kessler B (1999) Appl Environ Microbiol 65:540
68. Hein S, Sohling B, Gottschalk G, Steinbuchel A (1997) FEMS Microbiol Lett 153:411
69. Song S, Hein S, Steinbuchel A (1999) Biotechnol Lett 21:193
70. Valentin HE, Dennis D (1997) J Biotechnol 58:33
71. Fukui T, Yokomizo S, Kobayashi G, Doi Y (1999) FEMS Microbiol Lett 170:69
72. Byrom D (1987) Trends Biotechnol 5:246
73. Kim BS, Lee SC, Lee SY, Chang HN, Chang YK, Woo SI (1994) Biotecnol Bioeng 43:892
74. Ryu HW, Hahn SK, Chang YK, Chang HN (1997) Biotechnol Bioeng 55:28
75. Park HC, Park JS, Lee YH, Huh TL (1995) Biotechnol Lett 17:729
76. Park JS, Park HC, Huh TL, Lee YH (1995) Biotechnol Lett 17:735
77. Park JS, Huh TL, Lee YH (1997) Enzym Microbiol Technol 21:85
78. Choi J, Lee SY (1997) Hwahak Konghak 35:684
79 Lee YH, Park JS, Huh TL (1997) Biotechnol Lett 19:77
80. Steinbuchel A, Valentin HE, Schonebaum A (1994) J Environ Polymer Degrad 2:67
81. Valentin HE, Steinbuchel A (1995) J Environ Polymer Degrad 3:169
82. Park JS, Lee YH (1996) J Ferment Bioeng 81:197
83. Lee IY, Kim GJ, Choi DK, Yeon BK, Park YH (1996) J Ferment Bioeng 81:255
84. Haywood GW, Anderson AJ, Dawes EA (1989) FEMS Microbiol Lett 57:1
85. Gerngross TU, Snell KD, People OP, Sinskey AJ, Csuhai E, Masamune S, Stubbe J (1994) Biochemistry 33:9311
86. Dennis D, McCoy M, Stangl A, Valentin HE, Wu Z (1998) J Biotechnol 64:177
87. Kobayashi G, Shiotani T, Shima Y, Doi Y (1994) In: Doi Y, Fukuda K (eds) Biodegradable plastics and polymers. Elsevier Science, Netherland, p 410
88. Haywood GW, Anderson AJ, Williams GA, Dawes EA, Ewing DF (1991) Int J Biol Macromol 13:83
89. Kato M, Bao HJ, Kang CK, Fukui T, Doi Y (1996) Appl Microbiol Biotechnol 45:363
90. Steinbuchel A, Wiese S (1992) Appl Microbiol Biotechnol 37:601
91. Liebergesell M, Hustede E, Timm A, Steinbuchel A, Fuller RC, Lenz RW, Schlegel HG (1991) Arch Microbiol 155:415
92. Fukui T, Doi, Y (1997) J Bacteriol 179:4821
93. Fukui T, Doi Y (1998) Appl Microbiol Biotechnol 49:333
94. Fukui T, Kichise T, Yoshida Y, Doi Y (1997) Biotechnol Lett 19:1093
95. Matsusaki H, Manji S, Taguchi K, Kato M, Fukui T, Doi Y (1998) J Bacteriol 180:6459
96. Witholt B, Kessler B (1999) Curr Opin Biotechnol 10:279
97. Huijberts GNM, Eggink G, de Waard P, Huisman GW, Witholt B (1992) Appl Environ Microbiol 58:536
98. Huisman GW, Wonink E, de Koning G, Preusting H, Witholt B (1992) Appl Microbiol Biotechnol 38:1
99. Kraak M, Smits THM, Kessler B, Witholt B (1997) J Bacteriol 179:4985
100. Huisman GW, Wonink E, de Koning G, Preusting H, Witholt B (1992) Appl Microbiol Biotechnol 38:1
101. Steinbuchel A, Schubert P (1989) Arch Microbiol 153:101
102. Timm A, Byrom D, Steinbuchel A (1990) Appl Microbiol Biotechnol 33:296

103. Preusting H, Kingma J, Huisman G, Steinbuchel A, Witholt B (1993) J Environ Poly Degrad 1:11
104. Liebergesell M, Mayer F, Steinbuchel A (1993) Appl Microbiol Biotechnol 40:292
105. Prieto MA, Buhler B, Jung K, Witholt B, Kessler B (1999) J Bacteriol 181:858
106. Yagi K, Miyawaki I, Kayashita A, Kondo M, Kitano Y, Murakami Y, Maeda I, Umeda F, Miura Y, Kawase M, Mizoguchi T (1996) Appl Environ Microbiol 62:1004
107. Boynton ZL, Koon JJ, Brennan EM, Clouart JD, Horowitz DM, Gerngross TU, Huisman GW (1999) Appl Environ Microbiol 65:1524
108. Follner CG, Muller S, Steinbuchel A, Babel W (1995) J Basic Microbiol 35:179
109. Suzuki T, Miyake M, Tokiwa Y, Saegusa H, Saito T, Asada Y (1996) Biotechnol Lett 18:1047
110. Takahashi H, Miyake M, Tokiwa Y, Asada Y (1998) Biotechnol Lett 20:183
111. Maehara A, Ikai K, Ueda S, Yamane T (1998) Biotechnol Bioeng 60:61
112. Zhang H, Obias V, Gonyer K, Dennis D (1994) Appl Environ Microbiol 60:1198
113. Leaf TA, Peterson MS, Stoup SK, Somer D, Srienc F (1996) Microbiology 142:1169
114. Williams MD, Rahn JA, Sherman DH (1996) Appl Environ Microbiol 62:2540

Received: December 1999

Production of Polyesters in Transgenic Plants

Yves Poirier

Institut d'Écologie-Biologie et Physiologie Végétales, Université de Lausanne, 1015 Lausanne, Switzerland
E-mail: yves.poirier@ie-bpv.unil.ch

Polyhydroxyalkanoates (PHAs) are bacterial polyesters having the properties of biodegradable thermoplastics and elastomers. Synthesis of PHAs has been demonstrated in transgenic plants. Both polyhydroxybutyrate and the co-polymer poly(hydroxybutyrate-*co*-hydroxyvalerate) have been synthesized in the plastids of *Arabidopsis thaliana* and *Brassica napus*. Furthermore, a range of medium-chain-length PHAs has also been produced in plant peroxisomes. Development of agricultural crops to produce PHA on a large scale and at low cost will be a challenging task requiring a coordinated and stable expression of several genes. Novel extraction methods designed to maximize the use of harvested plants for PHA, oil, carbohydrate, and feed production will be needed. In addition to their use as plastics, PHAs can also be used to modify fiber properties in plants such as cotton. Furthermore, PHA can be exploited as a novel tool to study the carbon flux through various metabolic pathways, such as the fatty acid β-oxidation cycle.

Keywords. Polyhydroxyalkanoates, Polyhydroxybutyrate, Polyester, Transgenic plants, Metabolic engineering

Advances in Biochemical Engineering/
Biotechnology, Vol. 71
Managing Editor: Th. Scheper

List of Abbreviations

ACP	acyl carrier protein
ALS	acetolactate synthase
BCOADC	branched-chain 2-oxoacid dehydrogenase complex
CaMV 35S	cauliflower mosaic virus 35S
GC-MS	gas chromatography-mass spectrometry
HV	3-hydroxyvalerate
Km	Michaelis-Menten constant
poly(3HA_{MCL})	medium-chain-length polyhydroxyalkanoates
NMR	nuclear magnetic resonance
PHA	polyhydroxyalkanoates
PHAs	polyhydroxyalkanoates
poly(3HB)	polyhydroxybutyrate
poly(3HB-*co*-3HV)	poly(hydroxybutyrate-hydroxyvalerate)
PDC	pyruvate dehydrogenase complex
TD	threonine deaminase

1 Introduction

Agricultural crops have the capacity to produce millions of tons of chemicals at prices comparable to many petroleum-derived commodities. The major plant commodity chemicals, such as starch and oils, can be used for both food and non-food purposes. With recent progress in plant molecular biology and transformation, agricultural crops have increasingly been regarded as potential vehicles for the production of novel valuable products which are not naturally found in plants [1]. One class of such products is polyesters of the family of polyhydroxyalkanoates (PHAs). These bacterial polyesters have attracted considerable interest in the past 20 years as a potential renewable source of biodegradable polymers [2]. The main rationale for the synthesis of PHA in plants is the potential for producing the polyester in a large scale and at a cost which is more competitive than bacterial fermentation [3].

This review will focus on the synthesis of polyhydroxybutyrate (poly(3HB)) and PHA co-polymers in plants. Most of the knowledge on PHAs, including the genes and proteins involved in their synthesis, have been obtained from studies on bacteria. It would, therefore, be useful for the reader to refer to other chapters of this book as well to recent reviews on bacterial PHAs [2–7] to integrate fully the topic of PHA synthesis in plants within the larger topic of bacterial polyesters.

Synthesis of PHA in plants is a relatively young field of study. Consequently, there remains a number of areas which have been poorly worked on and for which there are more questions than answers. Nevertheless, recent studies have shown that the usefulness of PHA in plants can reach beyond the biotechnology of polyester synthesis in agricultural crops to include PHA synthesis as a novel tool in the study of basic plant biochemical pathways and metabolism.

2
Synthesis of PHA in Plants

Since the first demonstration of the synthesis of poly(3HB) in plants in 1992 [8], the range of PHAs which have been synthesized in plants has increased to include a number of co-polymers. This section will thus discuss in detail the synthesis of poly(3HB) in *Arabidopsis thaliana* and in other plants, as well as present the most recent data on synthesis of various PHA co-polymers in plants.

2.1
Synthesis of Poly(3HB) in *Arabidopsis thaliana*

Synthesis of PHA in plants was initially demonstrated in the plant *Arabidopsis thaliana* [8, 9]. This plant was chosen because it could easily be transformed with *Agrobacterium tumefaciens* and its short life cycle (approximately 2 months from seed to seed) enabled the rapidly create hybrid plants expressing the various genes necessary for PHA biosynthesis. *A. thaliana* also has the additional advantage of being a model organism for which a number of mutants affected in various aspects of metabolism were available and could be used to study PHA synthesis [10, 11]. Finally, *A. thaliana* is a plant which accumulates a large quantity of oil in its seeds, and thus can be used as a good model for the synthesis of PHA in oil crops, such as rapeseed or soybean. The first bacterial genes involved in the synthesis of PHA were cloned from *Ralstonia eutropha* (formerly *Alcaligenes eutrophus*) and were involved in the synthesis of poly (3HB) [12–15]. Thus, although poly(3HB) has physical properties which are not ideal as a commodity plastic, being too stiff and brittle for most uses in consumer products [16], it was the *R. eutropha* genes for poly(3HB) biosynthesis that were first expressed in plants.

Poly(3HB) is synthesized in bacteria from acetyl-CoA by a three-step reaction (Fig. 1). The first enzyme of the pathway, 3-ketothiolase, catalyzes the condensation of two molecules of acetyl-CoA to form acetoacetyl-CoA. Acetoacetyl-CoA reductase subsequently reduces acetoacetyl-CoA to R-3-hydroxybutyryl-CoA, which is then polymerized by the PHA synthase to produce poly(3HB). Since acetyl-CoA is present in plant cells in the cytosol, plastid, mitochondrion, and peroxisome, the synthesis of poly(3HB) in plants could, in

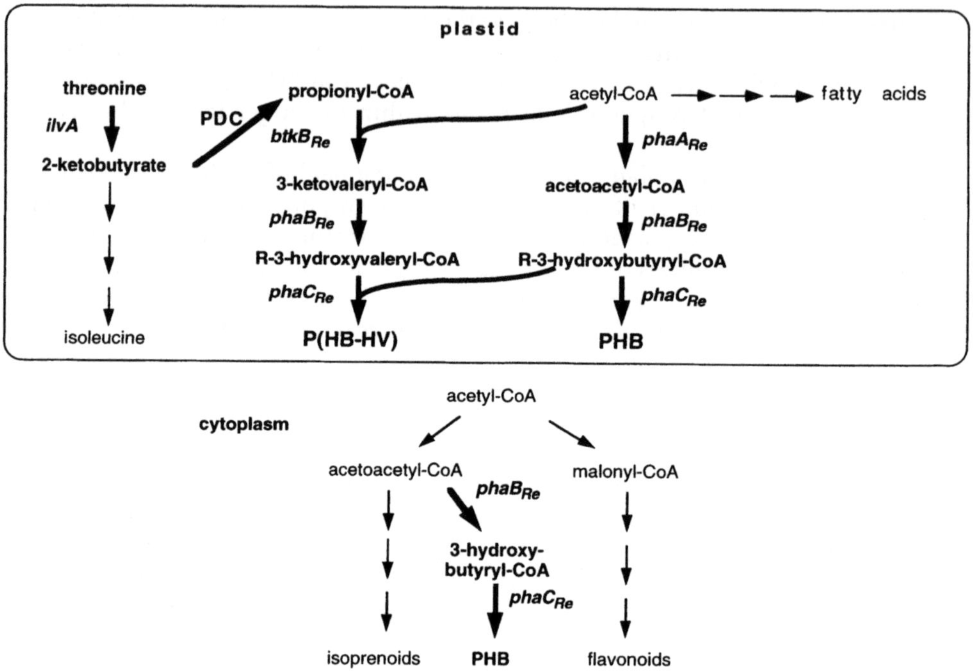

Fig. 1. Modification of plant metabolic pathways for the synthesis of poly(3HB) and poly(3HB-*co*-3HV). The pathways created or enhanced by the expression of transgenes are highlighted in *bold*, while endogenous plant pathways are in *plain letters*. The various transgenes expressed in plants are indicated in *italics*. The *ilvA* gene encodes a threonine deaminase from *E. coli*. The $phaA_{Re}$, $phaB_{Re}$, and $phaC_{Re}$ genes encode a 3-ketothiolase, an acetoacetyl-CoA reductase, and a PHA synthase from *R. eutropha*, respectively. The $btkB_{Re}$ gene encodes a second 3-ketothiolase isolated from *R. eutropha* which shows high affinity for both propionyl-CoA and acetyl-CoA [40]. PDC refers to the endogenous plant pyruvate dehydrogenase complex

theory, be achieved in any of these compartments. The cytoplasm was targeted as the first site for poly(3HB) synthesis because it had the advantage that the bacterial enzymes could be directly expressed in this compartment without any modification of the proteins. Furthermore, since the 3-ketothiolase is already normally available in the cytoplasm as part of the isoprenoid pathway, expression of the poly(3HB) biosynthetic pathway required only the expression of two additional enzymes, the reductase and synthase (Fig. 1). The *R. eutropha* genes encoding the acetoacetyl-CoA reductase (*phaB*) and PHA synthase (*phaC*) were expressed in plants under the control of the cauliflower mosaic virus (CaMV) 35S promoter, allowing a high constitutive expression of the enzymes [8]. Transgenic *A. thaliana* expressing acetoacetyl-CoA reductase or poly(3HB) synthase were cross-pollinated to obtain F1 hybrids having all the enzyme activities necessary for poly(3HB) production. The highest amount of poly(3HB) measured in these hybrids was approximately 0.1% of the dry weight of the plant [8]. Epifluorescence microscopy on tissues stained with Nile Blue A revealed the presence of poly(3HB) inclusions in all organs of the hybrid transgenic plants, including root, leaf, cotyledon, and seed. Transmission electron microscopy revealed that poly(3HB) accumulated in the form of agglomerations of electron-lucent inclusions surrounded by an electron-dense layer [8, 9]. The size (0.2 – 1 μm) and general appearance of these inclusions were similar to bacterial PHA inclusions. Surprisingly, even though the poly(3HB) biosynthetic enzymes were expressed in the cytoplasm, poly(3HB) agglomerations were located in several cellular compartments, i. e., cytosol, vacuole, and nucleus [8, 9]. No poly(3HB) inclusions were found in plastids and mitochondria . It is speculated that the hydrophobic nature of poly(3HB) inclusions allows them to penetrate through the single membrane of the vacuole but not the double membrane of organelles such as the plastid and mitochondrion. The localization of poly(3HB) granules in the nucleus may be explained by some affinity of the inclusions to nuclear constituents, leading to their entrapment in the nucleus during cell division.

The first experiments on the synthesis of poly(3HB) in plants established that the bacterial enzymes were active in plants. Detailed analysis of the polymer produced in these transgenic plants confirmed that the plant polymer was isotactic poly([*R*]-(–)-3-hydroxybutyrate) and that the thermal properties of plant poly(3HB) were similar to bacterial poly(3HB) [17]. The only difference noted was that plant poly(3HB) had a broader molecular weight distribution, ranging from 10^4 to 10^6, compared to bacterial poly(3HB) with a narrow molecular weight distribution near 10^6 [17]. These results demonstrated that although other proteins, such as phasins, may be found on the surface of PHA granules in bacteria [18, 19], expression of only the acetoacetyl-CoA reductase and PHA synthase were sufficient for the synthesis of high molecular weight PHA accumulating in the form of inclusions. It was notable that the size of the inclusions found in the nucleus was smaller than granules found in the cytoplasm or vacuole [8, 9]. It is likely that plant amphiphatic proteins can also be absorbed on the surface of inclusions and that these proteins, in a manner analogous to phasins, may affect the sizes of the granules through the promotion or prevention of inclusion fusion [20 – 22].

Plants synthesizing poly(3HB) in the cytoplasm accumulated the polymer to 0.1 % of the plant dry weight [8, 9], which is approximately 200 – 400 times lower than lipid accumulation in seeds of oil crops (20 – 40 % dry weight in soybean or rape, respectively) and 800 – 900 times lower than poly(3HB) accumulation in *R. eutropha*. In addition to low poly(3HB) synthesis, plants expressing both acetoacetyl-CoA reductase and poly(3HB) synthase were small in comparison to wild type plants [8, 9]. Although the reasons for the dwarf phenotype has not been unambiguously determined, it is thought that the diversion of cytoplasmic acetyl-CoA and acetoacetyl-CoA away from the endogenous isoprenoid and flavonoid pathways might lead to a depletion of essential metabolites which may affect growth. For examples, the plant isoprenoid pathway contributes to the synthesis of the three classes of plant hormones cytokinins, gibberelins, and brassinosteroids. It is thus possible that an imbalance in the synthesis of these hormones strongly affects plant growth. The fact that plants expressing high levels of acetoacetyl-CoA reductase have a white seed coat also indicates that the carbon flux through the flavonoid pathway, giving rise to the dark-colored antocyanins of the seed coat, is affected in these transgenics [9].

Since the limited supply of acetyl-CoA and the depletion of metabolites derived from it was thought to lead to low poly(3HB) accumulation and reduced growth of plants expressing the poly(3HB) pathway in the cytoplasm, expression of the poly(3HB) pathway in a compartment with a higher flux through acetyl-CoA was the next logical step. In plants, fatty acids are synthesized in the plastid using acetyl-CoA as precursor (Fig. 1). The plastid is therefore a site of high flux of carbon through acetyl-CoA. This is particularly true in tissues of plants accumulating triglycerides as the main carbon reserve. For example, *A. thaliana* synthesize up to 40% of the seed dry weight as triacylglycerides in 5–7 days. The large flux of acetyl-CoA in the plastids was thus hypothesized to allow a significantly higher production of poly(3HB) without deleterious effects on plant growth. Plastids are also the site of starch accumulation. Starch is synthesized as osmotically inert inclusions and can accumulate to high levels in photosynthetic chloroplasts and amyloplasts of storage tissues. The plastid can, therefore, accommodate inclusions without disruption of organelle function. In addition, the absence of poly(3HB) granules in plastids of transgenic plants expressing the poly(3HB) enzymes in the cytoplasm raised the possibility that the plastid envelope may be impervious to penetration by poly(3HB) inclusions. Expression of the poly(3HB) biosynthetic pathway in the plastid was hypothesized to lead to the accumulation of inclusions exclusively in the plastid, thus preventing potential disruption of other sub-cellular structures by the inclusions.

For the expression of the poly(3HB) biosynthetic pathway in plastids of *A. thaliana*, the bacterial *pha*A, *pha*B, and *pha*C proteins were modified by the addition of a plastid targeting sequence derived from the chloroplast small subunit of the ribulose bisphosphate carboxylase (Rubisco) from pea [23, 24]. The modified 3-ketothiolase, acetoacetyl-CoA reductase, and poly(3HB) synthase were independently expressed in *A. thaliana* under the control of the constitutive CaMV 35S promoter. Transgenic plants expressing a high level of plastid-targeted reductase were cross-pollinated with plants expressing a high level of plastid-targeted PHA synthase. The resulting double hybrids did not produce detectable poly(3HB) [24], since 3-ketothiolase activity is most likely not available in the plastid. Double hybrids were subsequently cross-pollinated with transgenic plants expressing the bacterial plastid-targeted 3-ketothiolase to obtain triple hybrids expressing all three PHA enzymes in the plastid. These triple hybrids produced poly(3HB) as detected by gas chromatography and mass spectrometry [24]. Transmission electron microscopy revealed that poly(3HB) inclusions accumulated exclusively in the plastids and were of size and appearance similar to bacterial PHA inclusions (Fig. 2) [24]. The poly(3HB) content gradually increased over the life span of the plant, with fully expanded pre-senescing leaves typically accumulating ten times more poly(3HB) than young expanding leaves of the same plant. The maximal amount of poly(3HB) detected in pre-senescing leaves was 10 mg/g fresh weight, representing approximately 14% dry weight. High level synthesis of poly(3HB) in plastids was not accompanied with a reduction in plant growth or seed yield. Only slight chlorosis of fully expanded leaves accumulating more than 3 mg/g fresh weight could be

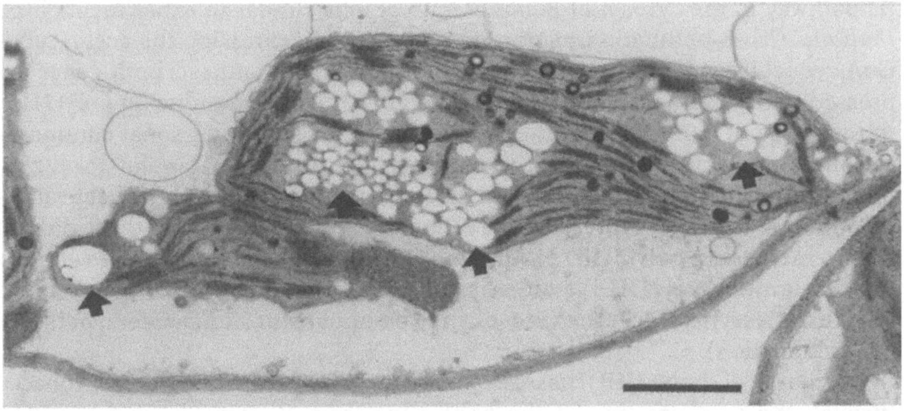

Fig. 2. Accumulation of PHA inclusions in *Arabidopsis* transgenic plants. Transgenic cell expressing the poly(3HB) pathway in the plastid showing accumulation of poly(3HB) inclusions (*arrows*) in the chloroplast of a leaf mesophyll cell. *Bar* represents 1 μm

detected [24]. Transmission electron micrographs indicated that in these leaves the majority of plastids were filled with poly(3HB) inclusions leaving little space for thylakoid membranes. In these plastids, it is possible that high poly(3HB) production interferes not only with photosynthesis but also with other pathways, such as starch accumulation or lipid biosynthesis. Thus, although plant growth was not affected under greenhouse or growth chamber conditions, leaf chlorosis indicates that the cell metabolism is affected to some extent in plants accumulating a large amount of poly(3HB) in the chloroplasts. It thus remains to be seen whether, under growth conditions such as encountered in the field, crop plants producing such high levels of poly(3HB) will show optimal growth behavior.

It has, thus, been demonstrated that redirecting the poly(3HB) biosynthetic pathway from the cytoplasm to the plastid resulted in an approximate 100-fold increase in poly(3HB) production [24]. However, it must be kept in mind that the rate of poly(3HB) biosynthesis in *A thaliana* leaves was relatively low, since poly(3HB) accumulated progressively over 40–60 days to reach 10–14% of the dry weight, whereas synthesis of starch can reach 17% dry weight for a 12 h photoperiod and seed storage lipids can reach 8% dry weight per day.

Production of poly(3HB) in the chloroplast of *A. thaliana* has also been independently demonstrated with similar experiments by the group of Monsanto. They have reported that by using the *phaA*, *phaB*, and *phaC* genes modified for plastid targeting and expressed under the CaMV35S promoter, poly(3HB) levels up to 12–13% dry weight were obtained in *A. thaliana* shoots.

2.2
Synthesis of Poly(3HB) in other Plants

Synthesis of poly(3HB) in the cytoplasm has also been successfully demonstrated in *Brassica napus* as well as tobacco. Expression of the poly(3HB) biosynthetic pathway in the cytosol of *B. napus* gave results similar to experiments in *A. thaliana*. Cross-pollination of transgenic rapeseed expressing the acetoacetyl-CoA reductase with plants expressing the poly(3HB) synthase, both genes expressed under the CaMV 35S promoter, led to F1 hybrids producing poly(3HB) in the range of 0.02–0.1% of the dry weight (P.A. Fentem, personal communication). Some transgenic rapeseed were stunted in growth, similar to *A. thaliana* producing poly(3HB) in the cytoplasm. Overexpression of the bacterial 3-ketothiolase in plants expressing the reductase and poly(3HB) synthase did not lead to increase in poly(3HB) production, indicating that 3-ketothiolase activity was not limiting poly(3HB) synthesis, but rather that other factors, such as the low flux of acetyl-CoA in the cytosol, may be important (P.A. Fentem, personal communication).

Synthesis of poly(3HB) has also recently been demonstrated in tobacco through the co-expression of the *phaB* gene from *R. eutropha* and the PHA synthase from *Aeromonas caviae* [25]. Although the bacterial genes were expressed under the strong constitutive CaMV35S promoter, expression of both proteins was relatively low. The maximal amount of poly(3HB) detected in leaves was 10 μg/g fresh weight. Analysis of the polymer revealed that the number-average molecular weights (M_n) and polydispersity (M_w/M_n) were 32,000 and 1.9, respectively [25].

The poly(3HB) biosynthetic pathway has also been successfully expressed in the leucoplasts of developing seeds of *B. napus*. In these experiments, the *phaA*, *phaB*, and *phaC* genes modified for plastid targeting were expressed under the control of the *Lesquerella* fatty acid hydroxylase seed specific promoter. Plants expressing the three enzymes were shown to accumulate poly(3HB) at levels up to 7.7% dry weight in mature seeds of heterozygous plants [26]. Electron-microscopy analysis indicated that leucoplasts from these plants were distorted, yet intact, and appeared to expand in response to polymer accumulation. These experiments were important in that they demonstrated for the first time that high level poly(3HB) production was possible in crop plants and was not restricted to *A. thaliana*. The potential effects of poly(3HB) synthesis in seeds on germination, seedling growth or triacylglyceride accumulation has not yet been reported.

2.3 Synthesis of Poly(3HB-co-3HV) Co-Polymer in Plants

The successful use of crop plants as a production method for biopolymer not only depends on the amount of PHA accumulated in plants but also on the type and quality of the PHA synthesized. Since poly(3HB) is a polymer with poor physical characteristics [16], it was important to engineer plants for the synthesis of PHA co-polymers with better physical characteristics. Poly(3-hydroxybutyrate-*co*-3-hydroxyvalerate) [poly(3HB-*co*-3HV)] is the best studied co-polymer. Poly(3HB-*co*-3HV) has lower crystallinity, and is more flexible and less brittle than poly(3HB) homopolymer [16]. Synthesis of poly(3HB-*co*-3HV) in bacteria was first achieved by fermentation of *R. eutropha* on glucose and propionic acid [2]. For a number of years, production of poly(3HB-*co*-3HV), known also under the trade name Biopol, has been central to the marketing and production strategies of PHA by Zeneca and strong intellectual property rights were obtained on this polymer and its synthesis. It was therefore natural that efforts of PHA co-polymer production in plants would also be focused on poly(3HB-*co*-3HV) synthesis.

Production of poly(3HB-*co*-3HV) co-polymer in plants has recently been demonstrated by the PHA group of Monsanto [27], which acquired the PHA business of Zeneca in 1996. In the commercial production of poly(3HB-*co*-3HV) from *R. eutropha*, propionate is added to the growth media in order to create an intracellular pool of propionyl-CoA which can be condensed to acetyl-CoA to form 3-ketovaleryl-CoA. The 3-ketovaleryl-CoA is then reduced by the acetoacetyl-CoA reductase to give 3-hydroxyvaleryl-CoA, which is co-polymerized with 3-hydroxybutyryl-CoA to synthesize poly(3HB-*co*-3HV) (Fig. 1). For the synthesis of poly(3HB-*co*-3HV) in plants, it was thus necessary to create an endogenous pool of propionyl-CoA which could be used by the PHA pathway.

Propionyl-CoA can potentially be generated by several pathways. Degradation of odd-chain fatty acids in the peroxisomes generates propionyl-CoA and acetyl-CoA [28]. The amount of odd-chain fatty acids in most plants is, however, very low. Propionyl-CoA can be generated in mammals and bacteria from succinyl-CoA via L-methylmalonyl-CoA and D-methylmalonyl-CoA [28]. The main problem with the creation of this biosynthetic route in plants is that the enzyme methylmalonyl-CoA mutase, which initiates this pathway, requires vitamin B12 as a cofactor, which is absent in plants. In *Clostridium propionicum*, propionyl-CoA is generated from pyruvate in four enzymatic steps which form part of the pathway of the conversion of lactate to propionate [29]. Synthesis of propionyl-CoA in *Rhodococcus rubrum* was proposed to be generated from methylmalonyl-CoA via either a methylmalonyl-CoA decarboxylase or a methylmalonyl-CoA:oxaloacetate transcarboxylase [30]. The pathway targeted by Zeneca and Monsanto was the conversion of 2-ketobutyrate to propionyl-CoA (Fig. 3). This enzymatic step is normally catalyzed by the branched-chain 2-oxoacid dehydrogenase complex (BCOADC).The BCOADC catalyses the irreversible oxidative decarboxylation of the branched-chain 2-ketoacids derived from valine, leucine, and isoleucine degradation pathways, as well as 2-ketobutyric derived from the methionine degradation pathway [28]. BCOADC is found in plants in the peroxisomes along with other enzymes of the degradation of amino acids [31]. Since high level poly(3HB) biosynthesis had been achieved through the expression of the *R. eutropha* enzymes in the plastid and that this pathway needs to be active for the synthesis of poly(3HB-*co*-3HV), the plastid was the favored site for poly(3HB-*co*-3HV) biosynthesis. In addition to the BCOADC, conversion of 2-ketobutyrate can also be done, albeit at low efficiency, by the pyruvate dehydrogenase complex (PDC) which is found in the plastid [32] (Fig. 3). Since 2-ketobutyrate is found in the plastid as an intermediate of the synthesis of isoleucine from threonine, both the substrate and the enzyme complex required for the generation of propionyl-CoA are present in the plastid.

In the synthesis of propionyl-CoA, the PDC competes with the enzyme of the isoleucine biosynthetic pathway for 2-ketobutyrate (Fig. 3). Since the PDC has a relatively low *Km* for 2-ketobutyrate (3 mmol/l for 2-ketobutyrate compared to 0.3 mmol/l for pyruvate), it was important to increase the concentration of 2-ketobutyrate in the plastid. This could potentially be done either by inhibiting the conversion of 2-ketobutyrate to 2-aceto-2-hydroxy-butyrate by the acetolactate synthase (ALS) or increasing the amount or activity of the threonine deaminase (TD) responsible for the synthesis of 2-ketobutyrate from threonine (Fig. 3). Sulfonylurea herbicides block the synthesis of branched-chain amino acids through the inhibition of the enzyme ALS. Studies in *Salmonella typhimurium* had shown that inhibition of ALS by herbicides leads to a large increase in the level of 2-ketobutyrate and propionyl-CoA, as well as of 2-aminobutyrate, which is the transaminated form of 2-ketobutyrate (Fig. 3) [33, 34]. Similarly, studies in *Lemna minor* [35] and corn [36, 37] have shown that inhibition of ALS lead to an increase in both 2-aminobutyrate and 2-ketobutyrate. Alternatively, an increase in threonine deaminase activity could also lead to an increase in 2-ketobutyrate concentration. Threonine deaminase is, however, feedback inhibited by isoleucine [38], the end product of the pathway, thus raising doubts as to whether simple overexpression of threonine deaminase gene would lead to an increase in 2-ketobutyrate availability. Mutant alleles of the TD of *S. typhimurium* [33] and *E. coli* [39] have been isolated which are insensitive to isoleucine. Overexpression of these mutant TD was shown to lead to an increased synthesis of isoleucine and in the accumulation of intermediates of the pathway [33, 39].

The functionality of the poly(3HB-*co*-3HV) biosynthetic pathway was first tested in bacteria before moving to plants. Although the *phaA* gene, encoding the 3-ketothiolase from *R. eutropha*, was initially thought to be the only enzyme responsible for the synthesis of poly(3HB-*co*-3HV) in this organism, studies indicated that this enzyme could not use propionyl-CoA for the synthesis of 3-ketovaleryl-CoA [40]. These results lead to the isolation of a second 3-ketothiolase from *R. eutropha*, designated *bktB*, which had a higher specific activity for propionyl-CoA than *phaA* [40]. Expression of the *bktB*, *phaA*, and *phaC* genes in *E. coli* lead to the synthesis of poly(3HB-*co*-3HV) containing 3.5 mol% HV when the media was supplemented with threonine, whereas only poly(3HB) was synthesized in the absence of threonine [41]. When the isoleucine insensitive TD gene from *E. coli*, named *ilvA 466*, was expressed along with the *bktB*, *phaA*, and *phaC* genes, synthesis of poly(3HB-*co*-3HV) containing 7.9 mol% HV was detected in the absence of any supplementary threonine in the media [41]. These results thus demonstrated that expression in *E. coli* of a mutant TD was essential for the creation of a pool of propionyl-CoA which could be used for the synthesis of poly(3HB-*co*-3HV).

The wild type *ilvA* gene was modified to target the protein to the plastid and expressed in *A. thaliana*. Transgenic plants showed a 20-fold increase in levels of 2-ketobutyrate as well as a large increase in 2-aminobutyrate, the transaminated product of 2-ketobutyrate [27, 41]. The levels of threonine remained stable whereas isoleucine concentration increased. Constitutive expression of the *ilvA* protein along with *bktB*, *phaA*, and *phaC* proteins in the plastids of *A. thaliana* led to the synthesis of poly(3HB-*co*-3HV) in the range of 0.2–0.8% dry weight, with a HV level between 4–17 mol % [27, 41]. Co-expression of the iso-

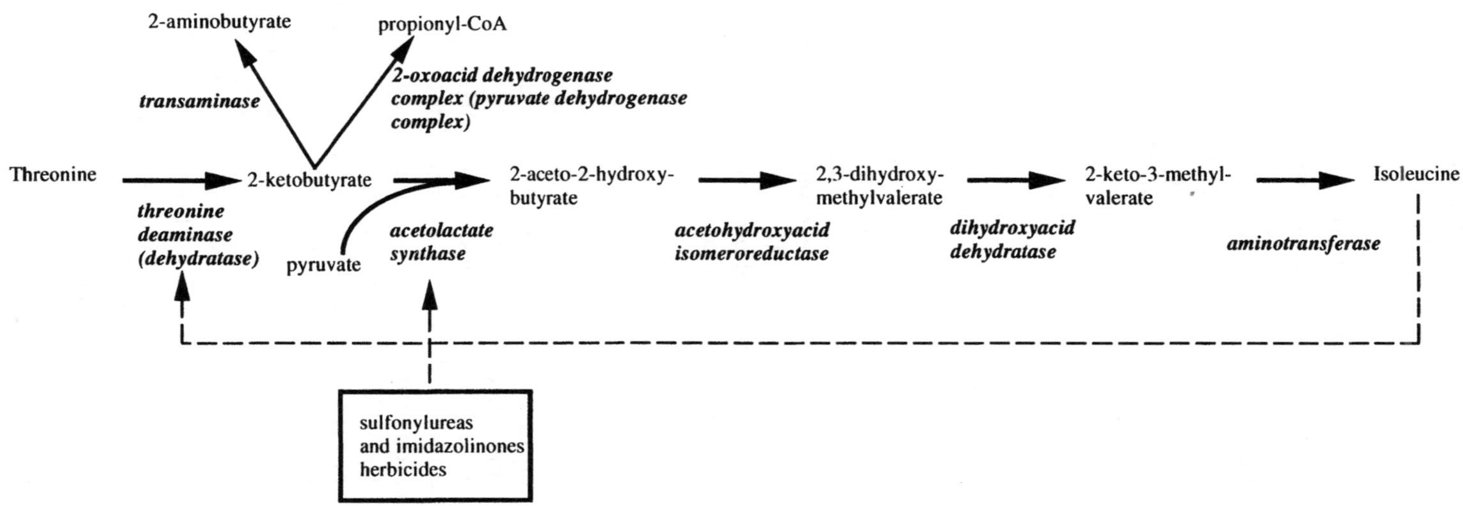

Fig. 3. Generation of propionyl-CoA from the isoleucine biosynthetic pathway. The intermediate 2-ketobutyrate can be decarboxylated by either the 2-oxoacid dehydrogenase complex or at low efficiency by the pyruvate dehydrogenase complex. Inhibition of the threonine deaminase by isoleucine and of the acetolactate synthase by herbicides are indicated with *dashed arrows*

leucine insensitive *ilvA466* TD allele along with the *bktB*, *phaA* and *phaC* genes in the developing seed of the oil crop *B. napus* led to the accumulation of poly(3HB-*co*-3HV) in the range of 0.7–2.3% dry weight, with a HV content of 2.3–6.4 mol % [27, 41]. In all these transgenic plants the presence of the HV units in the polymer was confirmed by GC-MS as well as NMR. The average molecular mass of the polymer produced ranged from 0.5×10^6 to 1.2×10^6 and polydispersity ranged from 1.8 to 2.4. Electron micrographs of transgenic *B. napus* seeds showed the presence of PHA inclusions in the leucoplasts. Furthermore, the size of the seed leucoplast accumulating PHA was significantly increased compared to negative controls, indicating that these organelles adjust their shape to accommodate PHA inclusions [27, 41].

Although the amount of poly(3HB-*co*-3HV) produced in transgenic plants is at present lower than poly(3HB), the demonstration of co-polymer synthesis in seeds of transgenic *B. napus* represent an important step in the development of crop plants for the production of PHA.

2.4 Synthesis of Poly(3HA$_{MCL}$) in Plants

There are more than 100 different hydroxyacid monomers which have been found to be included into PHAs [42]. One large group of PHAs, defined as medium-chain-length PHAs [poly(3HA$_{MCL}$)], represents polyesters containing 3-hydroxyacids ranging from 6 to 16 carbons. These PHAs are typically described as elastomers, although the actual physical properties of the polymers are dependent on the monomer composition and are very diverse [16]. Some monomers present in poly(3HA$_{MCL}$) may have functional groups, such as unsaturated bonds, epoxy, and hydroxy groups [42]. The presence of such reactive groups offers the opportunity to modify the structure and physical properties of PHAs after extraction. For example, the inclusion of unsaturated monomers into poly(3HA$_{MCL}$) enabled the formation of a cross-linked polymer after electron-beam irradiation, resulting in the formation of a true rubber with constant physical properties over a wide range of temperatures [43].

There are two main routes for the synthesis of poly(3HA$_{MCL}$) in bacteria [3, 4, 7]. The first is the synthesis of PHA using intermediates of fatty acid degradation. This pathway is found in bacteria, such as *Pseudomonas oleovorans*, which can synthesize poly(3HA$_{MCL}$) when fed on a source of alkanoic acid. In these bacteria, the type of PHAs produced is directly influenced by the carbon source added to the growth media, being composed of monomers which are 2n carbons shorter than the substrates used. For example, growth of *P. oleovorans* on octanoate (C8) generates a PHA co-polymer containing 89 mol% C8 and 11 mol% C6 monomers, whereas growth on dodecanoate (C12) generates a PHA containing 31 mol% C12, 36 mol% C10, 31 mol% C8, and 2 mol% C6 [44]. Alkanoic acids are thought to be converted to CoA esters by an acyl-CoA synthetase and then channeled to the β-oxidation pathway where a number of 3-hydroxyacyl-CoA intermediates can be synthesized. Since the PHA synthase is thought to accept only the R isomer of 3-hydroxyacyl-CoAs and β-oxidation of saturated fatty acids generates only the S isomer of 3-hydroxyacyl-CoAs [45,

46], the main question remains how are *R*-3-hydroxyacyl-CoAs intermediates generated. Possible pathways are reduction of 3-ketoacyl-CoA to *R*-3-hydroxyacyl-CoA by a novel ketoacyl-CoA reductase, conversion of *S*-3-hydroxyacyl-CoA to the R isomer by an epimerase, or the direct hydration of 2-*trans*-enoyl-CoA by an enoyl-CoA hydratase II [47].

While *P. oleovorans* can only synthesize poly(3HA_{MCL}) from related alkanoic acids present in the growth media, other pseudomonads, such as *Pseudomonas putida*, can synthesize poly(3HA_{MCL}) co-polymers when grown on unrelated substrates, such as glucose [48, 49]. The pathway leading to poly(3HA_{MCL}) synthesis from unrelated carbon sources has not been fully elucidated. Initial studies revealed the presence of mainly saturated 3-hydroxyacid monomers of 2n carbons in length, suggesting again the contribution of the β-oxidation pathway in generating substrates for the PHA synthase. However, detailed analysis of the composition of PHA produced by *P. putida* grown on glucose revealed the presence of the monomers 3-hydroxy-5-*cis*-dodecenoic acid and 3-hydroxy-7-*cis*-tetradecenoic acid [50]. Since these monomers are structurally identical to the acyl-moieties of the 3-hydroxyacyl-acyl carrier protein (ACP) intermediates of the *de novo* fatty acid biosynthesis, it was hypothesized that PHA synthesized from sugars can generate intermediates from the *de novo* fatty acid biosynthetic pathway, a conclusion also supported by studies using ^{13}C-labeled acetate [51, 52]. More recently, a gene encoding a protein having 3-hydroxyacyl-CoA-ACP transferase activity has been cloned from *P. putida* [53]. It is thought that this activity provides a link between fatty acid biosynthetic intermediates and poly(3HA_{MCL}) biosynthesis.

Synthesis of poly(3HA_{MCL}) in plants has recently been demonstrated in *A. thaliana* [54, 55]. The approach was to use intermediates of the β-oxidation of endogenous fatty acids for poly(3HA_{MCL}) production. Since in plants β-oxidation occurs principally in the peroxisomes, PHA biosynthetic proteins must be targeted to this organelle (Fig. 4). The PHAC1 synthase from *Pseudomonas aeruginosa* was thus modified for peroxisome targeting by the addition of the last 34 amino acids from the peroxisomal protein isocitrate lyase of *B. napus*. The modified gene was expressed under the control of the constitutive CaMV35S promoter. Immunolocalization of the PHA synthase demonstrated the appropriate targeting of the PHA synthase in plant peroxisomes [54]. Furthermore, plants expressing the PHA synthase showed the presence of electron-lucent inclusions within the peroxisomes. Synthesis of poly(3HA_{MCL}) in these transgenic plants was confirmed by GC-MS and NMR. The polymer had a weight-average molecular weight of 23,700, a number-average molecular weight of 5,500, and a polydispersity of 4.3, indicating that in comparison to poly(3HA_{MCL}) produced in *P. putida*, the plant polymer had a significantly lower molecular weight and a broader distribution [54]. The monomer composition was fairly complex, including saturated and unsaturated monomers ranging from 6 to 16 carbons. The production of peroxisomal poly(3HA_{MCL}) was low, reaching 0.4% dry weight in 7-day-old germinating seedlings and then decreasing as the plant matured to reach approximately 0.02% dry weight. Interestingly, there was a two- to threefold increase in PHA during leaf senescence. These data support the link between β-oxidation and PHA synthesis,

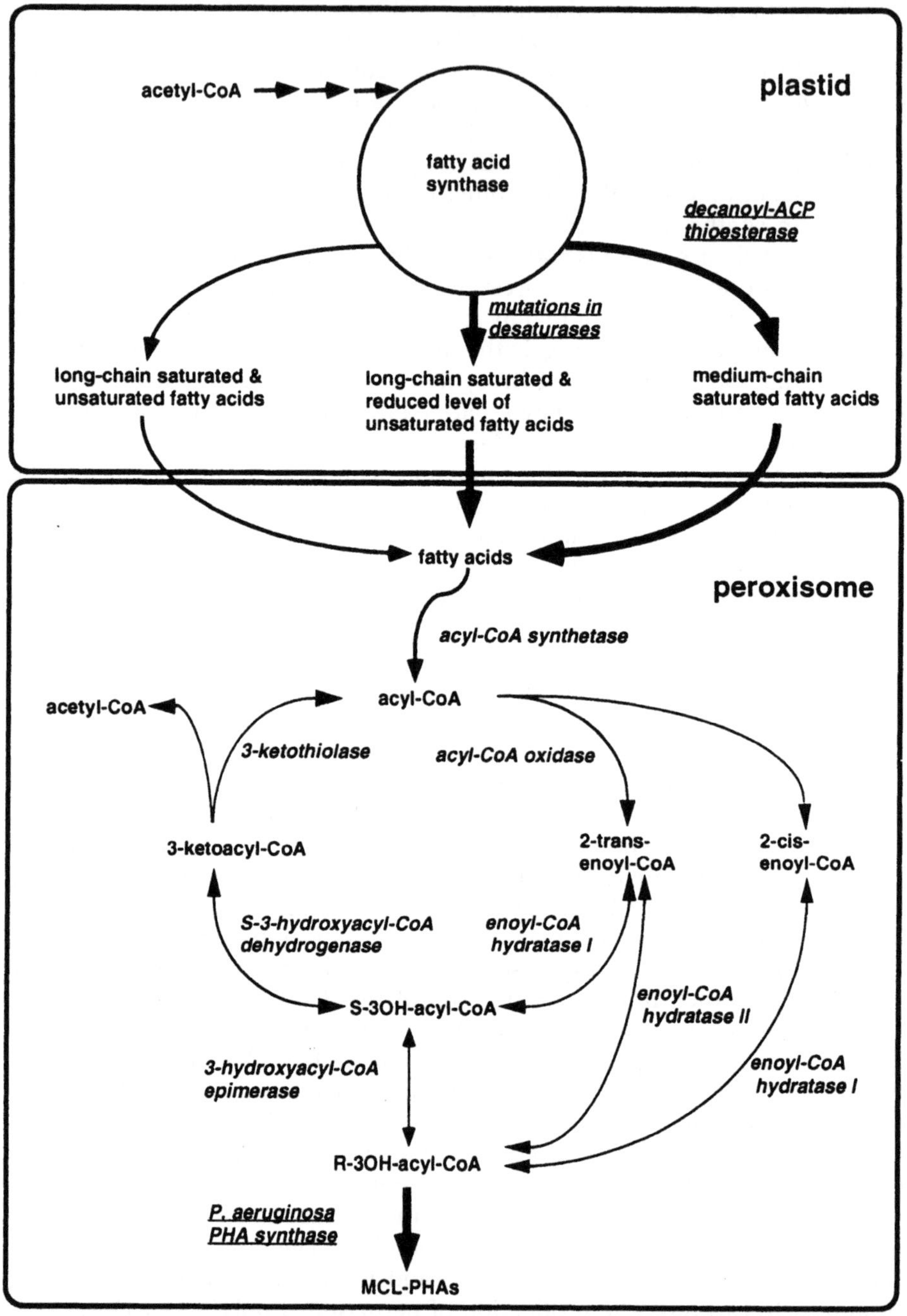

Fig. 4. Modification of plant metabolic pathways for the synthesis of poly($3HA_{MCL}$) in peroxisomes. The pathways created or enhanced by the expression of transgenes (*P. aeruginosa* PHA synthase and *C. lanceolata* decanoyl-ACP thioesterase) and of mutant alleles of plant fatty acid desaturase genes are highlighted by *bold arrows* and the enzymes involved *underlined*

since this degradative pathway is most active during germination and senescence, being involved in the conversion of fatty acids to carbohydrates. The wide range of monomers found in plant poly(3HA_{MCL}) suggests that, as with bacteria, plants also have enzymes capable of converting the β-oxidation intermediates *S*-3-hydroxyacyl-CoA to the R isomer (Fig. 4). Such enzymes could be either a 3-hydroxyacyl-CoA epimerase or an enoyl-CoA hydratase II activity which is specific for the generation of *R*-3-hydroxyacyl-CoA form 2-*trans*-enoyl-CoA [56, 57]. A third route for the synthesis of *R*-3-hydroxyacyl-CoA is the hydration of 2-*cis*-enoyl-CoA by the enoyl-CoA hydratase I activity [45]. The substrate 2-*cis*-enoyl-CoA is derived from the β-oxidation of unsaturated fatty acids having a double bond at an even position, such as found in linoleic and linonelic acid, two fatty acids which are abundant in plants.

Modulation of the quantity and quality of the poly(3HA_{MCL}) synthesized in plant peroxisomes has been demonstrated in various experiments aimed primarily at influencing the quantity and nature of the fatty acids targeted to the β-oxidation pathway [58]. Growth of PHAC1-expressing transgenic plants in liquid media supplemented with detergents containing various fatty acids resulted in an increased accumulation of poly(3HA_{MCL}) containing monomers derived from the β-oxidation of these fatty acids [58]. For example, addition to the growth media of the detergent polyoxyethylenesorbitan esterified to lauric acid (Tween-20) resulted in an 8- to 10-fold increase in the amount of PHA synthesized in 14-day-old plants compared to plants growing in the same media without detergent. Furthermore, the monomer composition of the poly(3HA_{MCL}) synthesized in the presence of Tween-20 showed a large increase in the proportion of saturated monomers lower than 14 carbons, and a corresponding decrease in the proportion of unsaturated monomers. This shift in monomer composition is accounted for by the fact that β-oxidation of lauric acid gives saturated 3-hydroxyacyl-CoA intermediates of 12 carbons and lower. Further experiments have shown that addition in the plant growth media of either tridecanoic acid, tridecenoic acid (C13:1, Δ12), or 8-methyl-nonanoic acid resulted in the production of poly(3HA_{MCL}) containing mainly saturated odd-chain, unsaturated odd-chain, or branched-chain 3-hydroxyacid monomers, respectively [58]. These studies demonstrated that the plant β-oxidation pathway was capable of generating from fatty acids a large spectrum of monomers which can be included in poly(3HA_{MCL}). Furthermore, the increase in the total amount of PHA produced in response to the addition of an exogenous source of fatty acids indicates that the amount of PHA synthase present in these transgenic plants was not limiting PHA production, but that more PHA could be synthesized if an additional flux of fatty acids is provided to the β-oxidation pathway.

Modulation of the quantity and/or quality of poly(3HA_{MCL}) synthesized in peroxisomes was also achieved by modifying the endogenous fatty acid biosynthetic pathway [58]. For example, expression of the peroxisomal PHA synthase in an *A. thaliana* mutant deficient in the synthesis of triunsaturated fatty acids [59] resulted in the synthesis of a PHA having an almost complete absence of triunsaturated 3-hydroxyacid monomers [58]. In a different strategy, expression of a fatty acyl-ACP thioesterase in the plastid was combined with the expression of a peroxisomal PHA synthase [58]. Fatty acyl-ACP thioesterases are responsible for stopping the elongation of fatty acids by catalyzing the release of the fatty acyl-ACP intermediate from the fatty acid synthase complex. Numerous thioesterases having different fatty acid chain-length specificities have been cloned in the last decade [60–62]. Typically, expression in the seeds of transgenic *B. napus* of a thioesterase specific for medium-chain fatty acyl-ACPs results in the accumulation of medium-chain fatty acids in the seed reserve lipids [60, 61]. However, expression of the same enzyme in leaves or roots does not result in the presence of measurable medium-chain fatty acids, suggesting that medium-chain fatty acids synthesized in vegetative tissues are channeled towards peroxisomal β-oxidation [63, 64]. These results indicated that expression of a thioesterase may be a way of increasing the carbon flux towards β-oxidation and peroxisomal PHA biosynthesis. This hypothesis was tested by combining the expression of a caproyl-ACP thioesterase from *Cuphea lanceolata* [65] in the plastid of *A. thaliana* with the expression of a peroxisomal PHA synthase. Expression of both enzymes lead to a 7- to 8-fold increase in the amount of poly($3HA_{MCL}$) synthesized in 40-day-old plants as compared to plants expressing only the PHA synthase [58]. Furthermore, the composition of the poly($3HA_{MCL}$) in the thioesterase/PHA synthase double transgenic plant was shifted towards saturated monomers containing 10 carbons and less, in agreement with an increase in the flux of decanoic acid towards β-oxidation triggered by the expression of the caproyl-ACP thioesterase [58]. From these results, a working model thus emerges where enzymes and genes involved in the synthesis of unusual fatty acids in plants can be used to modulate the quantity and quality of substrates channeled towards poly($3HA_{MCL}$).

The absolute amount of poly($3HA_{MCL}$) accumulating in mature shoots of transgenic plants expressing both thioesterase and PHA synthase remains relatively low ($\approx$ 0.1% dry weight) compared to poly(3HB) synthesis in the plastids ($\approx$ 10% dry weight). The reasons for this difference could be multiple, including different activities of bacterial enzymes in various plant subcellular compartments and the relative ability of the created and endogenous metabolic pathways to compete for the same substrates (acetyl-CoA for poly(3HB) and 3-hydroxyacyl-CoAs for poly($3HA_{MCL}$). There is, thus, a need for further metabolic engineering in order to increase the amount of poly($3HA_{MCL}$) synthesized in plants.

3 Novel Uses for PHA in Plants

Although for many decades the primary interest in the production of PHAs has been as a source of biodegradable plastics and elastomers, PHA synthesis in plants has opened novel avenues for the use of these polymers in both plant biotechnology and basic research.

3.1 PHA to Modify Fiber Properties

A novel perspective on the use of the synthesis of PHA in plants was achieved by the expression of the poly(3HB) biosynthetic pathway in the cytoplasm of cotton fiber cells [66, 67]. In this system, PHA is produced not as a source of polymer which must be extracted from the plant but rather as an intracellular agent which modifies the physical properties of the fiber. The genes from *R. eutropha* encoding the 3-ketothiolase, acetoacetyl-CoA reductase, and poly(3HB) synthase were expressed in transgenic cotton under the control of a fiber specific promoter [66]. Poly(3HB) was synthesized in the cytoplasm of cotton fibers and accumulated to 0.3% dry weight of the fiber at maturity. The polymer was in the molecular mass range of 0.6×10^6 to 1.8×10^6. Interestingly, even if the amount of poly(3HB) accumulated was relatively low, significant changes in the thermal properties of the fiber were measured [66]. The transgenic fibers conducted less heat, cooled down more slowly, and took up more heat than conventional fibers. It is intriguing to speculate what could be the changes in fiber properties if higher amount of poly(3HB) could be synthesized through expression of the pathway in the plastid, or if different PHA co-polymers, including poly($3HA_{MCL}$), could be accumulated. It is also tempting to speculate whether the properties of other fibers, such as flax, or of wood and rubber, could also be positively modified through the co-accumulation of PHAs.

3.2
PHA as a Tool to Study Plant Metabolic Pathways

Synthesis of PHAs in plants can not only be used directly in biotechnology for the creation of novel crop plants, but can also be a utilized as a unique novel tool in the basic studies of plant biochemistry. PHA synthesized in plants acts as a terminal carbon sink, since plants do not have enzymes, such as PHA depolymerases [68], required for degradation of the polymer. The quantity and composition of PHA can thus be used to monitor the quantity and quality of the carbon flux to different pathways.

Poly(3HB) synthesis in various subcellular compartments could be used to study how plants adjust their metabolism and gene expression to accommodate the production of a new sink, and how carbon flux through one pathway can affect carbon flux through another. For example, one could study how modifying the flux of carbon to starch or lipid biosynthesis in the plastid affects the flux of carbon to acetyl-CoA and poly(3HB). Alternatively, one could study how plants adjust the activity of genes and proteins involved in isoprenoid and flavonoid biosynthesis to the creation of the poly(3HB) biosynthetic pathway in the cytoplasm, since these three pathways compete for the same building block, i.e., acetyl-CoA.

With poly($3HA_{MCL}$) synthesis in the peroxisomes, the nature and proportion of monomers which are derived from β-oxidation intermediates could be used to elucidate the pathways involved in the degradation of unsaturated and unusual fatty acids. Such studies could also have an important aspect in lipid biotechnology since transgenic plants are created to synthesize novel fatty acids, such as ricinoleic acid and vernolic acid [69, 70]. In some cases, it is expected that an oilseed crop accumulating a novel fatty acid may show poor germination if the specialized enzymes required to handle the presence of novel groups in the fatty acids, such as epoxy or hydroxy groups, are missing in the transgenic plants. Thus, it may important to know what are the enzymes and genes required for the β-oxidation of unusual fatty acids. Similarly, the quantity of poly ($3HA_{MCL}$) can be used to study flux of fatty acids towards peroxisomal β-oxidation under different conditions. Poly($3HA_{MCL}$) has already been used to demonstrate that expression of a medium-chain fatty acyl-ACP thioesterase in leaves leads to an increase flux of medium-chain fatty acids towards β-oxidation in this tissue.

4
Extraction of PHA

Extraction of PHA from plants is likely to be a major factor affecting the production cost of PHA from crops and, therefore, the economic viability of this approach. In contrast to production of PHA from bacterial fermentation, where the production system is designed to produce only PHA, an agricultural production of PHA is likely to be most viable only through the recovery of not only PHA but also all other useful components of the harvested crop, i.e., oil, proteins, and carbohydrates. This fact, combined with the lower level of PHA accumulation in plants in comparison to micro-organism, is likely to make PHA recovery from plants a challenging task.

4.1
Extraction of PHA from Bacteria

The purification of PHA from plants is, in part, based on methods developed for the extraction of bacterial PHAs. It is, therefore, valuable to review briefly the two main approaches used for bacterial PHA purification. The first approach is based on the solubility of poly(3HB) in chlorinated hydrocarbons, such as chloroform, and its insolubility in methanol. Bacteria producing a high amount of PHA (typically $> 50\%$ dry weight) are harvested by conventional procedures such as centrifugation, refluxed in hot methanol to remove lipids and other lipophilic components, and then extracted with warm chloroform to solubilize PHA. The PHA can then be recovered from the chloroform by either solvent evaporation or precipitation by the addition of methanol. Repeated cycles chloroform solubilization and methanol precipitation can be used to obtain highly purified PHA. Although this protocol works very well for small laboratory scale PHA purification, it was found to be unsuitable for commercial production of PHA from fermentation because of the large amounts of solvents needed to avoid problems associated with the high viscosity of PHA in solution [71]. Furthermore, the use of hazardous solvents, such as chloroform, are unattractive in the context of a polymer which is marketed as biodegradable and environmentally friendly. An alternative extraction protocol was thus developed to avoid the use of organic solvents. Typically, the bacteria containing PHA are first heated to lyse the cells and then treated with a series of detergents, enzymes, or oxidizing chemicals aimed at degrading and/or solubilizing the non-PHA component while leaving the PHA intact [71]. Although sodium hypochlorite can be used as oxidizing and solubilizing agents, it causes a reduction in the molecular weight of the PHA, affecting the physical properties of the polymer [72, 73]. The extraction procedure developed by Zeneca in the 1980s for the production of PHA from *R. eutropha* uses a cocktail of enzymes, including proteases, nucleases, and lysozyme to remove proteins, nucleic acids, and bacterial cell wall [71]. Following washes with an anionic surfactant and water to remove solubilized non-PHA components, the PHA granules are harvested and dried.

4.2
Small-Scale Extraction of PHA from Transgenic Plants

The scientific literature on the purification of PHA from plants is essentially limited to small laboratory scale experiments. These experiments have been done with transgenic *A. thaliana*, producing either poly(3HB) or poly(3HA$_{MCL}$) [8, 17, 24, 54]. For poly(3HB) extraction, suspension cultures of microcalli derived from transgenic *A. thaliana* producing either poly(3HB) in the cytoplasm at 0.5% dry weight [17] or in the plastids at approximately 5% dry weight [C. Nawrath, Y. Poirier, unpublished results] were chosen as the starting material. For poly(3HA$_{MCL}$) synthesized in the peroxisomes, plants were grown in liquid cultures to facilitate the harvesting of germinating seedlings [54]. These plant cultures accumulated poly(3HA$_{MCL}$) at approximately 0.1% dry weight. Purification of PHA from digestion of the non-PHA components was found to be very difficult for transgenic plant material either due to the difficulties in removing the thick cell wall found in plant microcalli or the very low amount of PHA found in the plant material [Y. Poirier, unpublished results]. Purification of PHA was thus based on the use of chloroform [17, 54]. In a typical experiment, plant material is harvested, lyophilized, and ground to a fine powder. The dry powder is first extracted in a Soxhlet with methanol for 24 h to remove the lipids and chlorophylls, followed by a 24-h Soxhlet extraction in chloroform to solubilize the PHA. The chloroform solution is cleared from debris either by water extraction or filtering through glass wool, the volume is reduced by evaporation, and PHA is precipitated by the addition of 10 volumes of methanol. PHA isolated from transgenic microcalli culture was analyzed by gas chromatography, mass spectrometry, proton nuclear magnetic resonance spectroscopy, infra-red spectrometry, spectropolarimetry, differential scanning calorimetry, X-ray diffraction, and size exclusion chromatography. The results from all these analyses indicated that the polymer produced from transgenic plant had a chemical structure identical to that of bacterial poly(3HB), except for the molecular weight distribution of the polymer, which was broader for plant poly(3HB) [17].

4.3
Extraction of PHA from Transgenic Plants on a Commercial Scale

There are no published reports in the scientific literature on the extraction and purification of PHA from transgenic plant material on the medium- to large-scale necessary for commercial production of PHA in agricultural crops. There are, however, a number of issued patents and patent applications which describe various methods for the isolation of PHA from crops plants. It must be kept in mind, however, that for all these patent applications, experimental extraction of PHA from plants was made by mixing purified bacterial PHA with plant material. No direct experiments aimed at validating these extraction methods with PHA produced *in situ* in transgenic plants have been reported. It therefore remains to be determined how robust and efficient these various extraction methods will prove to be with PHA produced in plants. As with bacterial PHA, the extraction procedures described for PHA in crop plants can be broadly classified into two methods, either relying on organic solvents to solubilize PHA or on the isolation of the PHA granules.

4.3.1
Solubilization of PHA with Organic Solvents

The methods developed using organic solvents are typically designed to maximize the use of crop plants. For example, in the case of the synthesis of PHA in seeds of oil crops, such as rapeseed or soybean, the extraction methods have the advantage that the oil can be recovered from the seeds as well as PHA, and that the residual meal can be used in animal feed (Fig. 5) [74–78]. In this process, the seeds containing oil and PHA are crushed and the oil is obtained with a

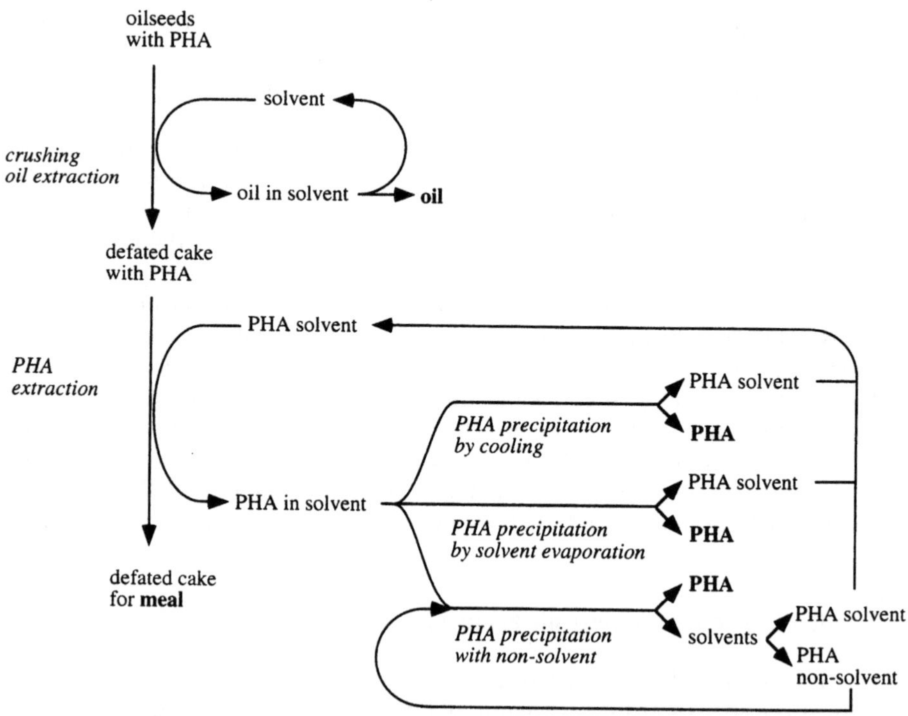

Fig. 5. Flow diagram of the extraction of PHA from oilseeds using PHA organic solvents. The final products obtained from the plant, i.e., PHA, oil and meal, are indicated in *bold*

combination of pressing and extraction with a solvent such as hexane, which does not solubilize PHA. The defatted cake containing PHA is then extracted with a solvent solubilizing PHA, leaving a cake which is rich in protein and cellulosic material. Although most of the examples cited used the seeds of oil-producing crops as the source material for PHA, the same extraction methods could also be adapted for the extraction of PHA from non-oily tissues, such as roots, tubers, or leaves.

An alternative to the extraction of intact PHA polymer is the isolation of PHA monomers, oligomers, or various derivatives such as esters [74]. PHAs are composed of stereo-chemically pure *R*-3-hydroxyacids, and therefore can be used as a source of optically pure organic substrates for the chemical and pharmaceutical industry [79]. In this protocol, the defatted cake containing PHA polymer would be chemically treated to obtain the PHA derivatives. For example, transesterification of the meal with methanol would give rise to methyl esters of 3-hydroxyalkanoic acids. The PHA derivatives would then be separated from the meal with appropriate solvents. One potential disadvantage of this method is the potential alteration of the quality of the residual meal if the harsh chemical treatments required for the production of PHA derivatives lead to protein or amino acid breakdown.

As previously discussed, the use of chlorinated hydrocarbons to solubilize PHA is not favorable because of its toxicity and negative environmental impact. It was thus important to find suitable alternative solvents for plant PHA extraction. This was facilitated by the finding that PHA co-polymers can be significantly more soluble in a range of non-chlorinated solvents than poly(3HB) homopolymer [75–78]. For example, whereas poly(3HB) is insoluble in butyl acetate, butyl propionate, 3-pentanol, or 1-hexanol, a co-polymer of poly(3HB-*co*-3HV) with a HV content ranging from 8–12 mol % is soluble in the same solvents [76]. Similarly, whereas poly(3HB) and poly(3HB-*co*-3HV) are not appreciably soluble in ethyl acetate or acetone, co-polymers containing longer-chain monomers, such as poly (3-hydroxybutyrate-3-hydroxyoctanoate), are readily dissolved in the same solvents [75]. The use of high temperature (typically above the boiling point) and high pressure considerably expand the range of solvents which can be used to extract PHA. For example, whereas poly(3HB-*co*-3HV) is insoluble in methanol at temperatures below boiling points, the co-polymer is soluble in the same solvent at 120 °C [78]. In this context, methanol is considered as a PHA-poor solvent, i. e., a solvent which does not appreciably dissolve PHA below its boiling point but which can dissolve the polymer at temperatures above its boiling point. Thus, by combining high temperature and pressure with the right choice of PHA co-polymer, a large spectrum of non-chlorinated solvents can be used to extract PHAs.

Following solubilization of the PHA from the defatted plant material, recovery of PHA from the solvent can be accomplished in various ways (Fig. 5) [74–78]. Addition of a PHA non-solvent to the solution would lead to PHA precipitation. If a solvent was used which dissolves PHA only under high temperature and pressure, cooling the solvent may be used to recover the polymer. Alternatively, evaporation of the solvent could also lead to polymer precipitation. Each of these methods have their disadvantages. Precipitation of PHA using a non-solvent involves the use and recovery of large quantities of a second organic solvent, in addition to the recovery of the solvent used to solubilize PHA. The principal problems associated with PHA precipitation by solvent evaporation or cooling is the formation of a highly viscous solution or gel which is very difficult to process to remove the residual solvent and harvest the solid PHA. Furthermore, PHA recovery by solvent evaporation leads to an accumulation of contaminants with the PHA which must be removed in a subsequent processing step.

A significant improvement to the recovery of PHA from solvents is the use of marginal non-solvents [77]. A PHA marginal non-solvent is a solvent which by itself cannot dissolve PHA appreciably, but when mixed with a PHA solvent, becomes capable of dissolving the polymer. Such marginal non-solvents can be alkanes or alcohols, but more interestingly can also be oil. Thus, in the case that PHA is synthesized in an oilseed crop, the oil present in the seed initially serves as a miscible co-solvent to promote the extraction of PHA. Alternatively, if the PHA is produced in leaves, vegetable oil can be added along with a PHA-solvent for polymer extraction [77]. An example of an extraction protocol using oil as a marginal non-solvent is shown in Fig. 6. The PHA-containing oilseeds are crushed and extracted with acetone to remove oil and PHA. Upon removal of the acetone, the oil becomes an effective suspending medium for the precipitating PHA due to its limiting PHA-solvating power. This leads to the absence of gel formation and increase processability of a solution with much lower viscosity than concentrated PHA solution in a solvent. The precipitate can be recovered from the oil and the evaporated acetone re-used for PHA and oil extraction (Fig. 6).

It is difficult, at present, to get a clear idea of the efficiency of the various plant PHA extraction protocols. In a patent application by Metabolix [74], it was reported that the efficiency of purifying poly(3HB) which was mixed with ground rapeseed was approximately 24% when using an oil-extraction step with hexane followed by refluxing in chloroform for 22 h and evaporating the chloroform to recover poly(3HB). The co-extraction of oil with a PHA contain-

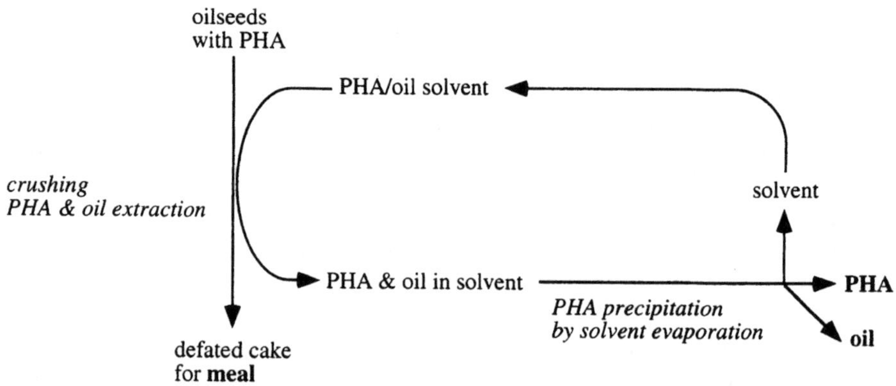

Fig. 6. Flow diagram of the extraction of PHA from oilseeds using a combination of PHA solvent and a marginal non-solvent (oil) for PHA

ing 94 mol % 3-hydroxyoctanoic acid and 6 mol % 3-hydroxyhexanoic acid using hot hexane followed by PHA recovery by solvent evaporation and purification by several washes with *n*-propanol and methanol gave an 84% yield. Recovery of PHA derivatives from the mixture of PHA and ground plant material was typically poor. For example, from a defatted meal containing bacterial poly(3HB) which was treated for 15 h with *n*-butanol and concentrated sulfuric acid, the recovery of butyl-3-hydroxybutyrate represented a 46% yield based on the initial amount of poly(3HB). Similar experiments with polyhydroxyoctanoate gave only a 25% yield of butyl-3-hydroxyoctanoate. From these data, it appears that significant optimization of the PHA purification methods based on organic solvents will be required before they can be applied to the production of PHA from transgenic plants.

4.3.2
Non-Solvent Extraction Procedures

Protocols have also been developed to avoid the use of organic solvents in the purification of PHAs from plant material. Similar to the extraction procedure developed by Zeneca for the purification of PHA from bacteria [71], the use of a cocktail of enzymes and oxidizing agents can be used to digest the non-PHA components of the vegetable matter [80]. Starting with an oilseed which accumulates PHA, the seeds would be crushed and extracted with hexane to recover the oil (Fig. 7). The defatted meal would then be grounded and wetted in order to make it more accessible to the enzymes and reagents involved in solubilizing non-PHA components. The carbohydrates, present mostly in the cell wall, would be stripped by a cocktail of carbohydratases containing arabanase, cellulase, β-glucanase, hemi-cellulase, pectinase, and xylanase. Solubilization of glycoprotein would be done enzymatically with lysozyme whereas the solubilization of nucleic acids and proteins would be done either enzymatically with nuclease and protease, or chemically using surfactants and oxidizing agents such as peroxide or hypochlorite. The PHA granules would then have to be recovered from the partially digested plant material using either decantation, filtration, and/or centrifugation. It is quite uncertain whether this extraction protocol would be really realistic in an industrial process of PHA purification from plants. One must consider the high cost of using enzymes and the fact that no meals valuable for animal feeding are recovered from the defatted cake. The patent application describing this procedure [80] experimentally demonstrates only the partial solubilization of the cell wall by a cocktail of enzymes and gives no data for the recovery of the PHA.

Recently, two novel procedures have been described for the isolation of PHA granules based on air classification [81] or centrifugal fractionation [82] (Fig. 8). These two processes are related to the well established wet and dry milling methods used in the corn industry for the fractionation of corn seeds into endosperm, germ, and hull and for the isolation of starch grains [83].

The air classification method separates dry solid particles according to weight and/or size by suspension in and settling from an air stream of appropriate velocity. Centrifugal fractionation separates particles suspended in a so-

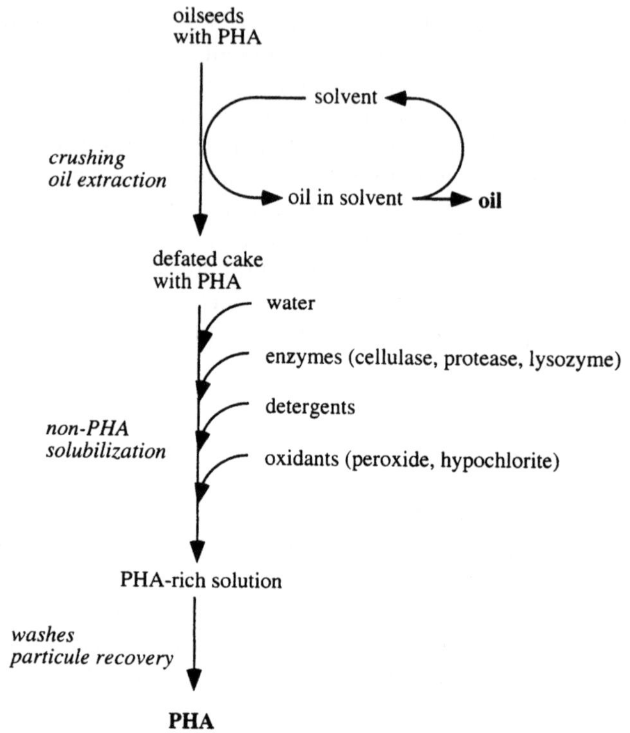

Fig. 7. Flow diagram of the extraction of PHA from oilseeds through the digestion of non-PHA constituents

lution based on size and/or density. PHA granules are typically 0.2–1 μm in diameter and thus expected to be one of the smallest particles present in plant cells, in comparison to starch grains or protein bodies. This size difference can thus be used to separate and purify PHA granules from the other plant components. In one example, grains of corn producing PHA are crushed and extracted with hexane to recover the oil (Fig. 8). The defatted meal can be washed with water and ethanol to remove soluble components such as sugars. The defatted/desugared meal is then pulverized to fine particles, creating a flour. In the air classification method, the particles present in the flour are separated into a PHA-poor coarse fraction and a PHA-rich fine fraction [81]. In the centrifugal fractionation method, the flour is mixed with water to create a fine suspension, and the particles separated into PHA-poor and PHA-rich fractions [82]. The PHA-rich fraction can then be processed further with enzymes, detergents, and oxidizing agents to solubilize the residual non-PHA components and recover the PHA granules.

As with most described methods for the purification of PHA from crops, no actual experimental data are available that would allow one to evaluate the efficiency of recovery of PHA based on air classification or centrifugal fractionation and the extra costs associated with the purification of PHA particles from the PHA-rich fractions.

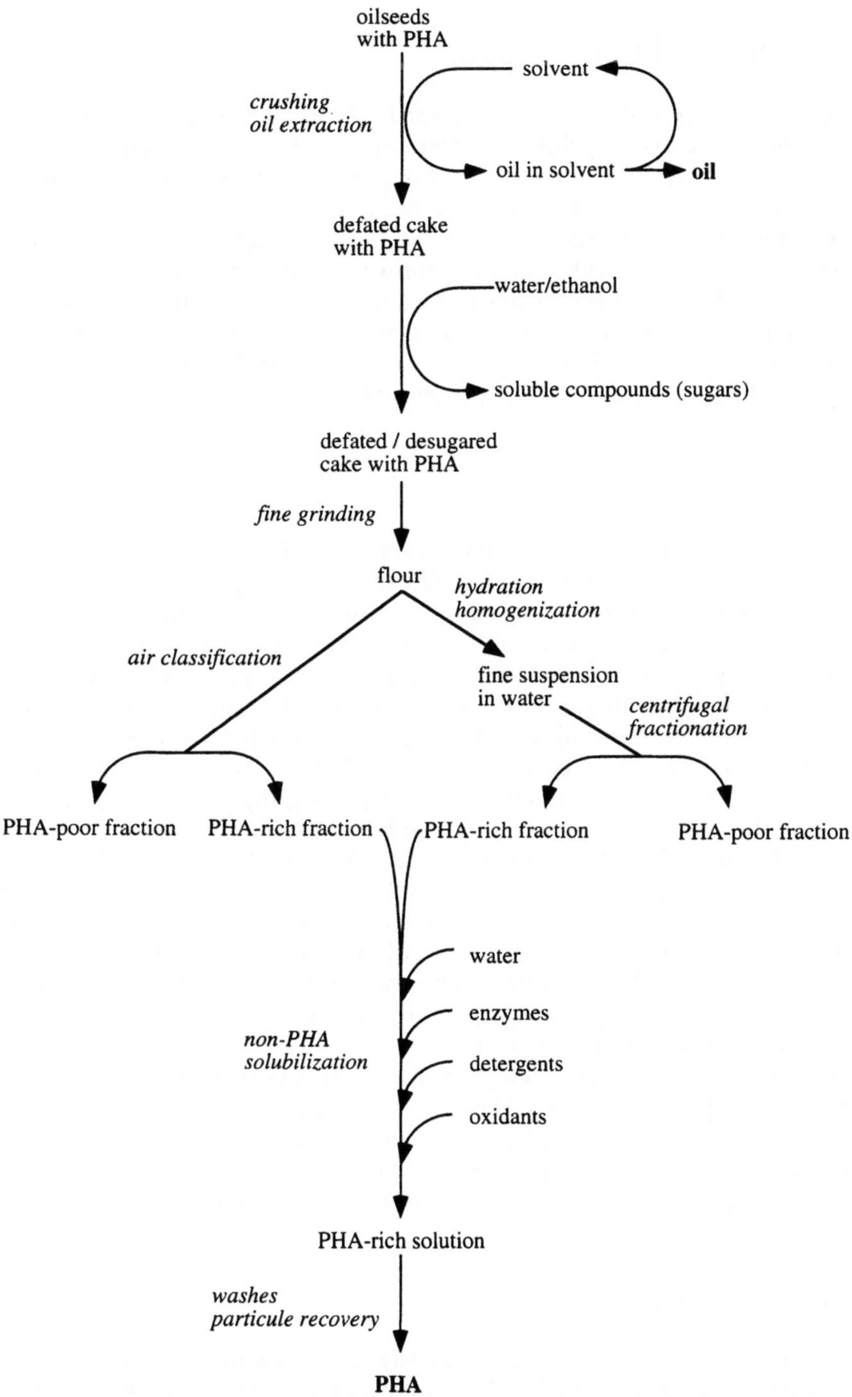

Fig. 8. Flow diagram of the extraction of PHA from oilseeds using air classification or centrifugal fractionation

5
PHA in Agricultural Crops

The plant of choice which can be used for PHA production will be influenced by a number of factors. Of prime importance is cost, i.e., in which crop will PHA production be cheapest. The answer to this question is likely to be different depending on the agricultural economics of each country. For example, if one considers oilseed crops, rapeseed may be the best crop for Northern European countries and Canada, sunflower for Southern European countries, and soybean for the USA. Other important factors which may influence the choice of target plant are the nature of the metabolic pathway that needs to modified for synthesis of a particular PHA, the procedure used for PHA purification, and the other uses of the crop besides PHA production.

5.1
Oilseed Crops

Oilseed crops have been the first plants targeted for PHA synthesis and have several advantages. First is that the metabolism of oilseed plants is well adapted to provide the right substrates in large quantities for PHA synthesis. Seeds accumulating oil have a large flux of carbon through acetyl-CoA, which is the precursor for the synthesis of fatty acids, poly(3HB), and poly(3HB-*co*-3HV). In addition to accumulating lipids, oilseeds also accumulate a substantial amount of storage protein. Thus, the pathway of branched-chain amino acid biosynthesis, required for the synthesis of poly(3HB-*co*-3HV), is also highly active (see Sect. 2.2). A strategy for producing poly($3HA_{MCL}$) using the intermediates of fatty biosynthesis (see Sect. 2.3) would also be well suited for oilseeds having a high flux through this pathway. As described in Sect. 4.3.1, the processing of oilseed crops using organic solvents could well be adapted for extraction of both the oil and PHA and for the production of animal feed. Although poly($3HA_{MCL}$) could be synthesized from germinating seedlings using the intermediates of the β-oxidation, it is less clear whether useful meal or fatty acids could also be recovered from germinating seedlings besides the PHA. One important factor that may negatively impact the use of oilseed crops for PHA is the effect of PHA synthesis on oil quantity and seed germination. It is possible that diversion of carbon towards poly(3HB) in seeds may result in a decrease in triacylglyceride accumulation. Although it is unlikely that a moderate decrease in triacylglycerides would be problematic for the viability or vigor of the germinating seedling, it could have an impact on the value of the crop as an oil producer.

5.2
Crops Storing Carbohydrates

Crops producing carbohydrates are typically more productive than oil crops. For example, whereas rapeseed and oil palm produce 1000 and 6000 kg of oil per hectare, respectively, sugar beet produces 9000 kg of sucrose per hectare and potato gives 19,000 kg of starch per hectare [84]. In this context, carbohydrate-producing crops may be good targets for PHA production. Furthermore, most potato cultivars are sterile and are propagated asexually, thus making the accidental spreading of the transgenes to other plant species very unlikely. The main challenge for these crops is that the metabolic pathways which are most active in these plants (sugar and starch biosynthesis) are not directly usable for PHA synthesis as in the case of oil crops. Since the flux of carbon through the fatty acid and amino acid biosynthetic pathways is typically low in the tissues that accumulate high amounts of carbohydrates (e.g., the potato tuber), it is likely that additional genetic modifications will be necessary to divert carbon from the sugar biosynthetic pathways into the lipid/PHA biosynthetic pathways. Thus, the metabolic genetic engineering required for PHA in these plants is likely to be more difficult and complex than for oilseed crops. Furthermore, the extraction of PHA may also be more challenging if both starch and PHA must be harvested from the same tissue. Although patent applications have described the theoretical aspects of adapting the dry and wet milling procedures normally used for starch isolation, for PHA purification, no experimentation has actually been reported (see Sect. 4.3.2). It is thus uncertain at this point if both PHA and starch could be extracted efficiently from the same tissue. One attractive alternative is to use crops that produce sucrose, such as sugar beet and sugar cane, for PHA synthesis. Following extraction of the water soluble sugars, PHA could be purified from the residual crushed material using an organic solvent extraction protocol (see Sect. 4.3.1).

5.3
Other Crops

A different strategy for the production of PHA in plants is to target its synthesis in tissue of the plant which is either not normally used in the production chain or which has low value. For example, in oilseed crops only the seeds have high value. Thus, by directing PHA synthesis to the leaves and stems of oilseed crops, the seed would be harvested and processed as usual, while the vegetative material would be harvested and extracted for PHAs, probably using organic solvents. Such a strategy may apply to the synthesis of a variety of PHAs, since the vegetative tissue has fairly active lipid and protein biosynthetic pathways. Even the synthesis of poly(3HA_{MCL}) could be envisaged in vegetative tissues since it was demonstrated that β-oxidation of fatty acids in leaves could be enhanced by the synthesis of unusual fatty acids such as medium-chain fatty acids (see Sect. 2.3). Synthesis of PHA in leaves could also be extended to fast growing grasses, such as switchgrass, which is considered a biomass energy crop.

5.4
Some Economic Considerations for the Production of PHA on an Agricultural Scale

One of the main barriers to the widespread use of biodegradable plastics is their higher production cost compared to petroleum plastics. For example, whereas the cost of most commodity plastics, such as polypropylene, is well below 1 US$/kg, the costs of some of the cheapest biodegradable plastics on the market, which are derived from starch, are 3–6 US$/kg [85, 86]. The economics of PHA synthesis by bacterial fermentation has been discussed in several papers [87–91]. The most optimistic production cost for poly(3HB) from bacteria is approximately 4 US$/kg, considering a production scale of one million tons per year. Production costs of bacterial PHA co-polymers are even higher, since either the yield of PHA co-polymer production in bacteria such as pseudomonads is lower, or more expensive feeding substrates must be used (e.g., propionic acid). The cost of bacterial PHA production can be divided into three main components: the starting material, fermentation, and polymer recovery. For bacterial fermentation, the cost of the starting material is mainly the cost of the carbon source. Considering that for the production of poly(3HB) from *R. eutropha* 3.3 g of glucose is used to synthesize 1 g poly(3HB) and that the price of glucose is ≈ 0.5 US$/kg, the cost of carbon is at least 1.65 US$/kg poly(3HB). Thus, even at this first stage in the production chain, the cost of poly(3HB) is significantly higher than petroleum-derived plastics. Although other cheaper carbon sources have been suggested, such as methanol or cane molasses [87], their uses do not necessarily translate into a lower final cost if, for example, the PHA production yield from bacteria using these carbon sources is lower or the use of more crude substrates leads to higher purification costs.

The main question is whether synthesis of PHA in plants can succeed in bringing the cost of the polymer down to the range of 0.5–1 US$/kg. Bacterial production of PHA typically relies on a carbon source, such as sucrose or glucose, which is produced from photosynthesis and extracted from plants. Synthesis of PHA directly in plants would, therefore, represent a saving in terms of the number of intermediary steps linking CO_2 fixation to PHA production. Furthermore, starch is one of the cheapest plant commodity product on the market, at about 0.25 US$/kg [86]. It is, thus, likely that the production cost of PHA in plants will be substantially cheaper than bacterial fermentation. The final cost of producing PHA in plants will depend on a number of factors.

One factor is the cost associated with the creation of a transgenic plant synthesizing PHA at high yield. Synthesis of PHA in crop plants is a challenging genetic engineering project. Most transgenic crops which are presently on the market, such as herbicide resistant plants or insect resistant plants expressing the *Bacillus thuringiensis* toxin, are created from the expression of one or two genes affecting a single metabolic pathway. In contrast, synthesis of poly(3HB-*co*-3HV) requires the expression of four novel genes affecting two independent metabolic pathways [27] (see Sect. 2.2). It is likely that optimization of the yield or monomer composition of PHAs in plants will require the expression of even more transgenes. Stable expression of transgenes over many generations in plants can be difficult when the number of transgenes increases, and particularly when transgenes share sequence homology between themselves, as is often the case when the same targeting sequence is used to direct the foreign proteins to the same organelle (e.g., the plastids) [92, 93]. This instability in gene expression has already been described in the case of transgenic *A. thaliana* expressing the three modified poly(3HB) biosynthetic genes for poly(3HB) biosynthesis in the plastid [24]. Rapid progress has, however, been made in understanding the mechanisms underlying transgene expression stability and novel genetic tools, including transformation vectors, are being developed to overcome these problems [94–97].

Probably the most important parameter that will influence the cost of producing PHA in transgenic plants will be the expenses associated with the extraction of the polymer. Cheap agricultural products, such as starch from corn or oil from soybean, are extracted using fairly simple physical and/or chemical separation procedures [83, 84]. Considering the investment required in the development of transgenic plants and that PHA must be a cheap polymer if it is to be successful on the market, it is unlikely that a transgenic plant will be developed to produce only PHA. Rather, the goal should be to add production of PHA onto a crop which is already used for the synthesis of other products, including oils and carbohydrates. It will, thus, be important that the extraction of PHA does not decrease the value of other useful components of the plant. As discussed in Sect. 4.2, very little is known about the efficiency and costs associated with the large scale extraction of PHA from plants. It is likely that the extraction costs will be higher for plants than for bacteria, if it is only for the fact that PHA synthesis in plants will probably be lower than the 80–85% dry weight achievable by bacterial fermentation.

All of these factors mean that production of PHA in plants will likely be more expensive than starch. However, considering that starch costs about 0.25 US$/kg, even tripling the production cost of PHA compared to starch would make PHA in plants at least five times cheaper than PHA obtained from bacterial fermentation and most likely the cheapest biodegradable plastic made from renewable resources.

The global market for biodegradable plastics has been evaluated at approximately 1.3 million tons per year. The demand and competitiveness of renewable biodegradable plastics over petroleum-derived polymers are likely to increase in the future as the expected decrease in the world production levels of petroleum [98] and potential CO_2 taxes imposed by governments make synthetic plastics more expensive. The benefits of producing cheap biodegradable polymers in crop plants would extend well beyond the polymer market to include the environment and agriculture. Production of novel industrially useful products in crop plants, such as PHA, would benefit the farmers and society by increasing the value of the agricultural products, thus contributing to alleviating problems of overproduction and large government subsidies associated with agriculture in the western world.

6 Conclusions

The synthesis of a number of different PHAs in plants has now been demonstrated. Synthesis of PHAs in plants is not only regarded as a promising way of producing biodegradable plastics on a large scale and at low cost, but can also be used to modify the physical properties of fibers and as a novel and useful tool in basic and applied research in plant biochemistry. The future of plant PHAs as a source of biodegradable plastics will rest on the development of transgenic crops capable of producing at high yields a number of key PHAs useful in large scale applications. This will likely require the expression of several transgenes affecting different plant metabolic pathways and the development of novel extraction processes aimed at purifying PHAs in a cost-effective manner. The production of PHAs in crop plants should beneficially impact the marketability of biodegradable plastics as well as the economic of agricultural production by increasing the value of crops.

Acknowledgment. The author thanks Christiane Nawrath for critical reading of the manuscript.

References

1. Goddijn OJM, Pen J (1995) Trends Biotechnol 13:379
2. Anderson AJ, Dawes EA (1990) Microbiol Rev 54:450
3. Poirier Y, Nawrath C, Somerville C (1995) Nature Biotechnol 13:142
4. Steinbüchel A (1991) Polyhydroxyalkanoic acids. In: Byrom D (ed) Novel biomaterials from biological sources. MacMillan, New York, p 123
5. Steinbüchel A, Schlegel HG (1991) Mol Microbiol 5:535
6. Doi K (1990) Microbial polyesters. VCH, New York
7. Steinbüchel A, Füchtenbusch B (1998) Trends Biotechnol 16:419
8. Poirier Y, Dennis DE, Klomparens K, Somerville C (1992) Science 256:520
9. Poirier Y, Dennis DE, Klomparens K, Nawrath C, Somerville C (1992) FEMS Microbiol Rev 103:237
10. Meyerowitz EM (1989) Cell 56:263
11. Meyerowitz EM, Somerville (eds) (1995) *Arabidopsis thaliana*. CSH, Cold Spring Harbor
12. Peoples OP, Sinskey AJ (1989) J Biol. Chem 264:15,293
13. Peoples OP, Sinskey AJ (1989) J Biol Chem 264:15,298
14. Slater SC, Voige WH, Dennis DE (1988) J Bacteriol 170:4431
15. Schubert P, Steinbüchel A, Schlegel HG (1988) J Bacteriol 170:5837
16. de Koning (1995) Can J Microbiol 41(suppl 1):303
17. Poirier Y, Schechtman LA, Satkowski MM, Noda I, Somerville C (1995) Int J Biol Macromol 17:7
18. Stuart ES, Lenz RW, Fuller RC (1995) Can J Microbiol 41(suppl 1):84
19. Steinbüchel A, Aerts K, Babel W, Föllner C, Liebergesell M, Madkour MH, Mayer F, Pieper-Fürst U, Pries A, Valentin HE, Wieczorek K (1995) Can J Microbiol 41(suppl 1):94
20. Pieper-Fürst U, Madkour MH, Mayer F, Steinbüchel A (1994) J Bacteriol 176:4328
21. Pieper-Fürst U, Madkour MH, Mayer F, Steinbüchel A (1995) J Bacteriol 17:2513
22. Wieczorek R, Pries A, Steinbüchel A, Mayer F (1995) J Bacteriol 177:2425
23. Nawrath C, Poirier Y, Somerville CR (1994) Plastid targeting of the enzymes required for the production of polyhydroxybutyrate in higher plants. In: Doi Y, Fukuda K (eds) Biodegradable plastics polymers. Elsevier, Amsterdam, p 136
24. Nawrath C, Poirier Y, Somerville CR (1994) Proc Nat Acad Sci USA 91:12,760
25. Nakashita H, Arai Y, Yoshioka K, Fukui T, Doi Y, Usami R, Horikoshi K, Yamaguchi I (1999) Biosci Biotechnol Biochem 63:870
26. Houmiel KL, Slater S, Broyles D, Casagrande L, Colburn S, Gonzalez K, Mitsky TA, Reiser SE, Shah D, Taylor NB, Tran M, Valentin HE, Gruys KJ (1999) Planta 209:547
27. Slater S, Mitsky TA, Houmiel KL, Hao M, Reiser SE, Taylor NB, Tran M, Valentin HE, Rodriguez DJ, Stone DA, Padgette SR, Kishore G, Gruys KJ (1999) Nature Biotechnol 17:1011
28. Lehninger AL, Nelson DL, Cox MM (1993) Principles of biochemistry. Worth Pub, New York
29. Schweiger G, Buckel W (1984) FEBS Lett 171:79
30. Williams DR, Anderson AJ, Dawes EA, Ewing DF (1994) Appl Microbiol Biotechnol 40:717
31. Gerbling H, Gerhart B (1989) Plant Physiol 91:1387

32. Camp PJ, Randall DD (1985) Plant Physiol 77:571
33. LaRossa RA, Van Dyk T, Smulski DR (1987) J Bacteriol 169:1372
34. Van Dyk TK, LaRossa RA (1987) Mol Gen Genet 207:435
35. Rhodes D, Hogan AL, Deal L, Jamieson GC, Haworth P (1987) Plant Physiol 84:775
36 Shaner DL, Singh BK (1993) Plant Physiol 103:1221
37. Singh BK, Shaner DL (1992) Carbon flow through branched-chain amino acid biosynthetic pathway: lessons from acetohydroxy acid synthase inhibitors. In: Singh BK, Flores HE, Shannon JC (eds) Biosynthesis and molecular regulation of amino acids in plants. American Society of Plant Physiologists, Maryland, p 354
38. Calhoun DH, Rimerman RA, Hatfield GW (1973) J Biol Chem 25:3511
39. Eisenstein E, Yu HD, Fisher KE, Iacuzio DA, Ducote DR, Schwarz RP (1995) Biochemistry 34:9403
40. Slater S, Houmiel KL, Tran M, Mitsky TA, Taylor NB, Padgette SR, Gruys KJ (1998) J Bact 180:197,941
41. Valentin HE, Broyles DL, Casagrande LA, Colburn SM, Creely WL, DeLaquil PA, Felton HM, Gonzalez KA, Houmiel KL, Lutke K, Mahadeo DA, Mitsky TA, Padgette SR, Reiser SE, Slater S, Stark DM, Stock RT, Stone DA, Taylor NB, Thorne GM, Tran M, Gruys K (1999) Int J Biol Macromol 25:30,341
42. Steinbüchel A, Valentin HE (1995) FEMS Microbiol Lett 128:219
43. De Koning GJM, Van Bilesen HMM, Lemstra PJ, Hazenberg W, Withold B, Preusting H, Van der Galiën JG, Schirmer A, Jendrossek D (1994) Polymer 35:2090
44. Lageveen RG, Huisman GW, Preusting H, Ketelaar P, Eggink G, Witholt B (1988) Appl Environ Microbiol 54:2924
45. Schulz H (1991) Biochim Biophys Acta 1081:109
46. Gerhard B (1993) Catabolism of fatty acid acids. In: Moore TS (ed) Lipid metabolism in plants. CRS Press, Bâton Rouge, p 527
47. Fukui T, Shiomi N, Doi Y (1998) J Bacteriol 180:667
48. Haywood GW, Anderson AJ, Ewing DF, Dawes EA (1990) Appl Environ Microbiol 56:3354
49. Timm A, Steinbüchel A (1990) Appl Environ Microbiol 56:3360
50. Huijberts GNM, Eggink G, de Waard P, Huisman GW, Witholt B (1992) Appl Environ Microbiol 58:536
51. Saito Y, Doi Y (1993) Int J Biol Macromol 15:287
52. Huijbert GNM, de Rijk TC, de Waard P, Eggink G (1994) J Bacteriol 176:1661
53. Rehm BHA, Krüger N, Steinbüchel A (1998) J Biol Chem 273:24,044
54. Mittendorf V, Robertson EJ, Leech RM, Krüger N, Steinbüchel A, Poirier Y (1998) Proc Natl Acad Sci USA 95:13,397
55. Mittendorf V, Krüger N, Steinbüchel A, Poirier Y (1998) Biosynthesis of medium-chain-length polyhydroxyalkanoates in transgenic *Arabidopsis* plants expressing the PhaC1 and PhaC2 synthases from *Pseudomonas aeruginosa*. In: Steinbüchel A (ed) Biochemical principles and mechanisms of biosynthesis and biodegradation of polymers. Wiley-VCH, Weinheim, p 368
56. Preisig-Müller R, Gühnemann-Schäfer KG, Kindl H (1994) J Biol Chem 269:20,475
57. Engeland KK, Kindl H (1991) Eur J Biochem 200:171
58. Mittendorf V, Bongcam V, Allenbach L, Coullerez G, Martini N, Poirier Y (1999) Plant J 20:45–55
59. McConn M, Browse J (1996) Plant Cell 8:403
60. Voelker TA, Worrell AC, Anderson L, Bleibaum J, Fan C, Hawkins DJ, Radke SE, Davies HM (1992) Science 257:72
61. Jones A, Davies HM, Voelker TA (1995) Plant Cell 7:359
62. Facciotti MT, Yuan L (1998) Fett/Lipid 4/5:167
63. Eccleston VS, Cranmer AM, Voelker TA, Ohlrogge JB (1996) Planta 198:46
64. Hooks MA, Fleming Y, Larson TR, Graham IA (1999) Planta 207:385
65. Martini N, Schell J, Abbadi A, Spener F, Töpfer R (1999) Vorträge für Pflanzenzüchtung 45:133
66. John ME, Keller G (1996) Proc Natl Acad Sci USA 93:12,768

67. John ME (1997) Crit Rev Biotechnol 17:185
68. Jendrossek D, Schirmer A, Schlegel HG (1996) Appl Microbiol Biotechnol 46:451
69. Van de Loo FJ, Broun P, Turner S, Somerville C (1995) Proc Natl Acad Sci USA 92:6743
70. Lee M, Lenman M, Banas A, Bafor M, Singh S, Schweizer M, Nilsson R, Liljenberg C, Dahlqvist A, Gummeson PO, Sjodahl S, Green A, Stymne S (1998) Science 280:915
71. Byrom D (1987) Trends Biotechnol 5:246
72. Berger E, Ramsay BA, Ramsay JA, Chavarie C, Braunegg G (1989) Biotechnol Tech 3:227
73. Ramsay JA, Berger E, Ramsay BA, Chavarie C (1990) Biotechnol Tech 4:221
74. Martin DP, Peoples OP, Williams SF (1997) PCT application WO 97/15,681
75. Noda I (1997) PCT application WO 97/07,230
76. Kurdikar DL, Strauser FE, Solodar AJ, Paster MD, Asrar J (1998) PCT application WO 98/46,782
77. Noda I (1998) US Pat 5,821,299
78. Kurdikar DL, Strauser FE, Solodar AJ, Paster MD, Asrar J (1998) PCT application WO 98/46,783
79. Seebach D, Roggo S, Zimmermann J (1987) Biological-chemical preparation of 3-hydroxycarboxylic acids and their use in EPC-synthesis. In: Bartmann W, Sharpless KB (eds) Stereochemistry of organic and bioorganic transformations. VCH Verlagsgesellschaft, Weinheim, p 85
80. Liddell (1997) PCT application WO 97/17,459
81. Noda I (1998) US Pat 5,849,854
82. Noda I (1999) US Pat 5,899,339
83. Whistler RL, BeMiller JN, Paschall EF (eds) (1984) Starch: chemistry and technology. Academic Press, Orlando
84. Röbbelen G, Downey RK, Ashri A (eds) (1989) Oil crops of the world. McGraw-Hill, New York
85. Moore JW (1992) Modern Plastics 69:58
86. Chemical Market Reporter (1998) 254
87. Page WJ (1997) Waste sources for polyhydroxyalkanoate production. In: Eggink G, Steinbüchel A, Poirier Y, Witholt B (eds) 1996 International Symposium on Bacterial Polyhydroxyalkanoates. NRC Research Press, Ottawa, p 56
88. Lee SY, Choi J, Chang HN (1997) Process development and economic evaluation for the production of polyhydroxyalkanoates by *Alcaligenes eutrophus*. In: Eggink G, Steinbüchel A, Poirier Y, Witholt B (eds) 1996 International Symposium on Bacterial Polyhydroxyalkanoates. NRC Research Press, Ottawa, p 127
89. de Koning G, Kellerhals M, van Meurs C, Witholt B (1997) A process for the production of bacterial medium-chain-length poly[(*R*)-3-hydroxyalkanoates]: reviewing the status quo. In: Eggink G, Steinbüchel A, Poirier Y, Witholt B (eds) 1996 International Symposium on Bacterial Polyhydroxyalkanoates. NRC Research Press, Ottawa, p 137
90. Lee SY (1996) Trends Biotechnol 14:431
91. de Koning G, Kellerhals M, van Meurs C, Witholt B (1997) Bioprocess Eng 17:15
92. Meyer P (1998) Stabilities and instabilities in transgene expression. In: Lindsey K (ed) Transgenic plant research. Harwood Academic Publishers, Amsterdam, p 263
93. Baulcombe DC (1999) Cur Opin Plant Biol 2:109
94. Ulker B, Allen GC, Thompson WF, Spiker S, Weissinger AK (1999) Plant J 18:25395. Vain P, Worland B, Kohli A, Snape JW, Christou P, Allen GC, Thompson WF (1999) Plant J 18:233
96. Vaucheret H, Beclin C, Elmayan T, Feuerbach F, Godon C, Morel JB, Mourrain P, Palauqui JC, Vernhettes S (1998) Plant J 16:651
97. Elmayan T, Balzergue S, Beon F, Bourdon V, Daubremet J, Guenet Y, Mourrain P, Palauqui JC, Vernhettes S, Vialle T, Wostrikoff K, Vaucheret H (1998) Plant Cell 10:1747
98. Kerr R (1998) Science 281:1128

Received: January 2000

In Vitro Biosynthesis of Polyesters

Shiro Kobayashi, Hiroshi Uyama

Department of Materials Chemistry, Graduate School of Engineering, Kyoto University, Kyoto 606-8501, Japan
E-mail: kobayasi@mat.polym.kyoto-u.ac.jp

In vitro synthesis of polyesters using isolated enzymes as catalyst via non-biosynthetic pathways is reviewed. In most cases, lipase was used as catalyst and various monomer combinations, typically oxyacids or their esters, dicarboxylic acids or their derivatives/glycols, and lactones, afforded the polyesters. The enzymatic polymerization often proceeded under mild reaction conditions in comparison with chemical processes. By utilizing characteristic properties of lipases, regio- and enantioselective polymerizations proceeded to give functional polymers, most of which are difficult to synthesize by conventional methodologies.

Keywords. Enzymatic polymerization, Polyester, Lipase, Enantioselectivity, Regioselectivity

Advances in Biochemical Engineering/
Biotechnology, Vol. 71
Managing Editor: Th. Scheper

List of Abbreviations

β-BL	β-butyrolactone
ε-CL	ε-caprolactone
CoA	coenzyme A
DDL	12-dodecanolide
DP	degree of polymerization
ee	enantiomeric excess
HDL	16-hecadecanolide
Lipase A	*Aspergillus niger* lipase
Lipase CA	*Candida antarctica* lipase
Lipase CC	*Candida cylindracea* lipase
Lipase MM	*Mucor miehei* lipase
Lipase PC	*Pseudomonas cepacia* lipase
Lipase PF	*Pseudomonas fluorescens* lipase
M_n	number-average molecular weight
M_w	weight-average molecular weight
8-OL	8-octanolide
PDL	15-pentadecanolide
PEG	poly(ethylene glycol)
PHA	poly(hydroxyalkanoate)
PHB	poly(hydroxybutylate)
β-PL	β-propiolactone
PPL	porcine pancrease lipase
UDL	11-undecanolide
δ-VL	δ-valerolactone

1 Introduction

Enzymes have several remarkable catalytic properties such as high catalytic power and high selectivities under mild reaction conditions, as compared with those of chemical catalysts. In the field of organic synthesis, enzymes have often been employed as catalyst; functional organic compounds were synthesized by the enzymatic selective reactions [1–5].

Production of all naturally occurring polymers in vivo is catalyzed by enzymes. Polymerizations catalyzed by an enzyme ("enzymatic polymerizations") have received much attention as new methodology [6–11], since in recent years structural variation of synthetic targets on polymers has begun to develop highly selective polymerizations for the increasing demands in the production of various functional polymers in material science. So far, in vitro syntheses of not only biopolymers but also non-natural synthetic polymers through enzymatic catalysis have been achieved [6–11].

Generally, there are three classes of polymer synthesis catalyzed by an enzyme:

1. Enzymatic synthesis in vivo (in living cells) via biosynthetic pathways
2. Enzymatic synthesis in vitro (outside cells) via biosynthetic pathways

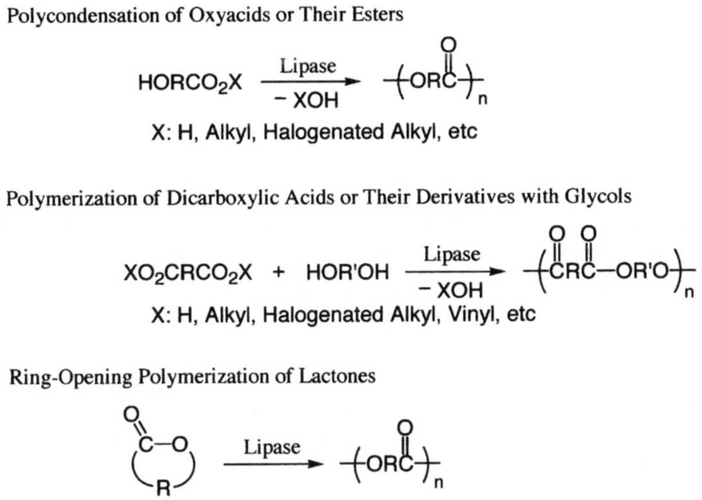

Fig. 1. Typical routes of polyester production using an isolated enzyme as catalyst

3. Chemical synthesis in vitro (in test tubes) via non-biosynthetic pathways catalyzed by an isolated enzyme

In the foregoing chapters, polyester syntheses belonging to classes 1 and 2 were described. The present chapter mainly reviews in vitro synthesis of polyesters through lipase catalysis (class 3). In vitro synthesis of poly(hydroxyalkanoate)s (PHAs) from hydroxyalkanoate coenzyme A (CoA) esters using PHA polymerase as catalyst is also briefly described. Lipids and polynucleotides (nucleic acids) are biogenic, naturally occurring esters. As to nucleic acids, a typical example of their enzymatic synthesis is the transcription of the genetic code from DNA to messenger RNA. The polymerization is catalyzed by RNA polymerase, which connects ribonucleotides by catalyzing the formation of the inter-nucleotide 3′-5′-phosphodiester bonds.

Figure 1 represents three major reaction types of lipase-catalyzed polymerization leading to polyesters [6–11].

2 Lipase-Catalyzed Polycondensation of Oxyacids and their Esters

Syntheses of aliphatic polyesters by fermentation and chemical processes have been extensively studied from the viewpoint of biodegradable materials science. Recently, another approach to their production has been made by using an isolated lipase or esterase as catalyst via non-biosynthetic pathways under mild reaction conditions. Lipase and esterase are enzymes which catalyze hydrolysis of esters in an aqueous environment in living systems. Some of them can act as catalyst for the reverse reactions, esterifications and transesterifications, in organic media [1–5]. These catalytic actions have been expanded to enzymatic synthesis of polyesters. First, enzymatic polymerization of oxyacids and their esters is described.

2.1 Oxyacids

In 1985, a lipase-catalyzed polymerization of 10-hydroxydecanoic acid was reported. The monomer was polymerized in benzene using poly(ethylene glycol) (PEG)-modified lipase soluble in the medium [12]. The degree of polymerization (DP) of the product was more than 5. PEG-modified esterase from hog liver and lipase from *Aspergillus niger* (lipase A) induced the oligomerization of glycolic acid [13].

The polymerization of ricinoleic acid proceeded using lipase from *Candida cylindracea* (lipase CC) or *Chromobacterium viscosum* as catalyst at 35 °C in water, hydrocarbons, or benzene to give the polymer with molecular weight of around 1×10^3 [14]. These lipases also induced the polymerization of 12-hydroxyoctadecanoic acid, 16-hydroxyhexadecanoic acid, and 12-hydroxydodecanoic acid. Oligoester from ricinoleic acid (estolide) possessing industrial applications in various fields was synthesized using lipase CC immobilized on ceramics [15]. The coloring of estolide improved by selecting the mild reaction conditions (40 °C).

Polyesters of relatively high molecular weight were enzymatically produced from 10-hydroxydecanoic acid [16] and 11-hydroxyundecanoic acid [17] using much lipase CC catalyst (tenfold the weight for the monomer). In case of 11-hydroxyundecanoic acid, the corresponding polymer with weight-average molecular weight (M_w) of 2.2×10^4 was obtained in the presence of activated molecular sieves. The enzymatic polymerization of 12-hydroxydodecanoic acid in the presence of 11-methacryloylaminoundecanoic acid produced the methacrylamide-type polyester macromonomer [18, 19]. Lipase CC and lipase derived from *Candida antarctica* (lipase CA) were active for the macromonomer synthesis.

By using characteristic catalysis of lipase, regio- and enantioselective polymerizations of oxyacids have been achieved. In the lipase CA-catalyzed polymerization of cholic acid, a hydroxy group at the 3-position was regioselectively acylated to give the oligoester with molecular weight of 920 (Fig. 2) [20]. Optically active oligoester was prepared by the enantioselective polymerization of racemic 10-hydroxyundecanoic acid catalyzed by lipase CC. The resulting polymer was enriched in the (S)-enantiomer to a level of 60% enantiomeric excess (ee) and the residual monomer was recovered with a 33% ee favoring the (*R*) enantiomer [21].

2.2 Oxyacid Esters

Porcine pancreatic lipase (PPL) catalyzed the polymerization of methyl 6-hydroxyhexanoate [22]. The polymer with DP up to 100 was synthesized by the polymerization in hexane at 69°C for more than 50 days. The PPL-catalyzed

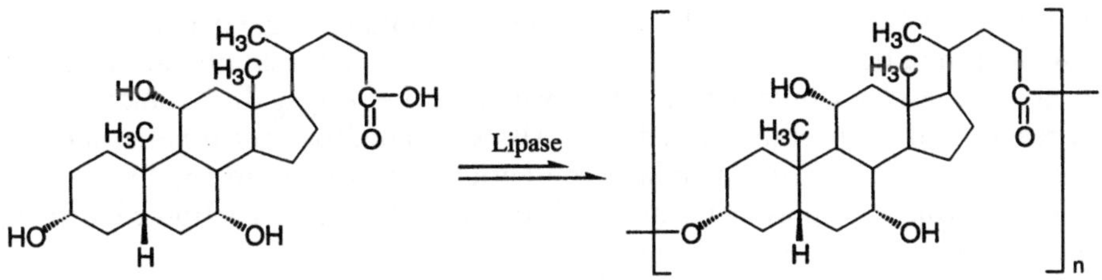

Fig. 2. Regioselective polymerization of cholic acid

polymerization of methyl 5-hydroxypentanoate for 60 days produced the polymer with DP = 29. Solvent effects were systematically investigated; hydrophobic solvents such as hydrocarbons and diisopropyl ether were suitable for the enzymatic production of high molecular weight polymer. Various hydroxyesters, ethyl esters of 3- and 4-hydroxybutyric acids, 5- and 6-hydroxyhexanoic acids, 5-hydroxydodecanoic acid, and 15-hydroxypentadecanoic acid, were polymerized by *Pseudomonas* sp. lipase at 45 °C to give the corresponding polyesters with molecular weights of several thousands [23].

Enzymatic enantioselective oligomerization of a symmetrical hydroxy diester, dimethyl β-hydroxyglutarate, produced a chiral oligomer (dimer or trimer) with 30–37% ee [24]. PPL catalyzed the enantioselective polymerization of ε-substituted-ε-hydroxy esters to produce optically active oligomers (DP < 6) [25]. The enantioselectivity increased with increasing bulkiness of the monomer substituent. Optically active polyesters with molecular weight of more than 1000 were obtained by the copolymerization of the racemic oxyacid esters with methyl 6-hydroxyhexanoate.

3
Lipase-Catalyzed Polymerization of Dicarboxylic Acids or their Derivatives with Glycols

Various combinations of dicarboxylic acid derivatives and glycols enzymatically afforded polyesters under mild reaction conditions. Dicarboxylic acids as well as derivatives, activated and non-activated esters, cyclic acid anhydride, and polyanhydrides, were found to be useful as monomer for the enzymatic synthesis of polyesters.

3.1
Dicarboxylic Acids

Immobilized *Mucor miehei* lipase (lipase MM) induced the polycondensation of adipic acid and 1,4-butanediol in ether solvents [26]. A horizontal two-chamber reactor was employed to facilitate the use of the molecular sieves. A low dispersity polyester with DP = 20 was obtained by two-stage polymerization.

The polymerization of dicarboxylic acids and glycols using lipase CA catalyst proceeded in a solvent-free system, despite the initial heterogeneous mixture of the substrates [27, 28]. The polymerization behaviors strongly depended on the chain length of both monomers. The detailed studies in the combination of adipic acid (A) and 1,4-butanediol (B) showed that the propagation took place by the reaction of the preliminary adduct (AB) with a hydroxy-terminated species. This solvent-free system has a large potential as an environmentally friendly synthetic process of polymeric materials owing to the mild reaction conditions and no use of organic solvents and toxic catalysts.

A dehydration reaction is generally realized in non-aqueous media. Since the water produced from the dehydration is in equilibrium with starting materials, a solvent water disfavors the dehydration to proceed in an aqueous medium due to the law of mass action. However, lipase catalysis achieved synthesis of aliphatic polyesters by dehydration polycondensation of dicarboxylic acids and glycols in water [29]. Various lipases such as lipases CA, CC, and MM were active for the polymerization of sebacic acid and 1,8-octanediol. In the polymerization of α,ω-dicarboxylic acids and glycols, the chain length of the monomers strongly affected the polymer yield and molecular weight; the highest yield was achieved from 1,12-docecanedioic acid and 1,10-decanediol, whereas no polymer formation was observed in using 1,6-hexanediol. In the above studies on the polyester synthesis from dicarboxylic acids and glycols, the molecular weight of the polymer was in the range of several thousands. High molecular weight polyester was obtained in the vacuum system, which removed the resulting water molecules during the polymerization [30, 31]. In the polymerization of sebacic acid and 1,4-butanediol using lipase MM catalyst in diphenyl ether at 37 °C for 7 days, the resulting polymer had an M_w of 4.2×10^4, which was much larger than that obtained under atmospheric pressure.

3.2
Dicarboxylic Acid Diesters

Unactivated esters, typically alkyl esters, often show low reactivity toward lipase catalyst for transesterifications. In the case of the lipase-catalyzed polycondensation of dialkyl esters with glycols, the polymer of high molecular weight was not obtained. The molecular weight improved when vacuum conditions were used; M_w reached more than 2×10^4 in the combination of diethyl sebacate and 1,4-butanediol catalyzed by lipase MM [30].

In the polymerization of dimethyl succinate and 1,6-hexanediol using lipase CA or MM as catalyst in toluene, equilibrium between the starting materials and polymer was observed [32]. Adsorption of methanol by molecular sieves or elimination of methanol by nitrogen bubbling shifted the thermodynamic equilibrium. Polyesters in the molecular weight range of several thousands were prepared from α,ω-alkylene dicarboxylic acid dialkyl esters and, whatever the monomer structure, cyclic oligomers were formed [33]. The yield of the cyclics depended on the monomer structure, initial concentration of the monomers, and reaction temperature. The ring-chain equilibrium was observed and the molar distribution of the cyclic species obeys the Jacobson-Stockmayer equation.

In lipase-catalyzed esterifications and transesterifications, esters of halogenated alcohols, typically 2-chloroethanol, 2,2,2-trifluoroethanol, and 2,2,2-

trichloroethanol, have often been used [1, 5], owing to increase of the electrophilicity (reactivity) of the acyl carbonyl and to avoiding significant alcoholysis of the products by decreasing the nucleophilicity of the leaving alcohols.

PPL catalyzed polycondensation of bis(2,2,2-trichloroethyl) alkanediaoates with glycols in anhydrous solvents of low polarity to produce the polyesters [34, 35]. In the polymerization of bis(2-chloroethyl) succinate and 1,4-butanediol using *Pseudomonas fluorescens* lipase (lipase PF) as catalyst, the polyester with low molecular weight was formed [36]. This may be due to the low enzymatic reactivity of the succinate substrate.

Vacuum was applied to shift the equilibrium forward by removal of the activated alcohol formed [30, 31, 37, 38]. In the enzymatic polycondensation of bis(2,2,2-trifluoroethyl) sebacate and aliphatic diols, the polymer with M_w of more than 1×10^4 was obtained using lipases CC, MM, PPL, and *Pseudomonas cepacia* lipase (lipase PC) as catalyst and lipase MM showed the highest catalytic activity [37]. Solvent screening indicated that diphenyl ether and veratrole were suitable for the production of the high molecular weight polyesters under vacuum. In the PPL-catalyzed reaction of bis(2,2,2-trifluoroethyl) glutarate with 1,4-butanediol in veratrole or 1,3-dimethoxybenzene, periodical vacuum method improved the molecular weight ($M_w \sim 4 \times 10^4$) [38].

In lipase-catalyzed transesterifications, frequent use of enol esters as acyl agents has been seen [1, 5], since the leaving unsaturated alcohol irreversibly tautomerizes to an aldehyde or a ketone, leading to the desired product in high yields. The polymerization of divinyl adipate and 1,4-butanediol proceeded in the presence of lipase PF at 45 °C [39]. Under similar reaction conditions, adipic acid and diethyl adipate did not afford the polymeric materials, indicating the high polymerizability of bis(enol ester) toward lipase catalyst.

Catalytic site of lipase is known to be a serine-residue and lipase-catalyzed reactions are considered to proceed via an acyl-enzyme intermediate. The mechanism of lipase-catalyzed polymerization of divinyl ester and glycol is proposed as follows (Fig. 3). First, the hydroxy group of the serine residue nucleophilically attacks the acyl-carbon of the divinyl ester monomer to produce an acyl-enzyme intermediate involving elimination of acetaldehyde. The reaction of the intermediate with the glycol produces 1:1 adduct of both

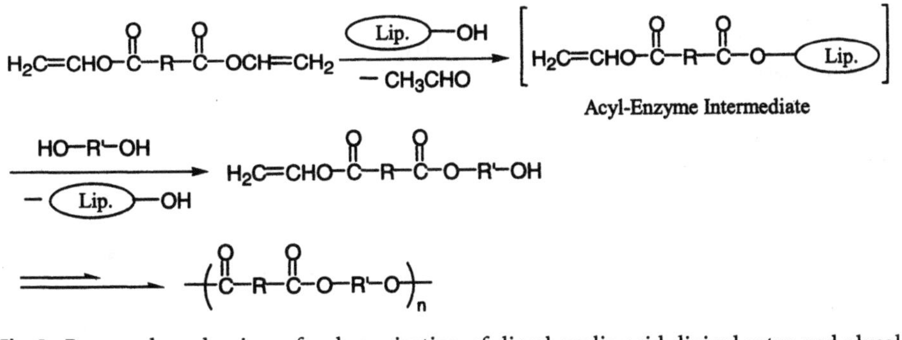

Fig. 3. Proposed mechanism of polymerization of dicarboxylic acid divinyl ester and glycol through lipase catalysis

monomers. In the propagation stage, the nucleophilic attack of the terminal hydroxy group to the acyl-enzyme intermediate formed from the vinyl ester group of the monomer, 1:1 adduct, and polymer takes place.

Lipase-catalyzed polymerization of divinyl adipate or divinyl sebacate with α,ω-glycols with different chain length has been reported [40]. Lipases CA, MM, PC, and PF showed high catalytic activity toward the polymerization. A combination of divinyl adipate, 1,4-butanediol, and lipase PC afforded the polymer with number-average molecular weight (M_n) of 2.1×10^4. The yield of the polymer from divinyl sebacate was higher than that from divinyl adipate, whereas the opposite tendency was observed in the polymer molecular weight.

Detailed studies on the lipase-catalyzed polymerization of divinyl adipate and 1,4-butanediol were performed [41–44]. Bulk polymerization increased the reaction rate and molecular weight of the polymer; however, the hydrolysis of the terminal vinyl ester significantly limited the formation of the polyester with high molecular weight. A mathematical model describing the kinetics of this polymerization was proposed, which effectively predicts the composition (terminal structure) of the polyester.

Another irreversible approach was performed by using bis(2,3-butanedione monoxime) alkanedioates as diester substrate [45]. Lipase MM efficiently produced the polymer with M_n up to 7.0×10^3.

Lipase catalysis induced the enantioselective polymerization, yielding optically active oligoesters and polyesters. The polymerization of racemic bis(2-chloroethyl) 2,5-dibromoadipate with excess of 1,6-hexanediol using lipase A catalyst produced optically active trimer and pentamer [46].

An optically pure polyester has been synthesized by PPL-catalyzed enantioselective polymerization of bis(2,2,2-trichloroethyl) *trans*-3,4-epoxyadipate with 1,4-butanediol in anhydrous diethyl ether (Fig. 4) [47]. The molar ratio of the diester to the diol was adjusted to 2:1 so as to produce the (–) polymer with enantiomeric purity of >96%. From end group analysis, the molecular weight was calculated to be 5.3×10^3.

Some proteases show an esterase activity, especially in their catalytic activity for regioselective acylation of sugars. By utilizing this property, enzymatic synthesis of polyester containing sugar group in the backbone was demonstrated

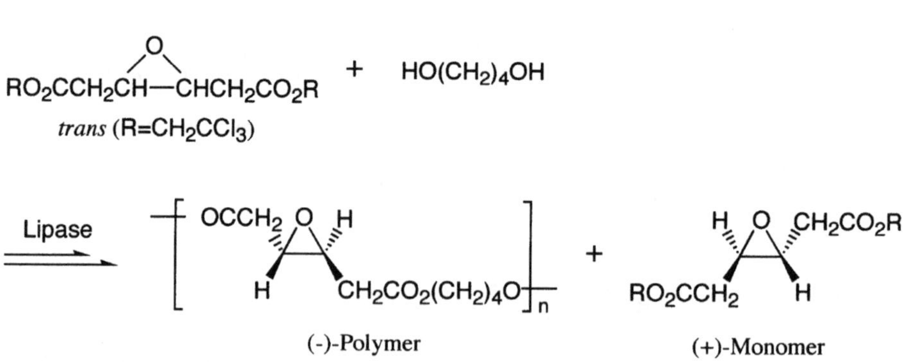

Fig. 4. Enantioselective polymerization of epoxy-containing diester with 1,4-butanediol

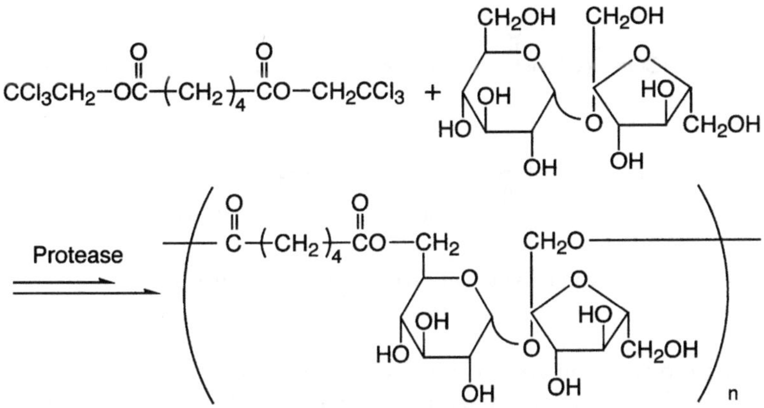

Fig. 5. Enzymatic synthesis of sucrose-containing polyester

[48]. Polycondensation of sucrose with bis(2,2,2-trifluoroethyl) adipate using an alkaline protease from *Bacillus* sp. as catalyst proceeds regioselectively to give an oligoester ($M_n = 1600$) having ester linkages at the C6 and C1′ positions on the sucrose (Fig. 5). The use of divinyl adipate as diester monomer gave an increase in the molecular weight of up to 1.1×10^4 [49].

Regioselective polymerization of divinyl esters with triols was achieved by using lipase CA as catalyst to give the soluble polymers with M_w of more than 1×10^4 [50, 51]. MALDI-TOF MS analysis confirmed the presence of a linear polyester with hydroxy substituents. NMR analysis of the product obtained from divinyl sebacate and glycerol in bulk at 60 °C showed that 1,3-diglyceride was a main unit and the branching unit (triglyceride) was contained in the resulting polymer. The regioselectivity of the acylation between primary and secondary hydroxy groups was 74:26. By choosing the reaction conditions, the regiospecific polymerization took place (Fig. 6).

Unsaturated ester oligomers have been synthesized by lipase-catalyzed polymerization of diesters of fumaric acid and 1,4-butanediol [52]. Isomerization of the double bond did not occur to give all-*trans* oligomers showing crystallinity, whereas the industrial unsaturated polyester having a mixture of *cis* and *trans* double bonds is amorphous [53] The enzymatic polymerization of bis(2-chloroethyl) fumarate with xylylene glycol produced the unsaturated oligoester containing aromaticity in its backbone [54].

The polymerization of dimethyl maleate and 1,6-hexanediol proceeded using lipase CA catalyst in toluene to give the polymer exhibiting exclusively *cis* structure [55]. During the polymerization, cyclic oligomers were formed. The cycles were semi-crystalline, whereas the linear polymer was amorphous. In the lipase CA-catalyzed copolymerization of dimethyl maleate and dimethyl fumarate with 1,6-hexanediol, the content of the cyclization was found to depend mainly on the configuration and concentration of the monomers [56].

Polyesters containing aromatic moiety in the main chain were synthesized by lipase CA-catalyzed polymerization of dicarboxylic acid divinyl esters and gly-

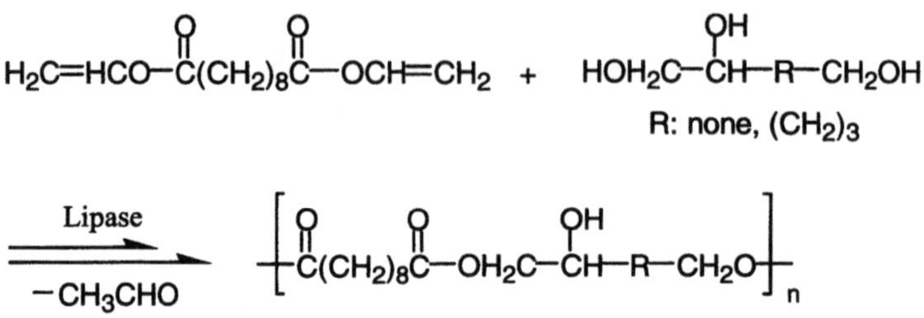

Fig. 6. Regiospecific polycondensation of divinyl sebacate and triols

cols under mild reaction conditions. Divinyl esters of isophthalic acid, terephthalic acid, and *p*-phenylene diacetic acid, were enzymatically polymerized in the presence of primary diols [57]. The highest M_n value (7.2×10^3) was obtained from a combination of divinyl isophthalate and 1,10-decanediol. Aromatic polyesters were also synthesized from methyl esters of terephthalic and isophthalic acids with 1,6-hexanediol in the presence of lipase CA [58]. In using methyl isophthalate as monomer, macrocyclic compounds were formed as by-product. Protease was also effective as catalyst for the aromatic polyester synthesis; *Bacillus licheniformis* protease catalyzed the oligomerization of esters of terephthalic acid with 1,4-butanediol [59].

In most enzymatic syntheses of polyesters, the polymerization was carried out in organic solvents or bulk. Using supercritical fluoroform as solvent, the polymerization of bis(2,2,2-trichloroethyl) adipate and 1,4-butanediol was demonstrated [60]. The molecular weight increased as a function of the pressure. By changing the pressure, the low-dispersity polymer fractions were separated.

3.3 Anhydrides

A new type of enzymatic polymerization involving lipase-catalyzed ring-opening poly(addition-condensation) of cyclic acid anhydride with glycols has been reported [61]. The polymerization of succinic anhydride with 1,8-octanediol proceeded using lipase PF catalyst at room temperature to produce the polyester with M_n of 3×10^3. In using glutaric anhydride as monomer, no polymer formation was observed, which suggests substrate selectivity for the acid anhydride.

Polyester synthesis was carried out by insertion-dehydration of glycols into polyanhydrides using lipase CA as catalyst (Fig. 7) [62]. The insertion of 1,8-octanediol into poly(azelaic anhydride) took place at 30–60 °C to give the corresponding polyester with a molecular weight of several thousands.

Polymerization of oxiranes with succinic anhydride proceeded in the presence of PPL at 60 °C or 80 °C [63]. The ring-opening of the oxirane might proceed by a carboxylic acid catalyst, which is formed by the reaction of succinic anhydride with serine residue of the lipase catalyst.

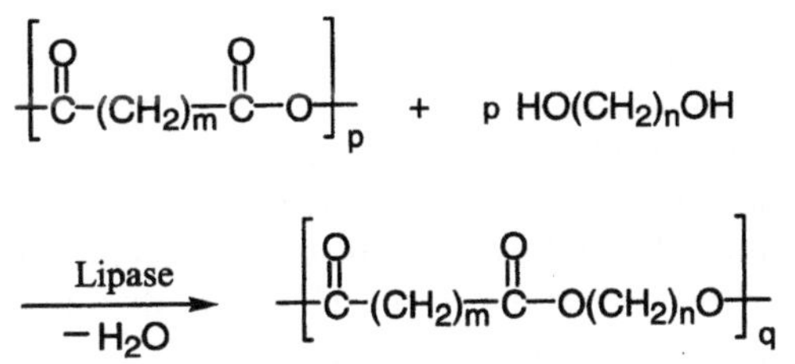

Fig. 7. Enzymatic insertion of glycols into polyanhydrides

4 Lipase-Catalyzed Ring-Opening Polymerization of Cyclic Monomers

Polyester syntheses have been achieved by enzymatic ring-opening polymerization of lactide and lactones with various ring-sizes. Here, we focus not only on these cyclic esters but also other cyclic monomers for lipase-catalyzed ring-opening polymerizations. Figure 8 summarizes cyclic monomers for providing polyesters via lipase catalysis.

4.1 Lactide and Lactones

Lactide was polymerized by lipase PC in bulk at high temperature (80–130 °C) to produce poly(lactic acid) with M_w up to 2.7×10^5 [64, 65]. The molecular weight of the polymer from the D,L-isomer was higher than that from the D,D- and L,L-ones. Protease (proteinase K) also induced the polymerization of lactide, however, the catalytic activity was relatively low.

Several small-size (four-membered) lactones have been reported to be polymerized through lipase catalysis. The polymerization of β-propiolactone (β-PL) by *Pseudomonas* family lipases as catalyst in bulk produced a mixture of linear

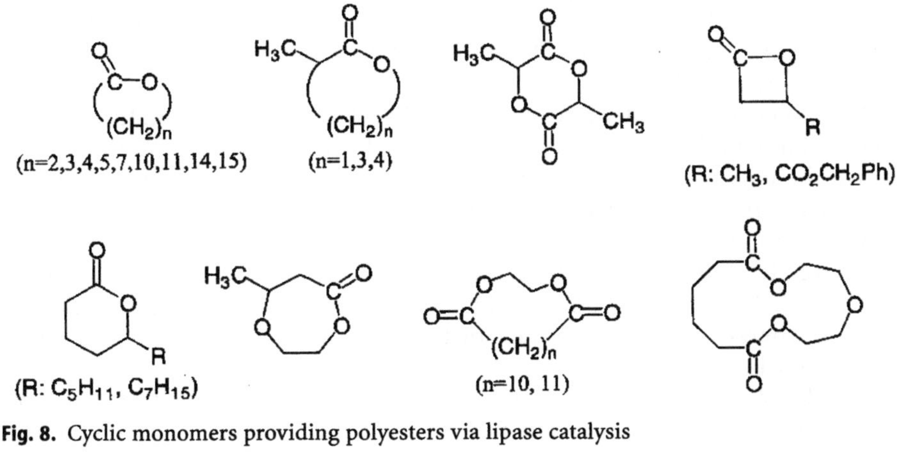

Fig. 8. Cyclic monomers providing polyesters via lipase catalysis

and cyclic oligomers with molecular weights of several hundreds [66]. In using a very small amount of lipase CC (0.5 wt % for the monomer), poly(β-PL) with M_w of 5×10^4 was obtained [67].

Ring-opening polymerization of racemic α-methyl-β-propiolactone using lipase PC catalyst proceeded enantioselectively to produce an optically active (*S*)-enriched polymer [68]. The highest ee value of the polymer was 0.50. NMR analysis of the product showed that the stereoselectivity during the propagation resulted from the catalyst enantiomorphic-site control.

β-Butyrolactone (β-BL) was also enzymatically polymerized to give poly(β-hydroxybutyrate) (PHB), which is a polyester produced in vitro by bacteria for an energy-storage substance. PPL-catalyzed polymerization of β-BL in bulk at room temperature produced PHB with molecular weight around 1×10^3 [69]. In the polymerization at high temperature (80 °C or 100 °C), lipase CC and PPL were very active for the polymerization to give PHB with a molecular weight of several thousands [70]. A significant amount of the cyclic PHB fraction was formed and the content of the cycles increased with increasing the monomer conversion. Enantioselective polymerization of β-BL was achieved by using thermophilic lipase to give (*R*)-enriched PHB with 20–37% ee [71].

Chemoenzymatic synthesis of biodegradable poly(malic acid) was performed by lipase-catalyzed polymerization of benzyl β-malolactone, followed by the debenzylation [72]. The addition of a small amount of β-PL (17 mol % for the monomer) increased M_w up to 3×10^4 [73].

Five-membered unsubstituted lactone, γ-butyrolactone, is not polymerized by conventional chemical catalysts. On the other hand, oligomer formation from γ-butyrolactone was observed by using PPL or *Pseudomonas* sp. lipase as catalyst [23, 69].

Lipase catalyzed the ring-opening polymerization of medium-size lactones, δ-valerolactone (δ-VL, six-membered) and ε-caprolactone (ε-CL, seven-membered). Lipases CC, PF and PPL showed high catalytic activity for the polymerization of δ-VL [74, 75]. The molecular weight of the polymer obtained in bulk at 60 °C was relatively low (less than 2000).

ε-CL was enzymatically polymerized by various lipases of different origin, lipases CA, CC, PC, PF, and PPL [22, 74–81]. Among the commercially available lipases, lipase CA was the most active toward the ε-CL polymerization; a very small amount of lipase CA (less than 1 wt % for ε-CL) was enough to induce the polymerization [76].

The molecular weight strongly depended on the enzyme origin. In the polymerization in bulk at 60 °C, *Pseudomonas* family lipases (lipases PC and PF) afforded the polymer of relatively high molecular weight. Under appropriate reaction conditions, the molecular weight reached more than 1×10^4 [75, 80]. The polymerization of ε-CL also proceeded in organic solvents. Properties of solvents greatly affected the polymerization behaviors [22, 75, 80]. In the lipase CA-catalyzed polymerization in organic solvents, cyclic oligomers were mainly formed, whereas the main product in the bulk polymerization was of linear structure [81].

The detailed kinetics of the ε-CL polymerization showed that termination and chain transfer did not occur and the monomer consumption followed a first-order rate law under appropriate conditions, indicating that the system provided controlled polymerizations where the molecular weight was a function of the monomer to initiator stoichiometry [78, 79].

Ring-opening polymerization of α-methyl-substituted medium-size lactones, α-methyl-γ-valerolactone and α-methyl-ε-caprolactone, proceeded by using lipase CA catalyst in bulk [82]. As to (*R*)- and (*S*)-3-methyl-4-oxa-6-hexanolides (MOHELs), lipase PC induced the polymerization of both isomers. The apparent initial rate of the S-isomer was seven times larger than that of the R-isomer, indicating that the enantioselective polymerization of MOHEL took place through lipase catalysis [83].

Nine-membered lactone, 8-octanolide (8-OL), was also enzymatically polymerized [84]. Lipases CA and PC showed the high catalytic activity for the polymerization.

Four macrolides, 11-undecanolide (12-membered, UDL) [85, 86], 12-dodecanolide (13-membered, DDL) [86, 87], 15-pentadecanolide (16-membered, PDL) [85, 86, 88, 89], and 16-hexadecanolide (17-membered, HDL) [90], were subjected to the lipase-catalyzed polymerization. For the polymerization of DDL, lipases CC, PC, PF, and PPL showed the high catalytic activity and the activity order in the bulk polymerization was as follows: lipase PC > lipase PF > lipase CC > PPL. These enzymes were also active for the polymerization of other macrolides. NMR analysis showed that the terminal structure of the polymer was of carboxylic acid at one end and of alcohol at the other terminal.

The bulk polymerization of PDL using lipase CA or MM as catalyst produced the corresponding polyester with high molecular weight ($1.5 \times 10^4 \sim 3.4 \times 10^4$) [89]. The polymerization behaviors (rate of the monomer consumption and molecular weight of the polymer) influenced the water content in the reaction system. Enzymatic ring-opening polymerization of macrolides (UDL, DDL, and PDL) proceeded even in an aqueous medium [91]. Lipases PC and PF produced the polymer with M_n of more than 1×10^3 in high yields.

The copolymerization of lactones took place through enzyme catalysis [92]. The copolymerization of ε-CL with δ-VL catalyzed by lipase PF affords the corresponding copolymer having a molecular weight of several thousand. From ^{13}C NMR analysis, the copolymer was found to be of random structure having both units, suggesting the frequent occurrence of transesterifications between the polyesters. In the copolymerization of 8-OL with ε-CL or DDL, random copolyesters were also formed [84], whereas the copolymer from ε-CL and PDL was not statistically random [88].

Enzyme activity for the polymerization of lactones was improved by the immobilization on Celite [93]. Immobilized lipase PF adsorbed on a Celite showed much higher catalytic activity than that before the immobilization. The catalytic activity was further enhanced by the addition of a sugar or poly(ethylene glycol) in the immobilization. Surfactant-coated lipase efficiently polymerized the ring-opening polymerization of lactones in organic solvents [94].

The enzymatic polymerization of lactones is explained by considering the following reactions as the principal reaction course (Fig. 9) [83, 85, 95, 96]. The key step is the reaction of the lactone with lipase involving the ring-opening of the lactone to give the acyl-enzyme intermediate (enzyme-activated monomer,

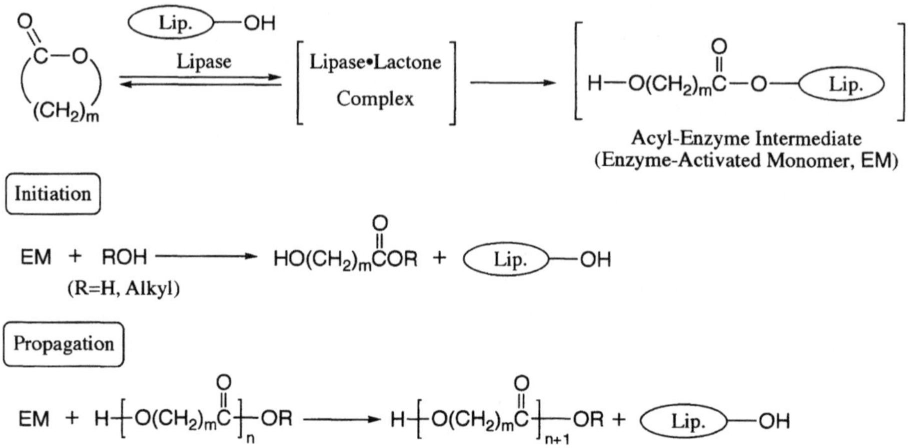

Fig. 9. Postulated mechanism of lactone polymerization catalyzed by lipase

EM). The initiation is a nucleophilic attack of water, which is probably contained in the enzyme, onto the acyl carbon of the intermediate to produce ω-hydroxycarboxylic acid (n = 1), the shortest propagating species. In the propagation stage, the intermediate is nucleophilically attacked by the terminal hydroxyl group of a propagating polymer to produce a one-unit-more elongated polymer chain.

Macrolides have virtually no ring strain, and hence show similar reactivities with acyclic fatty acid alkyl esters in alkaline hydrolysis and lower anionic ring-opening polymerizability compared with ε-CL (Table 1). On the other hand, the macrolides showed unusual reactivity in the lipase catalysis, that is, lipase PF-catalyzed polymerization of the macrolides proceeded much faster than that of ε-CL. The difference of the enzymatic polymerizability was quantitatively evaluated by Michaelis-Menten kinetics [83, 85, 95–97]. For all monomers, linearity was observed in the Hanes-Woolf plot, indicating that the polymerization followed Michaelis-Menten kinetics. $V_{max(lactone)}/K_{m(lactone)}$ of HDL was the largest, indicating that HDL had the largest enzymatic polymerizability [89]. The larger the ring size of lactone, the larger the $V_{max(lactone)}/K_{m(lactone)}$ value. $K_{m(lactone)}$ values were not so different with each other; on the other hand, $V_{max(lactone)}$ increased with increasing the ring size. These data imply that the enzymatic polymerizability increased as a function of the ring size, and the large polymerizability of macrolides through lipase catalysis is mainly due to the large reaction rate (V_{max}), but not to the binding abilities, i.e., the reaction process of the lipase-lactone complex to the acyl-enzyme intermediate is the key step of the polymerization.

Structural control of polymer terminal has been extensively studied since terminal-functionalized polymers, typically macromonomers and telechelics, are often used as prepolymers for synthesis of functional polymers. Various methodologies for synthesis of these polymers have been developed; however, most of them required elaborate and time-consuming procedures. By selecting

Table 1. Dipole moments and reactivities of lactones

Lactone[a]	Dipole moment (μ)	Rate constant		Michaelis-Menten kinetics[d]		
		Alkaline hydrolysis[b] ($l \cdot mol^{-1} \cdot s^{-1}$, $\times 10^4$)	Propagation[c] (s^{-1}, $\times 10^3$)	$K_{m(lactone)}$ ($mol \cdot l^{-1}$)	$V_{max(lactone)}$ ($mol \cdot l^{-1} \cdot h^{-1}$, $\times 10^2$)	$V_{max(lactone)}/K_{m(lactone)}$ (h^{-1}, $\times 10^2$)
δ-VL (*6*)	4.22	55000	–	–	–	–
ε-CL (7)	4.45	2550	120	0.61	0.66	1.1
UDL (12)	1.86	3.3	2.2	0.58	0.78	1.4
DDL (13)	1.86	6.0	15	1.1	2.3	2.1
PDL (16)	1.86	6.5	–	0.80	6.5	8.1
HDL (17)	–	–	–	0.63	7.2	11
Butyl caproate	1.75	8.4	–	–	–	–

[a] In parenthesis, ring-size of lactone.
[b] Alkaline: NaOH. Measured in 1,4-dioxane/water (60/40 vol. %) at 0 °C.
[c] Measured using NaOMe initiator (6 mol %) in THF at 0 °C.
[d] Kinetics of the polymerization was carried out using lipase PF (200 mg) as catalyst in the presence of 1-octanol (0.03 mol l^{-1}) in isopropyl ether (10 ml) at 60 °C.

the reaction conditions, facile synthesis of terminal-functionalized polyesters has been achieved using lipase catalyst.

An alcohol could initiate the ring-opening polymerization of lactones by lipase catalyst ("initiator method"). In the lipase CA-catalyzed polymerization of DDL using 2-hydroxyethyl methacrylate as initiator, the methacryloyl group was quantitatively introduced at the polymer terminal, yielding the methacryl-type polyester macromonomer [98]. This methodology was expanded to synthesis of ω-alkenyl- and alkynyl-type macromonomers by using 5-hexen-1-ol and 5-hexyn-1-ol as initiator.

Alkyl glucopyranosides also initiated the polymerization of ε-CL in the presence of lipase CA to give the polymer bearing the sugar moiety at the polymer terminal (Fig. 10) [99, 100]. In the initiation step, the primary hydroxy group of the glucopyranoside was regioselectively acylated. Poly(ε-CL) monosubstituted first generation dendrimer was synthesized by using lipase CA, in which the initiator was selectively monoacylated at the initial stage (Fig. 11) [101].

The enzymatic polymerization of lactones could be initiated at the hydroxy group of the polymer, which expanded to enzymatic synthesis of graft copolymers. The polymerization of ε-CL using thermophilic lipase as catalyst in the presence of hydroxyethyl cellulose (HEC) film produced HEC-*graft*-poly(ε-CL) with degree of substitution from 0.10 to 0.32 [102].

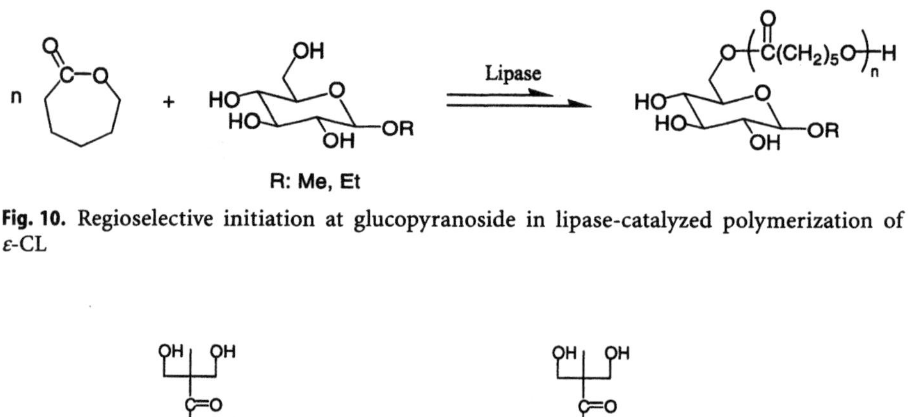

Fig. 10. Regioselective initiation at glucopyranoside in lipase-catalyzed polymerization of ε-CL

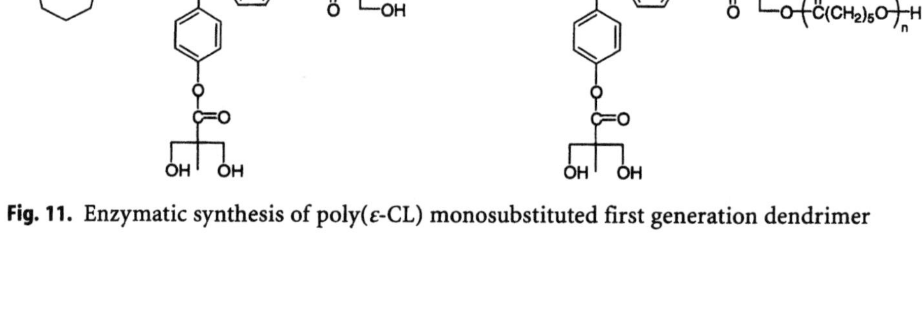

Fig. 11. Enzymatic synthesis of poly(ε-CL) monosubstituted first generation dendrimer

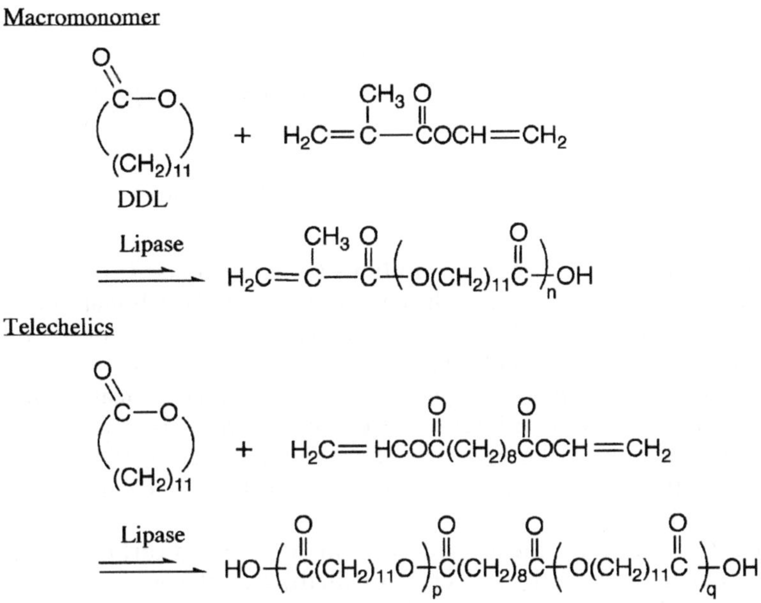

Fig. 12. Enzymatic synthesis of polyester macromonomer and telechelics by terminator method

End-functional polymers were also synthesized by lipase-catalyzed polymerization of DDL in the presence of vinyl esters [103, 104]. The vinyl ester acted as terminator ("terminator method"). In using vinyl methacrylate (12.5 mol % or 15 mol % based on DDL) and lipase PF as terminator and catalyst, respectively, the quantitative introduction of methacryloyl group at the polymer terminal was achieved to give the methacryl-type macromonomer (Fig. 12). By the addition of divinyl sebacate, the telechelic polyester having a carboxylic acid group at both ends was obtained.

Cyclic diesters were also subjected to the lipase-catalyzed ring-opening polymerization. The polymerization of ethylene dodecanoate and ethylene tridecanoate proceeded through lipase catalysis and lipases CA, PC, and PF were highly active for the polymerization [105]. The enzyme origin affected the polymerization behaviors; in using lipase PC catalyst, these bislactones were polymerized faster than ε-CL and DDL, whereas the reactivity of these cyclic diesters was in the intermediate between ε-CL and DDL in using lipase CA. The polymerization of 1,4,7-trioxa-cyclotridecane-8,13-dione catalyzed by lipase CA was also demonstrated [106].

4.2 Other Cyclic Monomers

Here, lipase-catalyzed ring-opening polymerization of cyclic compounds giving polymers other than polyesters is described. 1,3-Dioxan-2-one, six-membered cyclic carbonate, was polymerized in the presence of lipase catalysts (Fig. 13)

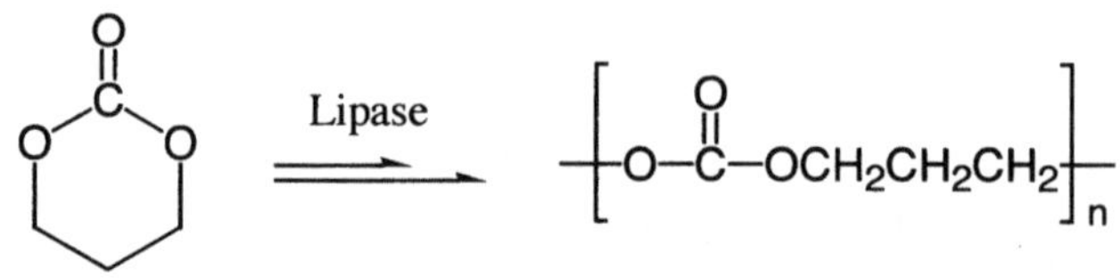

Fig. 13. Enzymatic polymerization of six-membered cyclic carbonate

[107–109]. Under mild reaction conditions (< 70 °C), lipase CA efficiently catalyzed the polymerization to give the corresponding polycarbonate with molecular weight more than 1×10^4. Lipases PC, PF, and PPL also induced the polymerization; however, the resulting polymer had low molecular weight. No ether bond was observed in the NMR spectrum of the product, indicating that elimination of carbon dioxide did not occur during the enzymatic polymerization. The polymerization in the presence of a small amount of PPL (0.1 wt% or 0.25 wt% for the monomer) at 100 °C produced the high molecular weight polymer ($M_w \sim 1.6 \times 10^5$).

Lipase CA catalyzed the polymerization of cyclic dicarbonates, cyclobis (hexamethylene carbonate) and cyclobis(diethylene glycol carbonate) to give the corresponding polycarbonates [105]. The enzymatic copolymerization of cyclobis(diethylene glycol carbonate) with DDL produced a random ester-carbonate copolymer. As to enzymatic synthesis of polycarbonates, reported were polycondensations of 1,3-propanediol divinyl dicarbonate with 1,3-propanediol [110], and of diphenyl carbonate with bisphenol-A [111].

3-(*S*)-Isopropylmorpholine-2,5-dione, six-membered depsipeptide, was polymerized by lipase PC and PPL catalysts [112]. High temperature (100 °C or 130 °C) was required for the polymerization, yielding biodegradable poly(depsipeptide). During the polymerization, the racemization of the valine residue took place. Demonstrated was PPL-catalyzed ring-opening polymerization of ethylene isopropyl phosphate, five-membered cyclic phosphate [113].

5
In Vitro PHA Polymerase-Catalyzed Polymerization to PHA

PHA is produced in *Alcaligenes eutrophus* from acetyl CoA in three steps and the last step is the chain growth polymerization of hydroxyalkanoate CoA esters catalyzed by PHA polymerase (synthase), yielding PHA of high molecular weight. Kinetics and mechanism of the polymerization of hydroxyalkanoyl CoA monomers with this bacterial polymerase have been investigated.

In the polymerization, the growing polymer chain was covalently attached to a highly conserved cysteine residue (Cy319) of the polymerase [114]. The granules of the precipitated polymer were quickly formed when the purified polymerase was exposed to (*R*)-hydroxybutyryl CoA [115]. The artificial PHB granules were spherical with diameters of up to 3 μm, significantly larger than the native ones, and the molecular weight was more than 10^7. In this polymerization, a lag time was observed and the enzyme on the granules was found to be much more active than the enzyme in solution.

This enzyme polymerized the CoA monomers of (*R*)-hydroxyalkanoates to give the high molecular weight homopolymers and copolymers [116, 117]. The addition of glycerol to the polymerase solution eliminated the lag phase of the polymerization. The molecular weight of the polymers was found to be inversely proportional to the molar ratio of monomer-to-enzyme and the obtained polymer had narrow molecular weight distribution, suggesting that the polymerization proceeds in a living fashion.

Random copolymers were obtained from the mixture of the two CoA esters in the presence of the polymerase, whereas block copolymers were synthesized when the two monomers were reacted sequentially with the enzyme. In the polymerization of racemic hydroxybutyryl CoA, only the (R)-monomer was polymerized. Furthermore, the presence of the (S)-monomer did not reduce the polymerization rate of the (R)-isomer. These data indicate that the (S)-monomer does not act as competitive inhibitor for the polymerase.

The polymerization behavior of hydroxybutyryl CoA by purified recombinant PHA synthase from *Chromatium vinosum* was different with that of *Alcaligenes eutrophus* [118]. This enzyme lost its activity during the polymerization and the yield and molecular weight were lower than those of *Alcaligenes eutrophus*. The molecular weight did not depend on the feed ratio of the monomer and enzyme.

In combination of this polymerase with purified propionyl-CoA transferase of *Clostridium propionicum*, a two-enzyme in vitro PHB biosynthesis system was established which allowed the PHB synthesis from (*R*)-hydroxybutyric acid as substrate [119]. In this way, the PHB synthesis was independent of the consumption of the expensive CoA, and hence PHA could be readily produced in a semipreparative-scale

6 Conclusion

In vitro polyester syntheses using an isolated enzyme as catalyst via non-biosynthetic pathways is reviewed. These enzymatic routes for production of biodegradable polyesters possess several advances in comparison with fermentation and chemical processes:

1. Structural variation of monomers and polymers.
2. Non-toxic catalyst and mild reaction conditions.
3. Enantio- and regioselective polymerizations to produce functional polymers, which are very difficult to be obtained by conventional methodologies.

By means of genetic engineering, including cloning and site-directed mutagenesis, it has become possible for modern synthetic chemists to utilize a sufficient amount of isolated enzyme catalysts and to modify the reactivity, stability, or even specificity of enzymes. Therefore, polymerizations catalyzed by isolated enzyme are expected to create a new area of precision polymer syntheses. Furthermore, enzymatic polymerizations have great potential as an environmentally friendly synthetic process of polymeric materials.

References

1. Wong C-H, Whitesides GM (1994) Enzymes in synthetic organic chemistry. Pergamon, Oxford
2. Whitesides GM, Wong C-H (1985) Angew Chem Int Ed Engl 24:617
3. Jones JB (1986) Tetrahedron 42:3351
4. Klibanov AM (1990) Acc Chem Res 23:114
5. Santaniello E, Ferraboschi P, Grisenti P, Manzocchi A (1992) Chem Rev 92:1071
6. Kobayashi S, Shoda S, Uyama H (1995) Adv Polym Sci 121:1
7. Kobayashi S, Shoda S, Uyama H (1997) Enzymatic catalysis. In: Kobayashi S (ed) Catalysis in precision polymerization. Wiley, Chichester, chap 8
8. Kobayashi S, Uyama H (1998) Biocatalytic routes to polymers. In: Schlüter A-D (ed) Materials science and technology – synthesis of polymers. Wiley-VCH, Weinheim, chap 16
9. Gross RA, Kaplan DL, Swift G (eds) (1998) Enzymes in polymer synthesis. ACS, Washington
10. Ritter H (1993) Trends Polym Sci 1:171
11. Ritter H (1996) Functionalized polymers via enzymatic synthesis. In: Arshady R (ed) Desk references of functional polymers: syntheses and applications. ACS, Washington, chap 1.7
12. Ajima A, Yoshimoto T, Takahashi K, Tamaura Y, Saito Y, Inada Y (1985) Biotechnol Lett 7:303
13. Ohya Y, Sugitou T, Ouchi T (1995) J Macromol Sci-Pure Appl Chem A32:179
14. Matsumura S, Takahashi J (1986) Makromol Chem Rapid Commun 7:369
15. Yoshida Y, Kawase M, Yamaguchi C, Yamane T (1995) Yukagaku 44:328
16. O'Hagan D, Zaidi NA (1993) J Chem Soc Perkin Trans I 2389
17. O'Hagan D, Zaidi NA (1994) Polymer 35:3576
18. Pavel K, Ritter H (1991) Makromol Chem 192:1941
19. Noll O, Ritter H (1997) Macromol Rapid Commun 18:53
20. Noll O, Ritter H (1996) Macromol Rapid Commun 17:553
21. O'Hagan D, Parker AH (1998) Polym Bull 41:519
22. Knani D, Gutman AL, Kohn DH (1993) J Polym Sci Polym Chem Ed 31:1221
23. Dong H, Wang H-D, Cao S-G, Shen J-C (1998) Biotechnol Lett 20:905
24. Gutman AL, Bravdo T (1989) J Org Chem 54:5645
25. Knani D, Kohn DH (1993) J Polym Sci Polym Chem Ed 31:2887
26. Binns F, Roberts SM, Taylor A, Williams CF (1993) J Chem Soc Perkin Trans I 899
27. Uyama H, Inada K, Kobayashi S (1998) Chem Lett 1285
28. Binns F, Harffey P, Roberts SM, Taylor A (1998) J Polym Sci Polym Chem Ed 36:2069
29. Kobayashi S, Uyama H, Suda S, Namekawa S (1997) Chem Lett 105
30. Linko Y-Y, Wang Z-L, Seppälä J (1995) J Biotechnol 40:133
31. Linko Y-Y, Seppälä J (1996) CHEMTECH 26(August):25
32. Mezoul G, Lalot T, Brigodiot M, Maréchal E (1995) J Polym Sci Polym Chem Ed 33:2691
33. Berkane C, Mezoul G, Lalot T, Brigodiot M, Maréchal E (1997) Macromolecules 30:7729
34. Wallace JS, Morrow CJ (1989) J Polym Sci Polym Chem Ed 27:3271
35. Jarvie AW, Samra BK, Wiggett AJ (1996) J Chem Research (S) 129
36. Linko Y-Y, Wang Z-L, Seppälä J (1994) Biocatalysis 8:269
37. Linko Y-Y, Wang Z-L, Seppälä J (1995) Enzyme Microb Technol 17:506
38. Brazwell EM, Filos DY, Morrow CJ (1995) J Polym Sci Polym Chem Ed 33:89
39. Uyama H, Kobayashi S (1994) Chem Lett 1687
40. Uyama H, Yaguchi S, Kobayashi S (1999) J Polym Sci Polym Chem Ed 37:2737
41. Chaundhary AK, Beckman EJ, Russell AJ (1997) Biotechnol Bioeng 55:227
42. Chaundhary AK, Lopez J, Beckman EJ, Russell AJ (1997) Biotechnol Prog 13:318
43. Chaundhary AK, Beckman EJ, Russell AJ (1998) Biotechnol Bioeng 59:428
44. Chaundhary AK, Beckman EJ, Russell AJ (1998) AIChE J 44:753
45. Athawale VD, Gaonkar SR (1994) Biotechnol Lett 16:149

46. Margolin AL, Crenne J-Y, Klibanov AM (1987) Tetrahedron Lett 28:1607
47. Wallace JS, Morrow CJ (1989) J Polym Sci Polym Chem Ed 27:2553
48. Patil DR, Rethwisch DG, Dordick JS (1991) Biotechnol Bioeng 37:639
49. Dordick JS (1992) Ann N Y Acad Sci 672:352
50. Kline BJ, Beckman EJ, Russell AJ (1998) J Am Chem Soc 120:9475
51. Uyama H, Inada K, Kobayashi S (1999) Macromol Rapid Commun 20:171
52. Geresh S, Gilboa Y (1990) Biotechnol Bioeng 36:270
53. Geresh S, Gilboa Y, Abrahami S, Bershadsky A (1993) Polym Eng Sci 33:311
54. Geresh S, Gilboa Y (1991) Biotechnol Bioeng 37:883
55. Mezoul G, Lalot T, Brigodiot M, Maréchal E (1995) Macromol Rapid Commun 16:613
56. Mezoul G, Lalot T, Brigodiot M, Maréchal E (1996) Macromol Chem Phys 197:3581
57. Uyama H, Yaguchi S, Kobayashi S (1999) Polym J 31:380
58. Mezoul G, Lalot T, Brigodiot M, Maréchal E (1996) Polym Bull 36:541
59. Park HG, Chang HN, Dordick JS (1994) Biocatalysis 11:263
60. Chaudhary AK, Beckman EJ, Russell AJ (1995) J Am Chem Soc 117:3728
61. Kobayashi S, Uyama H (1993) Makromol Chem Rapid Commun 14:841
62. Uyama H, Wada S, Kobayashi S (1999) Chem Lett 893
63 Matsumura S, Okamoto T, Tsukuda K, Toshima K (1998) Macromol Rapid Commun 19:295
64. Matsumura S, Mabuchi K, Toshima K (1997) Macromol Rapid Commun 18:477
65. Matsumura S, Mabuchi K, Toshima K (1997) Macromol Symp 130:285
66. Namekawa S, Uyama H, Kobayashi S (1996) Polym J 28:730
67. Matsumura S, Beppu H, Tsukuda K, Toshima K (1996) Biotechnol Lett 18:1041
68. Svirkin YY, Xu J, Gross RA, Kaplan DL, Swift G (1996) Macromolecules 29:4591
69. Nobes GAR, Kazlauskas RJ, Marchessault RH (1996) Macromolecules 29:4829
70. Matsumura S, Suzuki Y, Tsukuda K, Toshima K, Doi Y, Kasuya K (1998) Macromolecules 31:6444
71. Xie W, Li J, Chen D, Wang PG (1997) Macromolecules 30:6997
72. Matsumura S, Beppu H, Nakamura K, Osanai S, Toshima K (1996) Chem Lett 795
73. Matsumura S, Beppu H, Toshima K (1999) Chem Lett 249
74. Uyama H, Kobayashi S (1993) Chem Lett 1149
75. Kobayashi S, Takeya K, Suda S, Uyama H (1998) Macromol Chem Phys 199:1729
76. Uyama H, Suda S, Kikuchi H, Kobayashi S (1997) Chem Lett 1109
77. Kobayashi S, Shoda S, Uyama H (1994) J Syn Org Chem Jpn 52:754
78. MacDonald RT, Pulapura SK, Svirkin YY, Gross RA, Kaplan DL, Akkara JA, Swift G, Wolk S (1995) Macromolecules 28:73
79. Henderson LA, Svirkin YY, Gross RA, Kaplan DL, Swift G (1996) Macromolecules 29:7759
80. Dong H, Wang H-D, Cao S-G, Shen J-C (1999) J Polym Sci Polym Chem Ed 37:1265
81. Córdova A, Iversen T, Hult K, Martinelle M (1998) Polymer 39:6519
82. Küllmer K, Kikuchi H, Uyama H, Kobayashi S (1998) Macromol Rapid Commun 19:127
83. Kobayashi S, Uyama H, Namekawa S (1998) Polym Degrad Stab 59:195
84. Kobayashi S, Uyama H, Namekawa S, Hayakawa H (1998) Macromolecules 31:5655
85. Uyama H, Takeya K, Kobayashi S (1995) Bull Chem Soc Jpn 68:56
86. Uyama H, Kobayashi S (1996) Enzymatic ring-opening polymerization of macrolides to polyesters. In: Yalpani M (ed) Biomedical functions and biotechnology of natural and artificial polymers. ATL Press, Schrewsbury, p 5
87. Uyama H, Takeya K, Hoshi N, Kobayashi S (1995) Macromolecules 28:7046
88. Uyama H, Kikuchi H, Takeya K, Kobayashi S (1996) Acta Polymerica 47:357
89. Bisht KS, Henderson LA, Gross RA, Kaplan DL, Swift G (1997) Macromolecules 30:2705
90. Namekawa S, Uyama H, Kobayashi S (1998) Proc Jpn Acad 74B:65
91. Namekawa S, Uyama H, Kobayashi S (1998) Polym J 30:269
92. Uyama H, Takeya K, Kobayashi S (1993) Proc Acad Jpn 69B:203
93. Uyama H, Kikuchi H, Takeya K, Hoshi N, Kobayashi S (1996) Chem Lett 107
94. Noda S, Kamiya N, Goto M, Nakashio F (1997) Biotechnol Lett 19:307

95. Kobayashi S, Uyama H (1999) Macromol Symp 144:237
96. Namekawa S, Uyama H, Kobayashi S (1999) Int J Biol Macromol 25:145
97. Uyama H, Namekawa S, Kobayashi S (1997) Polym J 29:299
98. Uyama H, Suda S, Kobayashi S (1998) Acta Polymerica 49:700
99. Bisht KS, Deng F, Gross RA, Kaplan DL, Swift G (1998) J Am Chem Soc 120:1363
100. Córdova A, Iversen T, Hult K (1998) Macromolecules 31:1040
101. Córdova A, Hult A, Hult K, Ihre H, Iversen T, Malmström E (1998) J Am Chem Soc 120:13,521
102. Li J, Xie W, Cheng HN, Nickol RG, Wang PG (1998) Macromolecules 32:2789
103. Uyama H, Kikuchi H, Kobayashi S (1995) Chem Lett 1047
104. Uyama H, Kikuchi H, Kobayashi S (1997) Bull Chem Soc Jpn 70:1691
105. Müller S, Uyama H, Kobayashi S (1999) Chem Lett 1317
106. Binns F, Taylor A (1995) Tetrahedron 47:12,929
107. Matsumura S, Tsukuda K, Toshima K (1997) Macromolecules 30:3122
108. Kobayashi S, Kikuchi H, Uyama H (1997) Macromol Rapid Commun 18:575
109. Bisht KS, Svirkin YY, Henderson LA, Gross RA, Kaplan DL, Swift G (1997) Macromolecules 30:7735
110. Rodney RL, Stagno JL, Beckman EJ, Russell AJ (1999) Biotechnol Bioeng 62:259
111. Abramowicz DA, Keese CR (1989) Biotechnol Bioeng 33:149
112. Feng Y, Knüfermann J, Klee D, Höcker H (1999) Macromol Rapid Commun 20:88
113. Wen J, Zhuo R-X (1998) Macromol Rapid Commun 19:641
114. Wodzinska J, Snell KD, Rhomberg A, Sinskey AJ, Biemann K, Stubbe J (1996) J Am Chem Soc 118:6319
115. Gerngross TU, Martin DP (1995) Proc Natl Acad Sci USA 92:6279
116. Martin DP, Zhang S, Su L, Lenz RW (1999) Extracellular polymerization of 3-hydroxyalkanoate monomers by the synthase from *Alcaligenes eutrophus*. In: Steinbüchel A (ed) Biochemical principles and mechanisms of biosynthesis and biodegradation of polymers. Wiley-VCH, Weinheim, pp 168–175
117. Lenz RW, Farcet C, Dijkstra PJ, Goodwin S, Zhang S (1999) Int J Biol Macromol 25:55
118. Jossek R, Reichelt R, Steinbüchel A (1998) Appl Microbiol Biotechnol 49:258
119. Jossek R, Steinbüchel A (1998) FEMS Microbiol Lett 168:319

Received: January 2000

Properties, Modifications and Applications of Biopolyesters

G. A. M. van der Walle[1], G. J. M. de Koning[2], R. A. Weusthuis[1], G. Eggink[1]

[1] Agrotechnological Research Institute (ATO), Bornsesteeg 59, P.O. Box 17, 6700 AA, Wageningen, the Netherlands
E-mail: g.a.m.vanderwalle@ato.wag-ur.nl

[2] DSM Research, Polymeric construction materials, P.O. Box 18, 5150 MD Geleen, the Netherlands

Poly(hydroxyalkanoates) (PHAs), of which poly(hydroxybutyrate) (PHB) is the most common, can be accumulated by a large number of bacteria as energy and carbon reserve. Due to their biodegradability and biocompatibility these optically active biopolyesters may find industrial applications. A general overview of the physical and material properties of PHAs, alongside with accomplished applications and new developments in this field is presented in this chapter.

The properties of PHAs are dependent on their monomer composition and therefore it is of great interest that recent research has revealed that, in addition to PHB, a large variety of PHAs can be synthesized microbially. The monomer composition of PHAs depends on the nature of the carbon source and microorganism used. PHB is a typical highly crystalline thermoplastic whereas medium chain length PHAs are elastomers with low melting points and a relatively lower degree of crystallinity. By (chemical) modification of the PHAs, the ultimate properties of the materials can be adjusted even further, when necessary.

Applications that have been developed from PHB and related materials (e.g. Biopol) can be found in very different application areas and cover packaging, hygienic, agricultural and biomedical products. Recent application developments based on medium chain length PHAs range from high solid alkyd-like paints to pressure sensitive adhesives, biodegradable cheese coatings and biodegradable rubbers. Technically, the prospects for PHAs are very promising. When the price of these materials can be further reduced, application of biopolyesters will also become economically very attractive.

Keywords. Poly(hydroxyalkanoates), Biopolyesters, Properties, Modification, Crosslinking, Biodegradable, Environmentally friendly paints, Coatings, Rubbers, Adhesives

Advances in Biochemical Engineering/
Biotechnology, Vol. 71
Managing Editor: Th. Scheper

1 Introduction

Since poly(hydroxybutyrate), a naturally abundant poly(hydroxyalkanoate) (PHA, Fig. 1), was first isolated and characterized by Lemoigne in 1925, PHAs have been studied extensively by biochemists who referred to them as lipids [1, 2]. Further research on PHAs however, made it clear that these materials are in fact water-insoluble polyesters, which are synthesized and accumulated intracellularly as storage compounds by many different bacteria. Using various types of substrates, an almost infinite variety of PHAs can be synthesized [3–9].

Apart from being natural products, PHAs proved to be thermoplastic materials as well. In contrast to synthetic polymers, PHAs have the fundamental advantage of being based solely on renewable resources. Moreover, as described in the chapter about degradation of polyesters in this volume, due to the fact that PHAs are completely digested and metabolized by a wide variety of bacteria and fungi, they are genuinely biodegradable [10–15].

Most of the plastics and synthetic polymers that are used worldwide are produced from petrochemicals. Replacing petroleum-based feedstocks with materials derived from renewable resources is an attractive prospect for manufacturers of polymers and plastics, since the production of such polymers does not depend on the limited supply of fossil fuels [16]. Furthermore, synthetic materials are very persistent in the environment long after their intended use, and as a result their total volume in landfills is giving rise to serious waste management problems. In 1992, 20% of the volume and 8% of the weight of landfills in the US were plastic materials, while the annual disposal of plastics both in the US and EC has risen to over 10 million tons [17]. Because of the biodegradability of PHAs, they would be mostly composted and as such would be very valuable in reducing the amount of plastic waste.

In addition to the biodegradability of PHAs, they have another important feature, which is their hydrophobicity. This makes PHAs superior to their biodegradable competitors like starch and proteins in moisture resistance, despite the higher price of PHAs [18]. Apart from the unique combination of biodegradability and hydrophobicity, PHAs have other interesting and useful material characteristics. The combination of the various special material properties should be kept in mind for application development.

In this chapter, a general overview of the physical and material aspects of PHAs, such as their chemical structure and composition, as well as their thermal and mechanical properties is presented. To conclude, some recent developments in various application areas such as the development of high solid alkyd-like paints, pressure sensitive adhesives, biodegradable cheese coatings and biodegradable rubbers are discussed.

2 Physical and Material Properties

2.1 Chemical Structure and Composition

Since the chemical structure and monomer composition of a specific polymer are the most important factors in determining the polymer's physical and material properties, a short recapitulation of typical representatives of microbially synthesized poly(hydroxyalkanoates) is presented in this section. A more detailed overview on this issue is available from References [19–21], but is not within our scope here. The monomer composition of PHAs depends on the nature of the carbon source and the microorganisms used. This way, numerous monomers have been introduced into PHA chains [3–9]. PHAs have been divided roughly into two classes [19].

First, the short chain length PHAs, poly(HA_{SCL}), are composed of monomeric units containing up to 5 carbon atoms. The most well-known representatives are poly(3-hydroxybutyrate) (PHB), and its copolymers with hydroxyvalerate. Of all the PHAs, PHB is by far the most commonly encountered in nature [18]. It is the simplest PHA with respect to chemical structure, having a methylene (-CH_3) group as the pendent R-unit in Fig. 1. Owing to its enzymatic synthesis, PHB has an exceptional stereochemical regularity. The chains are linear and the chiral centers all are in the *R*-stereochemical conformation, which implies that this polymer is completely isotactic.

Comonomers, such as 3-hydroxyvalerate (3HV, ethylene R-unit (–CH_2–CH_3) in Fig. 1) and 4-hydroxybutyrate, have been incorporated in the PHB chains using specific additives in the growth medium of the bacteria [21–25]. It has been shown by nuclear magnetic resonance (NMR) studies that poly(3HB-*co*-3HV) has a statistically random distribution of the monomer units throughout a range of compositions varying from 0 to 90 mol % 3HV [23–26].

Second, some organisms are able to incorporate longer pendent chains yielding another class of PHA; medium chain length PHA, poly(HA_{MCL}). Poly (HA_{MCL}) is specifically accumulated by fluorescent pseudomonads. When aliphatic hydrocarbons like *n*-alkane, *n*-alkanoate, or *n*-alkanol serve as feedstocks for *Pseudomonas oleovorans* the resulting PHA is a random copolymer

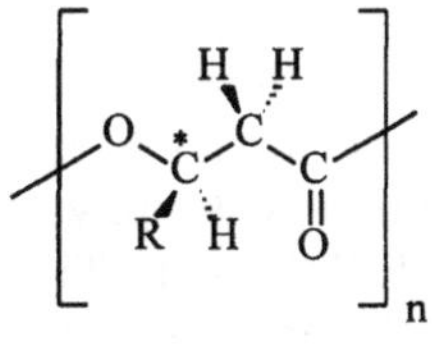

Fig. 1. Chemical structure of poly(hydroxyalkanoate) repeating unit

Table 1. Chemical Composition of Medium Chain Length PHAs Synthesized by *Pseudomonas putida* KT2442 During Growth on Different Types of Carbon Sources[a]

Substrate	$C_{6:0}$	$C_{8:0}$	$C_{8:1}$	$C_{10:0}$	$C_{10:1}$	$C_{12:0}$	$C_{12:1}$	$C_{12:2}$	$C_{14:0}$	$C_{14:1}$	$C_{14:2}$	$C_{14:3}$	$C_{16:3}$
Glucose	tr[b]	6.9	–	74.3	–	7.7	8.8	–	tr[b]	1.6	–	–	–
Fructose	0.5	12.6	–	70.8	–	5.7	8.5	–	0.3	1.6	–	–	–
Decanoic acid	5.3	52.3	–	42.3	–	–	–	–	–	–	–	–	–
Linoleic acid	5.6	38.9	–	22.7	–	–	15.9	–	–	–	16.9	–	–
Glycerol	1.7	21.4	–	63.6	–	3.8	8.6	–	0.1	0.8	–	–	–
Coconut fatty acids	3.8	38.1	–	37.8	–	18.1	1.0	–	1.1	–	–	–	–
Oleic acid	4.4	33.5	–	32.2	–	14.4	tr[b]	–	–	15.5	tr[b]	–	–
Tall oil fatty acids	4.1	22.9	5.2	26.6	–	5.9	11.4	–	–	5.5	17.7	–	–
Linseed oil fatty acids	2.5	11.6	7.9	13.3	7.8	5.6	3.5	9.1	1.1	5.0	6.8	19.1	6.7

[a] Relative amounts of monomers in purified PHA (%, wt/wt), determined by GC and GC-MS of the respective 3-hydroxy fatty acid methyl esters after hydrolysis of the poly(HA_{MCL})s.
[b] Trace amounts (< 0.1%, wt/wt).

containing a 3-hydroxyalkanoate unit as the major component, with a carbon chain length equivalent to that of the growth substrate [27–32]. When *P. oleovorans* is grown on mixtures of *n*-octane and 1-octene, a polymer is synthesized containing unsaturated and saturated pendent groups in a ratio of up to 45:46 [28, 33]. Like PHB, the poly(HA_{MCL}) is fully isotactic [27, 28].

P. putida is also capable of synthesizing poly(HA_{MCL}). Feedstocks for this microorganism range from sugars to glycerol and fatty acids. All these biopolyesters consist of randomly ordered 3-hydroxy fatty acid-based comonomers, with a chain length varying from 6 to 16 carbon atoms depending on the chemical structure of the starting material (Table 1).

An example of the large variety of monomer structures present in poly(HA_{MCL}) is given in Fig. 2. Also different degrees of unsaturation in poly(HA_{MCL}) can be established relatively easily [3–5, 34–39]. For example, the compositional data in Table 1 for the repeat units show that about 16% of the mono-unsaturated double bonds are incorporated when oleic acid is used as feedstock. When tall oil fatty acids are used, over 40% of the subunits of the resulting poly(HA_{MCL}) are mono- or di-unsaturated, while the total degree of unsaturation of the alkyl side chains of linseed oil-based PHA is even higher (> 65%). Moreover, a substantial part (about 30%) of these unsaturated linseed oil-based poly(HA_{MCL}) subunits have up to three double bonds present.

Unsaturated groups are very interesting for application development because this specific functionality opens up a broad range of possibilities for further (chemical) modification of the polymer structure, and therefore its physical and material properties. The direct microbial incorporation of other functional substituents to the polymer side chains, e.g. epoxy-, hydroxy-, aromatic-, and halogen functional groups, influences the physical and material properties of poly(HA_{MCL}) even further [28, 33, 35, 39–41]. This features many possibilities to produce tailor-made polymers, depending on the essential material properties that are needed for the development of a specific application.

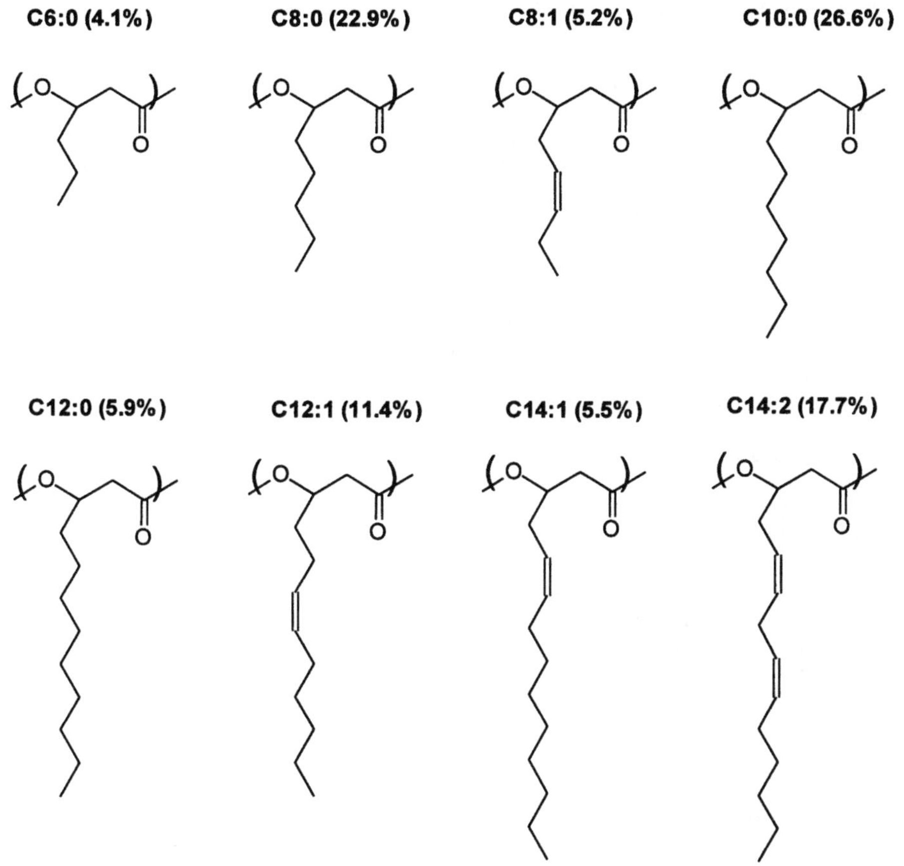

Fig. 2. Monomer units in medium chain length PHA, based on tall oil fatty acids

2.2
Molecular Weight and Molecular Weight Distribution

The molecular weight of the purified PHAs is a function of both biotechnological and down-stream processing conditions. The mechanisms that affect and determine the molecular weights of PHAs in the bacterial cells are not yet well understood. However, the molecular weight of the PHAs varies greatly depending on the microorganism and the feedstock used [21]. Also fermentation and cultivation conditions, such as the point during the growth cycle at which the cells are harvested and the production time seem, to play an important role in microbial polymer synthesis [42, 43].

The method of pure polymer recovery from the biomass prior to characterization can influence the molecular weight of the polymer significantly. Extraction of PHB-bacterial cells with organic solvents yields polymers with higher molecular weight compared to sodium hypochorite treatment [44–46]. Pretreatment of the biomass with a surfactant prior to hypochlorite digestion

Table 2. Polymer Properties of Some Saturated and Unsaturated Types of PHA

	M_w[a] (g mol^{-1})	M_w/M_n[b]	T_g (°C)[c]	T_m (°C)[d]
PHA from *A. eutrophus* [21, 52, 53]				
PHB	up to 1,400,000	1.9	9	177
Poly(3HB-*co*-3HV) (15 mol % 3HV)	982,000	2.5	-1	140
Poly(3HB-*co*-3HV) (28 mol % 3HV)	930,000	4.0	-8	102
PHA from *P. oleovorans* [30]				
grown on:				
n-hexane	330,000	1.8	-25.8	-
n-octane	178,000	1.8	-36.5	58.5
n-decane	225,000	2.0	-38.4	47.6
1-octene	242,000	2.4	-36.6	-
1-decene	260,000	2.2	-43.1	-
PHA from *P. putida* [4, 35, 36]				
grown on:				
coconut fatty acids	84,000	1.6	-43.7	42.4
oleic acid	112,000	1.6	-56.3	-
tall oil fatty acids	56,000	1.7	-59.8	-
linseed oil fatty acids	97,000	2.3	-61.7	-

[a] M_w = weight average molecular weight, determined by GPC.
[b] M_n = number average molecular weight, determined by GPC; M_w/M_n = molecular weight distribution (polydispersity).
[c] T_g = glass transition temperature, determined by DSC.
[d] T_m = melt temperature, determined by DSC.

yields polymers of higher purity and higher molecular weight, although the results obtained by extraction are better still [47–49]. Other polymer recovery methods, like enzymatic digestion of the biomass or treatment of the cells by ammonia solution have been reported [50, 51]. However, due to its ease and reproduceability extraction of the biomass is generally performed to obtain the pure materials for characterization.

Weight average molecular weights (M_w) ranging from 600,000 to over one million g mol^{-1} have been reported for PHB (Table 2) [21, 52, 53], which is very high compared to synthetic polyesters. Furthermore, the polydispersity of PHB has been reported to be around 1.9. Copolymerization of 3HV with HB yielded polymers with a somewhat lower molecular weight. At the same time, the molecular weight distribution of the poly(3HB-*co*-3HV) copolymers is higher compared to that of PHB, up to 4.0, although being unimodal.

Weight average molecular weights of poly(HA_{MCL}) with saturated or unsaturated pendent groups are relatively low, compared to M_w's of poly(HA_{SCL}), and in the range of 60,000 to 360,000 g mol^{-1}, as depicted in Table 2 [4, 30, 35, 36]. Also for the poly(HA_{MCL}) copolymers, the molecular weight distributions are unimodal. Their polydispersities are in the range of 1.6–2.4, which is narrower than the polydispersity of poly(3HB-*co*-3HV) copolymers, and close to the theoretical value of 2.0 for synthetic polycondensates such as chemically synthesized polyesters [54].

2.3 Thermal Properties

Most PHAs are partially crystalline polymers and therefore their thermal and mechanical properties are usually represented in terms of the glass-to-rubber transition temperature (T_g) of the amorphous phase and the melting temperature (T_m) of the crystalline phase of the material [55]. The melting temperature and glass transition temperature of several saturated and unsaturated PHAs have been summarized in Table 2.

Poly(HA_{SCL}) (i.e. PHB and poly(3HB-*co*-3HV)) are typical thermoplastic polymers, which become fluid and moldable above their melting points, whereas poly(HA_{MCL})s generally are elastomers with low melting points. PHB has been reported to show a relatively high melting temperature of about 180 °C, while the T_g of this polymer is approximately 9 °C [21, 52, 53]. Incorporation of 3HV-comonomer units into PHB decreases both the T_m as well as the T_g significantly. Poly(3HB-*co*-3HV), containing up to 28 mol% 3HV, has a T_m of about 102 °C and a T_g of –8 °C, respectively [21, 52, 53].

It has been stated that the lower T_m makes poly(3HB-*co*-3HV) series a more versatile range of materials than the homopolymer PHB, since a major disadvantage of PHB is its thermal instability in the melt [18]. Changes in molecular weight and the formation of crotonic acid have been reported at temperatures above 170 °C [56–58]. A drop in molecular weight to about half of its original value was reported when a sample of PHB was kept at 190 °C for one hour [59]. Thermogravimetric analysis showed complete weight loss in a single step between 225 and 300 °C corresponding to a quantitative conversion of the polymer to crotonic acid. The dominating degradation reaction is a β-elimination reaction, the kinetics of which fit a random scission model [58, 60]. In principle, this reaction even occurs under ambient conditions and its significance increases with temperature.

Although the decomposition rate of PHB is considerable at its melting temperature and melt-processing of pure PHB therefore may be critical, it is not considered impracticable, provided that both melt temperature and melt residence time are minimized. The incorporation of 3HV units decreases the melting point, whereas the thermal stability is not affected. This allows thermal processing of poly(3HB-*co*-3HV) as a melt, without the risk for thermal degradation. Consequently, processing will become less critical with increasing 3HV content. Therefore, the processability window of poly(3HB-*co*-3HV) is widened considerably, compared to that of the homopolymer PHB [18].

Poly(HA_{MCL})s are highly amorphous polymers with a glass transition temperature well below room temperature (T_g of –62 to –26 °C) and a relatively low melting point (42–58 °C) [4, 30, 35, 36]. As a consequence these materials can be classified as typical elastomers. The T_g values of poly(HA_{MCL}) decreases considerably as the average length of the side-chain groups increases, due to the increase in mobility of the polymer chains. From Table 2, a decrease in the T_g value of almost 20 °C is observed for poly(HA_{MCL}) produced from coconut fatty acids compared to that produced from linseed oil fatty acids. This corresponds to the increase in the average length of the predominant side-chain units, as shown in Table 1 of the previous section. An increase in the variety of side-chains within one polymer chain hampers its ability to crystallize. Therefore, poly(HA_{MCL}) crystallizes very slowly and some copolymers do not crystallize at all. For the latter, no T_m is observed (Table 2).

Because of its thermal nature, poly(HA_{MCL}) will become soft and lose its coherence even at relatively low temperatures. For many applications, this will make the use of common melt processing techniques such as injection molding and film blowing difficult [18]. Although some researchers have reported on the addition of nucleating agents such as talc or boron nitride in order to improve the applicability of conventional melt processing techniques, due to the low crystallization rates of these polymers, no significant improvement was found [18, 61].

2.4 Crystallinity and Morphology

PHB is an optically active polyester with completely linear chains, forming a compact right-handed helix with a twofold screw axis and a fiber repeat of 0.596 nm [62]. As a result of its stereochemical *R*-configuration, PHB can achieve a relatively high degree of crystallinity, usually between 55–80% [26]. PHB forms extremely thin lamellar crystals that are organized in oblong lath-like single crystals when grown from dilute solution, or in spherulites when it is grown from the melt [59]. The lamellar thicknesses in PHB spherulites are typically around 5 nm, and in crystals grown from solution they can be even smaller [59].

It has been reported that the overall rate of crystallization of pure PHB is relatively low compared to that of common synthetic polymers, showing a maximum in the temperature range of 55–60 °C [23]. The spherulite growth rate kinetics have been evaluated [59] in terms of the theory by Hoffmann et al. [63]. At about 90 °C, the spherulite growth rate displayed a maximum, which is not excessively low compared to that of common synthetic polymers. Therefore it was stated that the low overall crystallization rate of PHB centers on the nucleation process rather than the subsequent crystal growth. Indeed, it has been shown that PHB has an exceptionally low level of heterogeneous nuclei [18].

To minimize cycle times during melt processing of PHB, the initially low nucleation density was increased by the addition of nucleation agents such as boron nitride, saccharine, or ammonium chloride [23, 59, 64]. It was reported that adding nucleating agents not only reduced the processing cycle times, but also improved the mechanical properties to some extent by reducing the average spherulite size [18].

Copolyesters of poly(3HB-*co*-3HV) have approximately the same degree of crystallinity as the homopolymer PHB and all copolymers show similar conformation characteristics as those observed for PHB [24, 26, 65]. They show a minimum in their melting point versus composition curve at a 3HV content of approximately 40 mol%. The apparent ability of the two different monomeric units to cocrystallize might result from the fact that the copolymers are prone to show isodimorphic behavior [21, 26, 66–70]. However, the considerable reduction of the heat of fusion upon 3HV inclusion, as reported by Bluhm et al.

[26], clearly indicates that the 3HV units have to be regarded as energetically unfavorable defects in the PHB crystal structure [65, 71]. Therefore, if the equilibrium is approached by using higher crystallization temperatures, the expectation is that 3HV units are partially excluded from the crystalline phase. The concomitant variations in density and melting point were indeed observed [71].

According to Hoffman's crystallization theory, a drop in the heat of fusion corresponds to an exponential decrease in nucleation and crystal growth rates [63]. Implicitly, the rate of crystallization is severely retarded by the presence of 3HV comonomer [64, 69, 72]. These low crystallization rates can hamper the melt processing of these copolymers since they necessitate longer processing cycle times.

Since an increase in the variety of side-chains within one polymer chain hampers its ability to crystallize, some distinct differences in crystallinity for poly(HA_{MCL})s were found [29, 30]. Although the saturated poly(HA_{MCL})s are able to crystallize due to their isotactic configuration, they show a much lower degree of crystallinity than PHB or poly(3HB-*co*-3HV) due to their low crystallization rates. Other poly(HA_{MCL}) copolymers do not crystallize at all due to the presence of functional groups which disorder the regularity of the polymer structure and prevent crystallization. All PHAs, either having short or medium chain lengths, ultimately seem to crystallize in a similarly layered packing order, forming thin lamellar crystals [53].

Surprisingly, PHA granules are completely amorphous in vivo [73, 74]. Some authors suggested that the presence of water and/or lipids would prevent crystallization of the nascent polymer by acting as plasticizers. In addition, some reports ascribed the amorphous character of the PHA granules found in the microbes to the presence of some other, not defined, agents [75–79]. However, this has been contradicted by others. According to these studies, it is simply the result of the slow nucleation kinetics that are operative for small particles, implying that the PHA granules in the cells are simply too small to be able to crystallize with a viable time span [80, 81].

2.5
Mechanical Properties and Performance in Comparison with other Polymers

The mechanical properties of PHB and its copolymers have been studied extensively [82, 83]. As shown in Table 3, the material properties of PHAs can be readily controlled by adjusting the polymer composition during the fermentation [84].

The mass fraction crystallinity of molded PHB samples is typically around 60%. As shown in Table 3, PHB resembles isotactic polypropylene (iPP) with respect to melting temperature (175–180 °C), Young's modulus (3.5–4 GPa) and the tensile strength (40 MPa). In addition, the crystallinity of iPP is approximately 65% [18]. Accordingly, the fracture behavior of PHB may be anticipated to be tough at room temperature. Molded PHB samples do indeed show ductile behavior, but over a period of several days at ambient conditions, they slowly become more brittle [82, 85, 86]. Consequently, the elongation to break of the ultimate PHB (3–8%) is markedly lower than that of iPP (400%).

Table 3. Mechanical Properties of Various PHAs Compared to Conventional Industrial Polymers

Polymer	Melting temper-ature (°C)	Young's modulus (GPa)	Tensile strength (MPa)	Elongation to break (%)	Notched izod impact strength (J/m)
PHB [21,88,89]	175–180	3.5–4	40	3–8	35–60
Poly(3HB-*co*-3HV) [21,86]					
3 mol % 3HV	170	2.9	38	–[a]	60
9 mol % 3HV	162	1.9	37	–	95
14 mol % 3HV	150	1.5	35	–	120
20 mol % 3HV	145	1.2	32	50–100	200–350
25 mol % 3HV	137	0.7	30	–	400
Poly(3HB-*co*-4HV) [21,85,90]					
3 mol % 4HB	166	–	28	45	–
10 mol % 4HB	159	–	24	242	–
16 mol % 4HB	–	–	26	444	–
64 mol % 4HB	50	30	17	591	–
90 mol % 4HB	50	100	65	1080	–
100 mol % 4HB	53	149	104	1000	–
Polyhydroxyoctanoate (PHO) [61]	61	–	6–10	300–450	–
Poly(HA_{MCL}) grown on oconut fatty acids [4,35,36]	40–55	–	0.5–8	50–300	–
Conventional polymers [88,91]					
Isotactic Polypropylene (iPP)	170–176	1.0–1.7	29.3–38.6	500–900	45
High density polyethylene (HDPE)	112–132	0.4–1.0	17.9–33.1	12–700	32
Low density polyethylene (LDPE)	88–100	0.05–0.1	15.2–78.6	150–600	>36
Polyethylene terephtalate (PET)	250–265	2.2–2.9	56–70	7300–100	240
Polystyrene (PS)	80–110	3.0–3.1	50	3–4	21
Nylon-6,6	265	2.8	83	60	–

[a] Data not available.

This unfavorable aging process is a major drawback for the commercial use of the PHB homopolymer. Reducing the spherulite radius by means of a nucleating agent did result in a minor increase of the maximum elongation by only a few percent, which is still insufficient for "tough" applications [18].

It was demonstrated that the considerable embrittlement of PHB is an intrinsic phenomenon and is most probably not related to the presence of additives or its orientation [80]. In contrast to earlier claims [86], it has been reported that physical aging hardly contributes to the embrittlement but that it should be ascribed to progressive crystallization that tightly constrains the amorphous phase between the crystals [87, 88].

It has been reported that annealing can dramatically improve the mechanical properties of PHB by changing its lamellar morphology while subsequent aging is prevented to a large extent [87, 88]. Annealing induces a process of melting and recrystallization, yielding a much coarser lamellar texture. This gives rise to an improved fracture behavior: short annealing of PHB at 150 °C increases the elongation to break to 30%, while the Young's modulus, and therefore the flexibility, is retained [87, 88].

Incorporation of 3HV or 4HB comonomers produces remarkable changes in the mechanical properties: the stiffness (Young's modulus, Table 3) decreases while the toughness (impact strength) increases with increasing fraction of the respective comonomer [21, 85]. Also a decrease in tensile strength is observed when 3HV or 4HB units are incorporated. PHB-4HB copolymers containing more than 40 mol% 4HB units exhibit the mechanical properties of an elastic rubber (Table 3) [89]. These copolymers show some embrittlement as well [86], but the ultimate maximum elongation is considerably higher compared to PHB (Table 3). As for PHB, annealing of poly(3HB-*co*-3HV) did significantly increase the toughness of the materials, even after storage of the polymers [18].

In order to improve the mechanical properties of PHB or poly(3HB-*co*-3HV), many have reported on blending these biopolymers with other, both degradable as well as non-degradable, materials. However, due to the lack in compatibility between most polymers no substantial improvements in mechanical properties were reported upon, up to now [90].

Poly(HA_{MCL}), unlike PHB or its copolymers, behave as elastomers with crystals acting as physical crosslinks and therefore can be regarded as a class of its own with respect to mechanical properties [53, 61, 91]. Elongation to break of 250–350% has been reported, and a Young's modulus up to 17 MPa [53, 91].

3
Post-Biosynthetic Modifications

When grown on carbon sources having functional groups, these functionalities are often incorporated as such into the side chains during microbial PHA synthesis. A broad range of functional groups have been introduced into the bulk of the polymers this way, including both terminal and non-terminal olefin groups [3, 4, 28, 33, 38, 39, 92], methyl branches [92–94], esterified carboxyl groups [95, 96] and hydroxy and epoxy groups [39]. Functional groups introduced in a terminal position only, include acetoxy [97], bromine [97, 98], chlorine [97, 99], fluorine [40, 100], phenyl [101, 102], cyclohexyl [103], cyano [103], phenoxy [104, 105], and *p*-cyano and *p*-nitrophenoxy groups [104].

In addition to this product versatility, post-biosynthetic modifications can be performed at the various PHAs to further alter the material properties of these polymers, compared to the original properties of their precursors. These reactions can be performed both in the bulk of the material or at its surface only. For instance, an enhanced chemical reactivity, the attachment of bioactive substances or the shift of the hydrophobicity/hydrophilicity balance could be achieved this way. Although the (chemical) modification of polymers is a very powerful tool in tailoring ultimate material properties, only a limited number of such adjustments have been reported until now.

As an example, bulk modification by the organic reaction of unsaturated PHA with sodium permanganate resulted in the incorporation of dihydroxyl or carboxyl functional groups [106]. Due to the steric hindrance of the isotactic pendant chains, complete conversion could not be obtained. However, the solubility of the modified polymers was altered in such a way that they were now completely soluble in acetone/water and water/bicarbonate mixtures, respectively [106]. Solubility can play an important role in certain applications, for instance in hydrogels. Considering the biosynthetic pathways, the dihydroxyl or carboxyl functional groups are very difficult to incorporate by microbial synthesis and therefore organic chemistry actually has an added value to biochemistry.

The quantitative bulk conversion of unsaturated functional groups in PHAs to epoxides has been achieved by reaction with *m*-chloroperbenzoic acid as the chemical reagent [107]. No chain scission of the macromolecular chain was observed. Epoxy-modified PHAs are chemically even more reactive than unsaturated PHAs and therefore could be useful in further chemical reactions (e.g. grafting of therapeutic important substances) [108].

Another way of modifying unsaturated PHAs in the bulk is by crosslinking of the material. This has been accomplished by either chemical reaction with sulfur or peroxides [109, 110], or by radiation curing [91, 111]. In all cases, crosslinking altered the ultimate material properties drastically, yielding a true rubbery material. The advantages of applying rubbers from crosslinked PHAs over the use of current rubbers will be elaborated in Sect. 4.5.

Apart from modifications in the bulk, also surface modification of PHAs has been reported. Poly(3HB-*co*-3HV) film surfaces have been subjected to plasma treatments, using various (mixtures of) gases, water or allyl alcohol [112–114]. Compared to the non-treated polymer samples, the wettability of the surface modified poly(3HB-*co*-3HV) was increased significantly [112–114]. This yielded a material with improved biocompatibility, which is imperative in the development of biomedical devices.

4 Recent Developments in Various Application Areas

4.1 Applications from Short Chain Length PHAs

Since PHB and poly(3HB-*co*-3HV) have already been available for several years to polymer and material scientists in sufficient quantities, product development from these polymers is more evolved at the moment than for applications based on poly(HA_{MCL}). The variation in physical and mechanical properties of the different PHA-types offers a wide range of applications. Since reports on applications based on poly(HA_{MCL}) are relatively scarce, new developments in this field are described in more detail in this chapter than for applications based on poly(HA_{SCL}).

Being a thermoplast, PHB and especially its derivatives are very suitable for polypropylene-type processing techniques such as molding. By introduction of other monomer units (e.g. 3-hydroxyvalerate, 3HV), the material properties and the processing properties of PHB are greatly enhanced. Therefore, many different applications including packaging, hygienic, agricultural and biomedical products have already been developed from poly(HA_{SCL}) (Fig. 3).

The most well known application of PHB and poly(3HB-*co*-3HV) is as substitute for conventional, non-biodegradable plastics used for packaging purposes and derived products [21, 115, 116]. Single-use bottles for shampoos, cosmetics and biodegradable motor-oil have been manufactured from these biopolyesters by common molding techniques. Containers and cups for food products were developed similarly, and bags have been produced from blown films of the material.

It is also possible to use PHB and poly(3HB-*co*-3HV) in the form of an aqueous latex, a dispersion of polymer granules in water, for coating fibrous materials such as paper or cardboard [115, 116]. From these PHA-coated materials paper cups and trays for holding foods and drinks were developed. Due to its outstanding water resistance, the hydrophobic poly(HA_{SCL}) coating protects the hydrophilic paper or cardboard against damages and deterioration caused by moisture of the packaged food or by the environment. A particular property of these films that could be of potential interest for its use in food packaging is their relatively low oxygen diffusivity [117]. However, due to a lack of particle coalescence of the polymer granules, the mechanical strength and coherence of the poly(HA_{SCL}) coatings were very poor when they were dried at room temperature. Only by additional treatment of the dried films (e.g. hot-pressing or solvent treatment), the coating properties could be improved significantly to yield stronger films [115, 116].

Various other disposable products have been developed from poly(HA_{SCL}). Using common techniques, fibers and non-woven fabrics have been manufactured. These materials would be very suitable for use in sanitary napkins and diapers [21, 115, 116]. Other typical single-use products include disposable razors and cutlery and were produced from molded poly(HA_{SCL}).

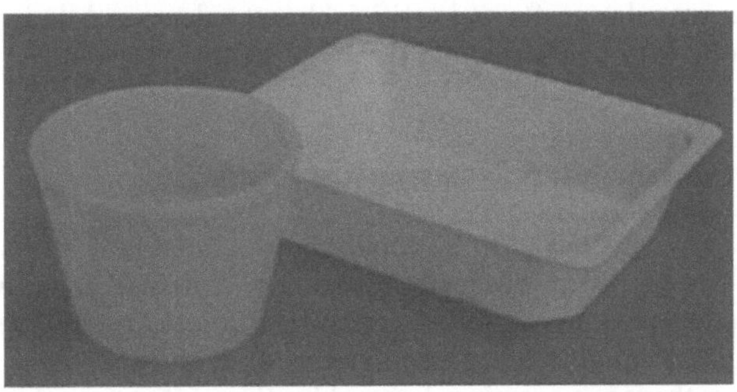

Fig. 3. Biodegradable flower pot and food container from molded biopolyesters

Other reported applications of poly(HA_{SCL}) are in agriculture, i.e. mulching films, biodegradable flower pots, one-season irrigation tubes and biodegradable matrices for the controlled release of plant growth factors (nutrients and fertilizers) or pesticides and herbicides [118]. Due to the very nature of the agricultural applications most probably there is no demand for highly purified polymers, which might facilitate the production process of the base material. Economically, this could be very interesting.

Hot melt adhesives based on poly(3HB-*co*-3HV) have also been described [119]. Hot melts are commonly used in bookbinding, bag ending and case and carton sealing and are mostly based on synthetic materials such as polyethylene, polypropylene ethylene-vinyl acetate and styrene block copolymers [119]. Hot melts based on PHAs alleviate the dependence on petroleum based materials and allow the development of biodegradable alternatives based on natural raw materials.

Very attractive, both from a technical and an economical point of view, are biomedical and related applications. PHB is biocompatible with mammalian tissue and is resorbed at a slow rate. When implanted in the human body it is hydrolyzed to naturally occurring mammalian metabolites [21, 117]. Due to these properties poly(HA_{SCL}) may find biomedical applications, such as in multifilament surgical sutures, wound dressings, pericardial substitutes and slow release drug delivery systems [120, 121]. Hydrolyzed PHAs could be a feedstock for the stereospecific drug industry [122, 123]. Also the application of poly(HA_{SCL}) in biomedical related products such as surgical swabs, lubricants for surgeon's gloves, blister packs and strips, and disposable syringes has been proposed.

PHB has been claimed to have piezo-electric properties similar to those of natural bone, giving it potential as biodegradable fixative plates that could actually stimulate bone formation and consequently promote the healing of the patient [117]. Furthermore, PHB has been used to produce non-woven patches for pericardium repair following open-heart surgery.

The development of tissue engineering scaffolds from poly(HA_{SCL}) has recently been evaluated [114]. Tissue engineering scaffolds are designed to provide a temporary three-dimensional support to the engineered tissue and must be at least biocompatible. It should first support cell growth and then degrade away, leaving viable tissue. poly(3HB-*co*-3HV) of biomedical grade has been processed into vascular grafts and artificial heart valves, and their performance is expected to be very promising [114].

4.2
Pressure Sensitive Adhesives from Unsaturated Poly(HA_{MCL})

A product is only considered to be totally biodegradable if all its single components can be degraded naturally. Currently, pressure sensitive adhesives (PSA) are mostly based on non-biodegradable synthetic polymers such as polyacrylates, ethylene-vinyl acetate copolymers and styrene block copolymers [124]. Therefore there is a growing demand for the application of biodegradable PSAs on naturally degradable products like paper and cardboard.

PSA compositions must have a good balance of different properties, depending on its specific application. PSAs for adhesive tapes for instance, must have a good adhesion, cohesion, stretchiness and elasticity. Furthermore the base polymers must be elastomers at room temperature [124]. Due to its elastomeric behavior and biodegradability, poly(HA_{MCL}) seems to be a promising base polymer for the development of a completely biodegradable PSA.

Babu et al. described the development of a biodegradable PSA formulation based on poly(HA_{MCL}) [124]. Different PHAs were tested, produced by cultivating *P. oleovorans* on octanoic acid (PHO) or decanoic acid (PHN), mixtures of octanoic and nonanoic acids (PHON) or mixtures of octanoic and 11-undecenoic acids (PHOU). Addition of different types and amounts of tackifiers to the various PHAs improved the tack of the PSA formulations. The strength properties of the PSAs were increased by UV-radiation crosslinking using a photosensitizer. Of the different poly(HA_{MCL})s tested PHN, PHON and PHOU gave PSAs with good properties. Biodegradation studies indicated that the (partially crosslinked) PSA formulations were still biodegradable [124].

4.3 High Solid Alkyd-Like Paints from Unsaturated poly(HA_{MCL})

In the manufacture of coatings and paints, resins consisting of synthetic polymers such as polyacrylates, polyurethanes and acid- or epoxy-functionalized polyesters are commonly used for many different applications, e.g. in primers, automotive coatings, printing inks, and for home decoration. Frequently, vegetable oil-based binders are also being used for this purpose. These vegetable oils are polycondensated with polyols and poly(carboxylic acids) to yield alkyd resins which are applied as fast-drying paints. It is estimated that about 1 million metric tons of the total world production of vegetable oils, approximately 63 million metric tons in 1992, nowadays is used in the manufacture of lubricants and coatings [125].

Most of the aforementioned binders in paints are diluted with organic solvents to afford an optimal viscosity, applicability, performance and stability to the final paint formulation. However, the use of these volatile organic solvents (VOCs) in paints has several important disadvantages. First, most of the applied organic solvents are highly inflammable. Furthermore, the emission of VOCs is believed to have a strong negative effect on the environment, since VOCs catalyze the photochemical synthesis of ozone and smog. Another drawback of the emission of VOCs is their huge impact on humans when exposed to organic solvents for prolonged periods, which is assumed to cause the psycho-neurological disease known as Organo Psycho Syndrome (OPS). Furthermore, there is evidence that the fertility of humans is most probably affected by exposure to organic solvent vapors. Therefore, a substantial reduction in the use of organic solvents in paints and coatings is urgently required [126].

Decreasing the viscosity of the currently applied synthetic alkyd resins reduces the amount of organic solvent that is needed in these paints for optimal performance. This could either be accomplished by decreasing the molecular weight of the applied alkyd resin, or by using polymers having a narrower molecular weight distribution [127, 128]. These conditions could be fulfilled by the application of poly(HA_{MCL}) as the polymer binder in paints.

Recently, the development of environmentally friendly binders produced from renewable agricultural resources, e.g. linseed and tall oil fatty acids, has been described [36]. These new poly(HA_{MCL}) resins were applied in high solid alkyd-like coatings and paints.

Coating properties and performance of pigmented poly(HA_{MCL}) alkyd-like paints. Poly(HA_{MCL}) pigmented binder solutions (30 to 70% of solid dry weight in organic solvent) have been applied to various substrates (Table 4). The different PHA coatings showed distinct drying characteristics depending on the type of PHA that was used in the formulation of the paints. Pigmented PHA coatings based on tall oil fatty acids (TOFA) were no longer tacky after curing for 40 hours at 50°C. For comparable PHA paints based on linseed oil fatty acids (LOFA), a drying time of only 16 hours was sufficient to yield tack-free coatings under the same conditions due to the relatively higher degree of unsaturation in the LOFA-PHA alkyl side chains. Using UV radiation the curing process of the wet films was accelerated to a mere 30 minutes (Table 4). The time of drying was independent of the type of substrate on which the coating was applied.

The initial drying of currently applied alkyd paints is accomplished by evaporation of solvent (physical drying). Subsequently, the eventual curing of the alkyd paint is completed by the formation of a polymer network, which is mainly formed by chemical crosslinks (oxidative drying) but in some cases also physical interactions between the fatty acid side chains occur, such as crystallization or proton-bridge formation [129]. Efficient network formation is crucial in the formation of dry films with good mechanical properties. Due to the presence of unsaturated units in the investigated LOFA- and TOFA-PHA bin-

Table 4. Formulation and Curing Conditions of Poly(HA_{MCL}) Alkyd Paints

No.	Substrate[a]	PHA-type[b]	PVC[c]	Pigment	Wet film thickness[d]	Curing conditions
1	glass	TOFA	0.25	TiO_2	90	UV, RT, 30 min
2	glass	TOFA	0.25	TiO_2	60	Oven, 50 °C 40 hours
3	glass	TOFA	0.25	TiO_2	120	Oven, 50 °C, 40 hours
4	glass	LOFA	0.25	TiO_2	60	Oven, 50 °C, 16 hours
5	glass	LOFA	0.25	TiO_2	120	Oven, 50 °C, 16 hours
6	glass	LOFA	0.15	Iriodin	60	Oven, 50 °C, 16 hours
7	glass	LOFA	0.15	Iriodin	120	Oven, 50 °C, 16 hours
8	glass	LOFA	0.25	Iriodin	60	Oven, 50 °C, 16 hours
9	glass	LOFA	0.25	Iriodin	120	Oven, 50 °C, 16 hours
10	PET	TOFA	0.25	TiO_2	90	Oven, 50 °C, 40 hours
11	PET	LOFA	0.25	TiO_2	90	Oven, 50 °C, 16 hours
12	PET	LOFA	0.25	Iriodin	90	Oven, 50 °C, 16 hours

[a] Glass = glass panels, PET = acrylate coated poly(ethylene terphthalate) transparencies
[b] PHAs based on tall oil fatty acids (TOFA) and linseed oil fatty acids (LOFA).
[c] Pigment volume concentration.
[d] Wet film thickness in μm.

ders, the drying and curing mechanism of the paints in this study will be a combination of physical and chemical processes, as described for conventional alkyd paints.

Recently, the application of a totally saturated biopolyester with short aliphatic side chains as an additive (15% w/w poly(3HB-*co*-3HV)) to conventional synthetic binders in paint formulations has been reported, in order to improve the curing and drying properties of these paints [130, 131]. The addition of poly(3HB-*co*-3HV) increased the speed of curing of the paints, and resulted in coatings that were faster tack-free, due to the specific physical properties of this copolyester. It was assumed that, due to the rapid crystallization of poly(3HB-*co*-3HV) in the presence of the conventional binder, "quasi-crosslinking" of the coating occurred, which enhanced the physical drying of the paint [130, 131]. However, it is not possible to accomplish crosslinking by oxidative processes in this type of saturated material. In contrast to the poly(HA_{MCL})s in this study, cohesive film formation of poly(3HB-*co*-3HV) will not occur when applied solely as the polymer binder in drying paint systems.

After curing of the various poly(HA_{MCL}) coatings smooth and homogeneous films were obtained, indicating a good leveling of the paints, with a dry film thickness in the range of 30 to 50 μm. The dispersion of commercially available inorganic pigments in the miscellaneous PHA paints was excellent. Depending on the pigment volume ratio and the type of pigment that was used, this resulted in high-gloss coatings (Table 5).

No significant variation in the color coordinates measured at different spots of the investigated PHA coatings was observed (Table 5). This demonstrated clearly that all coatings were uniform in color and that the different pigments

Table 5. Coating Properties of Pigmented Poly(HA_{MCL}) Alkyd Paints

No.[a]	Dry film thickness[b]	Hardness[c]	Adhesion[d]	Gloss[e]		
				20°	60°	85°
1	30	14		79.9	87.5	96.8
2	30	20	3.1	76.9	88.0	93.5
3	50	16		75.6	88.7	95.8
4	30	36	3.2	71.7	85.9	94.8
5	50	33		72.8	86.2	96.3
6	30	41	2.9	51.9	83.3	81.0
7	50	42		63.6	85.8	87.9
8	30	33		6.8	17.7	8.2
9	50	47		6.5	16.5	7.7
10	30	23	0.6	76.8	89.6	95.0
11	30	33	0.7	45.2	70.4	83.3
12	30	47		4.4	12.1	4.5

[a] Sample numbers correspond to the samples described in Table 4.
[b] Dry film thickness in μm.
[c] König hardness, number of deflections.
[d] Adhesion in N/mm^2.
[e] Gloss measured at three different angles.

used in the formulation of the paints were distributed homogeneously in the coatings.

The adhesion of the various investigated PHA coatings on glass substrates was very good, with a value of about 2.9–3.2 N/mm^2 for the force of adhesion (Table 5). A similar force of adhesion was observed for PHA coatings applied to aluminium test panels. A reference commercial powder coating, based on an acid functionalised synthetic polyester resin with the trade name Uralac, showed an adhesive force of 3.0 N/mm^2 to glass or aluminum, which is comparable to the adhesion of PHA coatings to these same substrates. On the other hand, the adhesion of the PHA paints to a PET surface was somewhat less satisfactory.

The flexibility of the pigmented TOFA and LOFA-PHA coatings was tested on two different types of substrate: flexible PET transparencies and transformable aluminium test panels. Heavy bending of the coated PET samples did not initiate cracking or delamination of the investigated coatings due to the rubbery character of the PHA binders. Furthermore, the PHA coatings showed an excellent resistance to chipping, as demonstrated by the conical mandrel test (Fig. 4). In contrast to a 85-μm thick reference Uralac coating, which showed the formation of cracks at 2 cm, no microscopically small cracks could be detected at the small end of the aluminium test cones when the PHA-coated mandrels were subjected to a bend test.

In conclusion, the application of homogeneously pigmented poly(HA$_{MCL}$) resins with a high solids content resulted in high-gloss, smooth and strong films upon curing. Furthermore, they showed an excellent flexibility, a good adhesion to different substrates, cohesive film properties and resistance to chipping. In

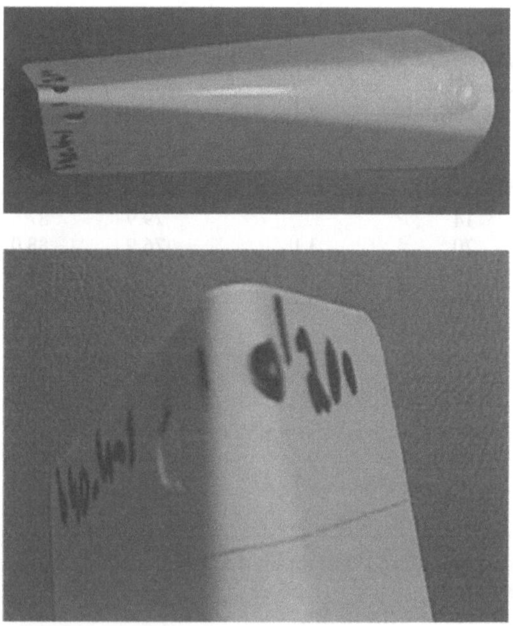

a

b

Fig. 4. Flexibility of poly(HA$_{MCL}$) alkyd paints, visualized by conical mandrel bending test

contrast to the commercial synthetic alkyd resins, the poly(HA_{MCL}) binders are produced biologically from relatively cheap renewable resources such as vegetable oils and fatty acids. Since there is a wide variation of such natural raw materials available, all with their own specific fatty acid composition, many different poly(HA_{MCL})s can be manufactured. This way, the polymer properties and therefore also the coating performance of poly(HA_{MCL}) based paint systems can be tailored easily, depending on the coating specifications that have to be met.

4.4 Biodegradable Cheese Coatings Based on Poly(HA_{MCL}) Latex

At present, during cheese making many cheeses are coated before or after the actual ripening and/or storage of the cheeses. This protects the cheeses from microbial attack and mechanical damage [132, 133]. Cheeses are generally coated by a non-biodegradable, synthetic plastic based latex, typically a copolymer of polyvinyl acetate and dibutyl maleic acid [134]. This material is better known as a wood adhesive. Because cheese rinds are mostly disposed of as kitchen and garden waste fractions, and therefore cannot be recovered or recycled, they should be biodegradable to avoid a build-up of waste.

Although alternative cheese coatings have been developed based on natural materials (e.g. starch and proteins) they were not commercialized, mainly because of technical reasons [133]. In the search for a new biodegradable cheese coating formula, the development of products based on poly(HA_{MCL}) latex was started. The fatty acid-based PHAs are completely biodegradable, while the physical properties of these materials (e.g. thermal properties, mechanical strength, water resistance and flexibility) were very likely to meet the technical demands. As a further advantage of the new PHA-based cheese coatings, being manufactured from hydrophobic polymers, the water permeability can be tailored to optimize ripening conditions and to protect against mold growth. Conventional cheese coatings have a fixed permeability, necessitating rigorous storage and ripening conditions of the cheeses.

General consumer acceptance is swiftly moving towards utilizing less packaging materials and to environmentally friendly, biodegradable materials. This also applies to traditional industries like the dairy industry, although the economics of the products are one of the most important factors. It was estimated that, using PHA-based cheese coatings, the required ripening and storage time would be reduced by at least 10% compared to conventionally coated cheeses [16]. Since cheese ripening times range from about six weeks for younger cheeses to up to three years for very old (Dutch) cheeses, this is a significant saving in time and will therefore be very profitable to the cheese sector. Furthermore, due to the biodegradability of the new cheese coating no landfilling or incineration of the waste will be necessary. This would save an estimated 20 million Euro a year [16].

The technical demands for a cheese coating are very comprehensive since it has to fulfill a large number of functions [132, 135]. First of all, it has to offer both mechanical and hygienic protection over a long period of time. Therefore, the coating should be homogeneous and crack free and adhere well to the wet and greasy cheese surface. After drying of the coating, the surface must be tack free to prevent sticking to boards or other cheeses during transportation. The coating should be semi permeable for water, but CO_2 and certain other flavoring components should also be able to diffuse through the coating in order not to affect maturing and taste development of the cheese. On the other hand, coating components should not diffuse into the cheese. Finally, the coating should be easy to apply, by hand as well as mechanically, and it has to be easy to rinse equipment of remaining coating after use. In addition, the coating itself should be stable for an extended period of time in order to be able to guarantee an acceptable shelf-life to the customer.

The new biodegradable cheese coatings (Fig. 5) are produced from poly (HA_{MCL}) latex based on saturated fatty acids and are easily composted.

An extensive test program showed that the functional aspects of the PHA-based cheese coatings, like ripening control and mechanical and bacterial protection, are equivalent to the current generation of plastic coatings. In Fig. 6, a comparison of the percentage average weight loss of Dutch Gouda cheeses as a function of time is presented. For this specific coating formulation, no significant differences in weight loss were observed compared to a commercial plastic cheese coating. Obviously, since it is in direct contact with food the safety of this material has to be examined first before commercializing this new cheese coating. Although the food contact approval of poly(HA_{MCL}) has not been assigned yet, Biopol has already been approved by both the Scientific Committee for

Fig. 5. Dutch Gouda cheeses coated with poly(HA_{MCL}) latex

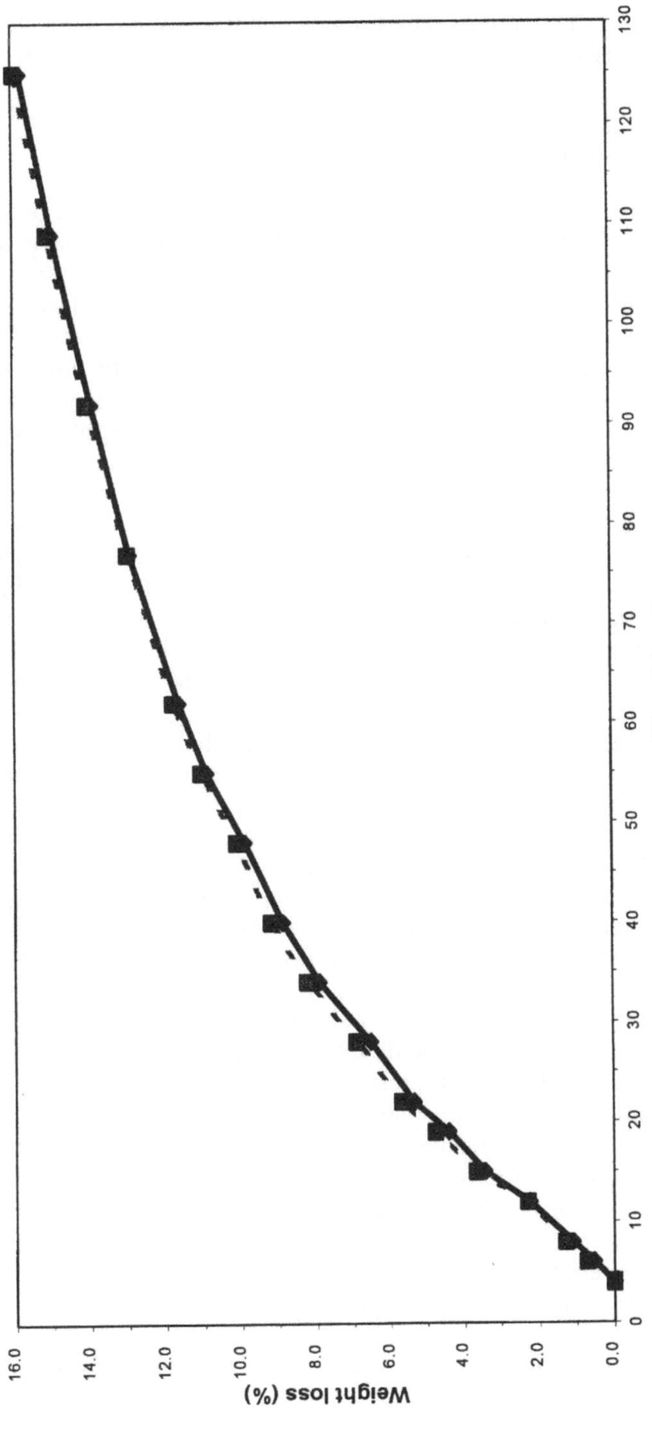

Fig. 6. Comparison of the percentage average weight loss of Dutch Gouda cheeses treated with either a poly(HA_{MCL}) latex coating (*solid line*, 12 cheeses in total) or a commercial cheese coating (*dashed line*, 6 cheeses) as a function of time

Food in the EU as well as by the Ministry of Health and Welfare in Japan to be used in food contact applications. Therefore, it is not expected that a clearance for the use of poly(HA_{MCL}) latex in cheese coatings will be a bottleneck.

4.5 Biodegradable Rubbers from Crosslinked Poly(HA_{MCL})

Both natural rubber and synthetic rubbers are currently used in a large number of articles that wind up as waste at the end of the life cycle of these products. Many of these rubbers are used in applications that are difficult to recycle, such as consumables, diapers, carpet backings, construction molds, and surgical devices and gloves [136–139]. Re-using the rubbers in these products is very difficult and since they are not biodegradable they build up in the environment, resulting in an increase in waste. Biodegradable rubbers are not yet commercially available but these would be a good alternative for such applications.

In 1994, the worldwide consumption of rubber was approximately 14.5 million tons a year, of which about 40% consisted of natural rubber. Natural rubber is produced as latex by tropical rubber trees (*Hevea brasiliensis*). It is processed locally and therefore the quality of natural rubber fluctuates remarkably [140]. Due to increasing demand for rubbers, combined with a decreasing production capacity in Asia and a vast increase in labor costs, the price of natural rubber is still rising sharply. In 1990–1994, the average price of natural rubber was about 0.38 $/lb, while in 1996 it was already over 0.80 $/lb. The remaining 60% of the articles were manufactured from synthetic petroleum-based rubbers such as isoprene rubber, styrene-butadiene rubber, chloroprene rubber and polyurethanes. The quality of synthetic rubbers is constant, and their price varies between 2 and 5 US$ per kilogram [137–140].

In current research, biodegradable rubbers have been manufactured from unsaturated PHAs by crosslinking of the biopolyesters. This has been accomplished by either chemical reaction (with sulfur or peroxides [109, 110]), or by radiation curing (using UV or an electron-beam source [91, 111]). By choosing different types of starting material and varying the crosslinking conditions, material properties like mechanical strength, tear resistance, tensile set and flexibility of the biorubbers were readily adjusted [91, 109–111]. An increase of the Young's modulus of the crosslinked material by approximately 130%, and a 75% increase of the tensile strength, respectively, has been reported [111].

Crosslinking of the biopolyesters mainly occurs by intra- and intermolecular reactions at the side chains of the material. In contrast to other existing rubbers, rubbers based on these biopolyesters are intrinsically biodegradable because the polymer main chain is still susceptible to degradation after crosslinking. The crosslinked polymer was degraded completely by enzymatic hydrolysis and subsequent surface erosion, similar to the degradation of the non-modified PHA [91]. The rate of degradation is regulated by changing the crosslink density [91].

Recently, PHAs based on fatty acids have been crosslinked. By varying the type of fatty acid in the fermentation process, the type and amount of double bonds in the pendant chains was easily adjusted (Table 6). PHA based on coconut

Table 6. Total Amount of Double Bonds in Monomer Units of Poly(HA_{MCL})

	0	1	2	3
poly(HA_{MCL}) grown on:				
coconut fatty acids (COFA)	98.9	< 1	–	–
oleic acid (OA)	84.0	15.5	0.5	–
tall oil fatty acids (TOFA)	58.2	22.1	19.7	–
linseed oil fatty acids (LOFA)	34.1	24.6	21.4	19.9

fatty acids was almost totally saturated, whereas PHA from linseed oil fatty acids contained about 65% of unsaturated monomer (divided in 25% mono-, 21% di- and 20% tri-unsaturated double bonds). PHA based on oleic acid or tall oil fatty acids contained a total number of double bonds of 15% and 40%, respectively (Table 6). The biopolyesters were crosslinked chemically by both sulfur vulcanization (Fig. 7) or using peroxides (Fig. 8) to yield a rubbery material.

This can be accomplished either from the pure material or in a latex form, in which polymer granules are dispersed in water. The gel formation of peroxide crosslinked poly(HA_{MCL}), based on coconut fatty acids (COFA), oleic acid (OA), tall oil fatty acids (TOFA) or linseed oil fatty acids (LOFA) is presented in Fig. 8. For all PHAs, in 4 to 5 hours the plateau value in peroxide crosslinking with benzoyl peroxide at 90 °C is reached. LOFA-PHA, containing most double bonds (Table 6) and the longest alkyl-side chains (Table 1) has the highest gel formation when crosslinked with benzoyl peroxide, indicating that the crosslink density of LOFA-PHA is the highest. The degree of crosslinking of benzoyl peroxide treated COFA-PHA is the lowest, compared to the other crosslinked fatty acid-based PHAs. Depending on the type of starting material and crosslinking conditions, the material properties also diversify to a great extent. Tensile strengths varying from 0.1 – 5.6 MPa and an elongation of break in the range of 5 – 1200% have been found. Peroxide crosslinked materials were much more susceptible to tearing than sulfur vulcanized PHAs.

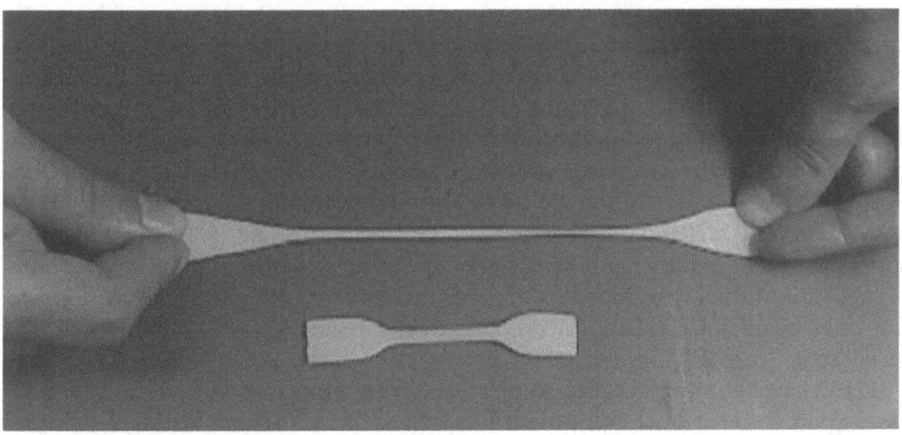

Fig. 7. Test specimens of vulcanized poly(HA_{MCL}), based on tall oil fatty acids (TOFA)

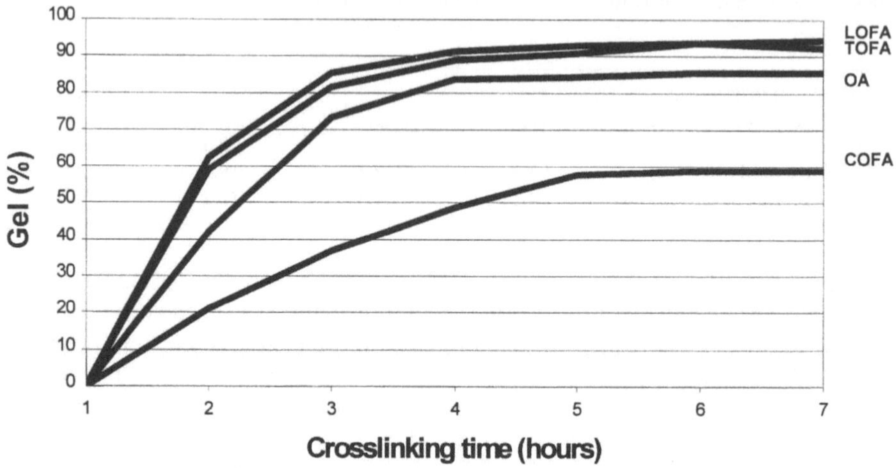

Fig. 8. Gel formation of peroxide crosslinked poly(HA_{MCL}), based on coconut fatty acids (*COFA*), oleic acid (*OA*), tall oil fatty acids (*TOFA*) or linseed oil fatty acids (*LOFA*)

Biologically degradable rubbers can be produced from relatively cheap renewable resources such as vegetable oils and fatty acids. Due to the abundance in fatty acid compositions, many different biorubbers can be manufactured, having their own specific properties. Based on these types of rubber, several commercially interesting applications, e.g. consumables, are now being developed.

5
Commercial Products: Current State and Future

In this section, it will be demonstrated that by using a single biotechnological production process, a wide variety of commercially available fatty acids can be converted into natural polymers with very divergent properties. The material properties of the biopolyesters are strongly related to the chemical characteristics (i.e. monomer composition) of the various polymers. Since the polyester structure can be tailored quite simply, the polymer properties can therefore be readily adjusted to meet the specific demands for a particular application. Properties such as flexibility can be modified by varying the length of the fatty acid chain. The melting temperature can also be controlled in a simple fashion: short chain length fatty acids result in final products with a higher melting temperature than those produced from medium chain length fatty acids. Moreover, the unsaturated poly(HA_{MCL})s are chemically reactive and completely amorphous.

PHAs can be manufactured to many different materials and shapes, by processing on conventional equipment for polyolefins or other plastics, e.g. injection molding, extrusion, film blowing and fiber-spray molding [141]. Furthermore, they can be processed in latex (granules in water), or in solution with several different solvents. Together with the material properties of PHAs this opens up a whole field of feasible commercial applications to be explored and exploited.

So far, commercial applications have been developed only for poly(3HB-*co*-3HV) (Biopol) by ICI [52, 142]. This material has been processed into bottles for hair care products (Wella) and biodegradable motor oil [84]. Moreover, various containers, disposable razors, and food trays for holding portions of fish and meat in the refrigerated section of supermarkets were all manufactured from Biopol and sold in Japan [84, 143]. In these cases, the rather expensive Biopol is used solely for its "green" image in order to increase product sales and to open up the market for other applications, which require biodegradability for functional reasons [18].

In general, due to their biodegradability, water resistance, and oxygen impermeability PHAs can be used for all sorts of biodegradable packaging materials, including composting bags and food packaging. Also the use of PHAs in single-use sanitary articles like diapers is considered as being economically feasible. Besides, in marine environments (fishing nets and other discarded objects that cause severe damage when made from non-degradable materials), construction materials (adhesives, laminates, foams and rubbers) and in agricultural industries there is a promising market potential for new biodegradable materials [144].

The potential for biomedical applications is very promising, since the added value to these special products is remarkably high [114, 117, 120, 121]. Although research in this field is of unique complexity, it is both technically and economically very compelling to succeed.

Recently, the development of environmentally friendly paints and coatings based on unsaturated poly(HA_{MCL}) has been reported [36]. This could have a significant potential, since organic solvents in DIY paints will be, and in some EU countries already are, further restricted by future legislation. Further studies are focussed on the application of poly(HA_{MCL}) latexes in totally organic solvent-free paints. The application of such water-borne paint systems is a promising perspective in further reducing the use of organic solvents in paints and coatings [36].

Technically, the prospects for PHAs are very promising. If the price of these materials can be further reduced, application of biopolyesters will also become economically very attractive. At the moment, a worldwide effort is being made to produce PHAs from major crop plants, such as corn, potatoes and rape seed [9, 145–148]. This should ultimately provide cheaper technology for PHA production, leading to the implementation of PHA-based products in everyday life.

References

1. Lemoigne M (1927) Ann Inst Pasteur (Paris) 41:148
2. Lemoigne M (1925) Ann Inst Pasteur (Paris) 39:144
3. Huijberts GNM, Eggink G, de Waard P, Huisman GW, Witholt B (1992) Appl and Environm Microbiol 58:536

4. Eggink G, van der Wal H, Huijberts GNM, de Waard P (1992) Industrial Crops and Products 1:157
5. Eggink G, de Waard P, Huijberts GNM (1992) FEMS Microbiology Reviews 103:159
6. de Waard P, van der Wal H, Huijberts GNM, Eggink G (1993) J Biol Chem 268:315
7. Huijberts GNM, de Rijk TC, de Waard P, Eggink G (1994) J Bacteriol 176:1661
8. Huijberts GNM, van der Wal H, Wilkinson C, Eggink G (1994) Biotech Techniques 8:187
9. Poirier Y, Nawrath C, Sommerville C (1995) Bio/Technology 13:142
10. Barak P, Coquet Y, Halbach TR, Molina JAE (1991) J Environ Qual 20:173
11. Gilmore DF, Antoun S, Lenz RW, Goodwin S, Austin R, Fuller RC (1992) J Ind Microbiol 10:199
12. Krupp LR, Jewell WJ (1992) Environ Sci Technol 26:193
13. Budwill K, Fedorak PM, Page WJ (1996) J of Environmental Polymer Degradation 4:91
14. Molitoris HP, Moss ST, de Koning GJM, Jendrossek D (1996) Appl Microbiol Biotechnol 46:570
15. Imam SH, Chen L, Gordon SH, Shogren RL, Weisleder D, Greene RV (1998) J of Environmental Polymer Degradation 6:91
16. Kim I, Ondrey G, Kamiya T (1998) Chem Eng July :43
17. Stein RS (1992) Proc Natl Acad Sci USA 89:835
18. de Koning G (1995) Can J Microbiol 41:303
19. Steinbüchel A, Hustede E, Liebergesell M, Pieper U, Timm A, Valentin H (1992) FEMS Microbiol Rev 103:217
20. Steinbüchel A, Valentin H (1995) FEMS Microbiol Rev 128:219
21. Doi Y (1990) Microbial Polyesters. VCH Publishers, Inc, NY
22. Doi Y (1995) Macromolecules 28:4822
23. Bloembergen S, Holden DA, Hamer GK, Bluhm TL, Marchessault RH (1986) Macromolecules 19:2865
24. Doi Y, Kunioka M, Nakamura Y, Soga K (1988) Makromol Chem 189:1077
25. Doi Y, Kunioka M, Nakamura Y, Soga K (1986) Macromolecules 19:2860
26. Bluhm TL, Hamer GK, Marchessault RH, Fyfe CA, Veregin RP (1986) Macromolecules 19:2871
27. de Smet MJ, Eggink G, Witholt B, Kingma J (1983) J Bacteriol 154:870
28. Lageveen RG, Huisman GW, Preusting H, Ketelaar P, Eggink G, Witholt B (1988) Appl Environm Microbiol 54:2924
29. Gross RA, DeMello C, Lenz RW, Brandl H, Fuller RC (1989) Macromolecules 22:1106
30. Preusting H, Nijenhuis A, Witholt B (1990) Macromolecules 23:4220
31. Brandl H, Gross RA, Lenz RW, Fuller RC (1988) Appl Environ Microbiol 54:1977
32. Haywood GW, Anderson AJ, Dawes EA (1989) Biotech Lett 11:471
33. Fritzsche K, Lenz RW, Fuller RC (1990) Int J Biol Macromol 12:85
34. Huijberts GNM, Eggink G (1996) Appl Microbiol Biotechnol 46:233
35. Casini E, de Rijk TC, de Waard P, Eggink G (1997) J of Environmental Polymer Degradation 5:153
36. van der Walle GAM, Buisman GJH, Weusthuis RA, Eggink G (1999) Int J Biol Macromol 25:123
37. Kim O, Gross RA, Rutherford DR (1995) Can J Microbiol 41:32
38. Kim YB, Rhee YH, Han S-H, Heo GS, Kim JS (1996) Macromolecules 29:3432
39. Eggink G, de Waard P, Huijberts GNM (1995) Can J Microbiol 41:14
40. Hori K, Soga K, Doi Y (1994) Biotech Lett 16:501
41. Lenz RW (1993) Advances in Polymer Science. Springer, Berlin Heidelberg New York
42. Shimizu H, Tamura S, Shioya S, Suga K-I (1993) J Ferm Bioeng 76:465
43. Birrer GA, Cromwick A-M, Gross RA (1994) Int J Biol macromol 16:265
44. Dawes EA, Senior PJ (1973) Adv Microb Physiol 10:135
45. Berger E, Ramsay BA, Ramsay JA, Chavarie C (1989) Biotechnol Tech 3:227
46. Hahn SK, Chang YK, Kim BS, Lee KM, Chang HN (1993) Biotechnol Tech 7:209
47. Ramsay JA, Berger E, Ramsay BA, Chavarie C (1990) Biotechnol Tech 4:221

48. Hahn SK, Chang YK, Kim BS, Chang HN (1994) Biotech Bioeng 44:256
49. Ramsay JA, Berger E, Voyer R, Chavarie C, Ramsay BA (1994) Biotechnol Tech 8:589
50. Holmes PA, Lim GB (1990) US Patent 4,910,145
51. Page WJ, Cornish A (1993) Appl Environ Microbiol 59:4236
52. Byrom D (1987) Tibtech 5:246
53. Marchessault RH, Monasterios CJ, Morin FG, Sundararajan PR (1990) Int J Biol Macromol 12:158
54. Hiemenz PC (1984) Polymer Chemistry. M Dekker Inc, NY
55. Anderson AJ, Dawes EA (1990) Microbiol Rev 54:450
56. Grassie N, Murray EJ, Holmes PA (1984) Polym Degrad Stab 6:47
57. Grassie N, Murray EJ, Holmes PA (1984) Polym Degrad Stab 6:95
58. Grassie N, Murray EJ, Holmes PA (1984) Polym Degrad Stab 6:127
59. Barham PJ, Keller A, Otun EL, Holmes PA (1984) J Mater Sci 19:2781
60. Lehrle R, Williams R, French C, Hammond T (1995) Macromolecules 25:4408
61. Gagnon KD, Lenz RW, Farris RJ, Fuller RC (1992) Macromolecules 25:3723
62. Cornibert J, Marchessault RH (1972) J Mol Biol 71:735
63. Hoffman JD, Davis GT, Lauritzen JI (1976) Treatise on Solid State Chemistry, Volume 3. Plenum Press, New York
64. Organ SJ, Barham PJ (1991) J Mater Sci 26:1368
65. Orts WJ, Romansky M, Guillet JE (1992) Macromolecules 25:949
66. Orts WJ, Marchessault RH, Bluhm TL (1991) Macromolecules 24:6435
67. Kamiya N, Sakurai M, Inoue I, Chujo R (1991) Macromolecules 24:3888
68. Yoshie N, Sakurai M, Inoue Y, Chujo R (1992) Macromolecules 25:2046
69. Scandola M, Ceccorulli G, Pizzoli M, Gazzano M (1992) Macromolecules 25:1405
70. Pearce RP, Marchessault RH (1994) Macromolecules 27:3869
71. Barker PA, Mason F, Barham PJ (1990) J Mater Sci 25:1952
72. Owen AJ, Heinzel J, Skrbic Z, Divjakovic V (1992) Polym Commun 33:1563
73. Calvert P (1992) Nature 360:535
74. Song JJ, Yoon SC, Yu SM, Lenz RW (1998) Int J Biol Macromol 23:165
75. Harrison STL, Chase HA, Amor SR, Bonthorne KM, Sanders JKm (1992) Int J Biol Macromol 12:112
76. Barnard GN, Sanders JKM (1988) J Biol Chem 264:241
77. Kawaguchi Y, Doi Y (1990) FEMS Microbiol Lett 70:151
78. Lauzier C, Marchessault RH, Smith P, Chanzy H (1992) Polymer 33:823
79. Hobbs JK, Barham PJ (1997) Polymer 38:3879
80. de Koning GJM, Lemstra PJ (1992) Polymer 33:3292
81. Horowitz DM, Clauss J, Hunter BK, Sanders JKM (1993) Nature 363:23
82. Scandola M, Cecorilli G, Doi Y (1990) Int J Biol Macromol 12
83. Nahamura S, Doi Y, Scandola M (1992) Macromolecules 25:3723
84. Lee SY (1996) Biotech Bioeng 49:1
85. Holmes P (1988) Biologically Produced (*R*)-3-hydroxyalkanoate Polymers and Copolymers. In: Development in Crystalline Polymers, Volume 2. Elsevier, London
86. Scandola M, Ceccorulli G, Pizzoli M (1989) Makromol Chem Rapid Commun 10:47
87. de Koning GJM, Lemstra PJ (1993) Polymer 34:4089
88. de Koning GJM, Scheeren AHC, Lemstra PJ, Peeters M, Reynaers H (1994) Polymer 35:4598
89. Kunioka M, Kawaguchi Y, Doi Y (1989) Polym Commun 29:174
90. Verhoogt H, Ramsay BA, Favis BD (1994) Polymer 35:5155
91. de Koning GJM, van Bilsen HMM, Lemstra PJ, Hazenberg W, Witholt B, Preusting H, van der Galiën JG, Schirmer A, Jendrossek D (1994) Polymer 35:2090
92. Choi MH, Yoon SC (1994) Appl Environm Microbiol 60:3245
93. Fritzsche K, Lenz RW, Fuller RC (1990) Int J Biol Macromol 12:92
94. Scholz C, Wolk S, Lenz RW, Fuller RC (1994) Macromolecules 27:6358
95. Scholz C, Fuller RC, Lenz RW (1994) Macromolecules 27:2886
96. Scholz C, Fuller RC, Lenz RW (1994) Macromol Chem Phys 195:1405

97. Lenz RW, Kim BW, Ulmer HW, Fritzsche K, Knee E, Fuller RC (1990) Novel Biodegradable Polymers. Kluwer Academic, Dordrecht, The Netherlands
98. Kim YB, Lenz RW, Fuller RC (1992) Macromolecules 25:1852
99. Doi Y, Abe C (1990) Macromolecules 23:3705
100. Abe C, Taima Y, Nakamura Y, Doi Y (1990) Polym Comm 31:404
101. Fritzsche K, Lenz RW, Fuller RC (1990) Makromol Chem 191:1957
102. Kim YB, Lenz RW, Fuller RC (1991) Macromolecules 24:5256
103. Lenz RW, Kim YB, Fuller RC (1992) FEMS Microbiol Rev 103:207
104. Kim OY, Gross RA, Rutherford DR (1995) Can J Microbiol 41:32
105. Ritter H, Vonspee AG (1994) Chem Phys 195:1665
106. Lee MY, Park WH, Lenz RW (1998) International Symposium on Biological Polyhydroxyalkanoates '98 :P77
107. Bear MM, Leboucher-Durant MA, Langlois V, Lenz RW, Goodwin S, Guerin P (1997) Reactive & Functional Pol 34:65
108. Dubois MC, Baltaze JP, Langlois V, Guerin P (1996) IUPAC Meeting on Macromolecules, Prague
109. Gagnon KD, Lenz RW, Farris RJ, Fuller RC (1994) Polymer 35:4358
110. Gagnon KD, Lenz RW, Farris RJ, Fuller RC (1994) Polymer 35:4368
111. Ashby RD, Cromwick A-M, Foglia TA (1998) Int J Biochem Macromol 23:61
112. Mas A, Jaaba H, Schue F, Belu AM, Kassis CM, Linton RW, Desimone JM (1997) JMS-Pure Appl Chem A34:67
113. Mas A, Jaaba H, Schue F, Belu AM, Kassis CM, Linton RW, Desimone JM (1997) Eur Pol J 33:331
114. Williams SF, Martin DP, Horowitz DM, Peoples OP (1999) Int J Biol Macromol 25:111
115. Lauzier CA, Monasterios CJ, Saracovan I, Marchessault RH, Ramsay BA (1993) Tappi Journal 76:71
116. Marchessault RH, Lepoutre PF, Peter E (1991) WO 91/13207
117. Lafferty RM, Korsatko B, Korstako W (1988) Biotechnology, Volume 6b. VCH Verlagsgesellschaft, Weinheim
118. White BG (1983) Brit Pat Appl 8221567
119. Kaufmann T, Brady FX, Puletti PP, Raykovitz G (1992) US 5,169,889
120. Hocking PJ, Marchessault RH (1994) Chemistry and Technology of Biodegradable Polymers, Chapter 4. Chapman and Hall, London
121. Duvernoy O, Maml T, Ramstrom J, Bowald S (1995) Thorac Cardiovasc Surg 43:271
122. Mueller HM, Seebach D (1993) Angew Chem 105:483
123. Coulombe S (1978) Macromolecules 11:279
124. Babu GN (1997) International symposium on bacterial polyhydroxyalkanoates :48
125. Derksen JTP, Cuperus FP, Kolster P (1995) Industrial Crops and Products 3:225
126. Wicks ZW, Jones FN, Pappas SP (1992) Organic Coatings: Science and Technology, Vol I, Chapter XV: Solvent Properties. John Wiley, New York
127. Holberg K (1987) High Solids Alkyd Resins. Marcel Dekker, New York
128. Hare CH (1994) Protective Coatings, Fundamentals of Chemistry and Composition. Techn Publishing Company, Lancaster, Basel
129. Wicks ZW, Jones FN, Pappas SP (1992) Organic Coatings: Science and Technology, Vol I, Chapter IX: Drying Oils. John Wiley, New York
130. Taylor PL UK Patent Application GB 2 291 648 (A)
131. Noda I (1997) Patent WO 97/07229
132. Fradin M (1989) Cheesemaking: Science and Technology. Lovoisier, New York
133. Kester JJ, Fennema OR (1986) Food Technology :47
134. Salame M Plastic Film Technology: High Barrier Plastic Films for Packaging. Technomic, Lancaster, Basel
135. Castle L, Kelly M, Gilbert J (1993) Food Additives and Contaminants 10:175
136. Hager T, MacArthur A, McIntyre D, Seeger R (1979) Rubber Chemistry and Technology 52:693
137. Brown RP (1987) Physical Testing of Rubber. Elsevier, Barking, Essex, England

138. Hirata Y (1996) Inform 7:506
139. Evans CW (1986) Practical Rubber Compounding and Processing. Elsevier, Barking, Essex, England
140. Buchanan RA, Cull IM, Otey FH, Russell CR (1978) Economic Botany 32:146
141. Zeneca (1993) Brochure: Biopol Resin: Nature's Plastic, The Natural Choice Zeneca BioProducts, Billingham, UK
142. Holmes PA (1985) Phys Technol 16:32
143. Byrom D (1994) Plastics From Microbes: Microbial Synthesis of Polymers and Polymer Precursors. Hansen, New York
144. (1995) Report by the study committee for the practical use of biodegradable plastics: The age of new plastics March 1995
145. Vanderleij FR, Witholt B (1995) Can J Microbiol 41:222
146. Nawrath C, Poirier Y, Somerville C (1995) Molec Breeding 1:105
147. Williams SF, Peoples OP (1996) Chemtech 26:38
148. see also the referring chapter of this book, and references herein

Received: March 2000

Microbial Degradation of Polyesters

Dieter Jendrossek

Institut für Mikrobiologie der Universität Stuttgart, Allmandring 31, 70569 Stuttgart, Germany
E-mail: dieter.jendrossek@po.uni-stuttgart.de

Dedicated on occasion of the 75th birthday of Hans Günter Schlegel in recognition of his outstanding ability to teach and to confer his fascination for microbiology to students.

Polyesters, such as microbially produced poly[(*R*)-3-hydroxybutyric acid] [poly(3HB)], other poly[(*R*)-hydroxyalkanoic acids] [poly(HA)] and related biosynthetic or chemosynthetic polyesters are a class of polymers that have potential applications as thermoplastic elastomers. In contrast to poly(ethylene) and similar polymers with saturated, non-functionalized carbon backbones, poly(HA) can be biodegraded to water, methane, and/or carbon dioxide. This review provides an overview of the microbiology, biochemistry and molecular biology of poly(HA) biodegradation. In particular, the properties of extracellular and intracellular poly(HA) hydrolyzing enzymes [poly(HA) depolymerases] are described.

Advances in Biochemical Engineering/
Biotechnology, Vol. 71
Managing Editor: Th. Scheper

1 Introduction

Poly[(*R*)-hydroxyalkanoic acids] [poly(HA)] are a class of bacterial storage compounds that are synthesized during unbalanced growth by many Gram-negative and Gram-positive bacteria. Poly(HA) are deposited intracellularly in form of inclusion bodies ("granules") to levels up to 90% of the cellular dry weight. For reviews see [1–11] and the chapters by Kim and Lenz, Steinbüchel and Hein, and Babel et al. in this book. Poly[(*R*)-3-hydroxybutyric acid] [poly(3HB)] was the first poly(HA) discovered by Lemoigne [12] and is the most abundant polyester in bacteria. However, in the last two decades more than 100 hydroxyalkanoic acids have been identified as constituents of poly(HA) (most of them summarized by Steinbüchel and Valentin [13]). The monomeric composition of poly(HA) is highly variable. It depends on the bacterial strain as well as on the carbon source applied during the accumulation of the polymer and determines its physical and chemical properties.

Poly(HA) can be biodegraded to water and carbon dioxide or methane by a large variety of ubiquitous microorganisms present in many ecosystems (Fig. 1). This fairly easy biodegradability came as a surprise given the inertness of the water-insoluble, hydrophobic, and (partially) crystalline polymers.

Any review on the biodegradation of poly(HA) should clearly distinguish between (i) the intracellular and (ii) the extracellular poly(HA) biodegradation. Intracellular biodegradation is the active mobilization of an endogenous carbon/energy storage reservoir by the accumulating bacterium itself. Extracellular degradation is the utilization of an exogenous carbon/energy source by

Fig. 1. Biodegradation of poly(HA_{SCL}). The bottles consist of the Biopol material and were incubated in aerobic sewage sludge for 0, 2, 4, 6, and 8 weeks, respectively (left to right)

a microorganism that must not necessarily accumulate poly(HA). Sources of this extracellular polymer are poly(HA) released by dying, accumulating cells.

The key enzymes of both processes are poly(HA) depolymerases, and this chapter will focus on the biochemical and molecular properties of these intriguing enzymes. The mechanism of poly(HA) degradation by extracellular poly(HA) depolymerases has been extensively studied over the last decade [14–18] and will be summarized in the first part of this chapter. However, the study of intracellular poly(HA) degradation and mobilization, after being the subject of detailed physiological studies in the 1960s [19–21], only very recently came back to our attention, and accordingly its molecular mechanism is only poorly understood. Our current knowledge on poly(HA) mobilization will be surveyed in the second part of this chapter.

2
Extracellular Degradation of Poly(HA)

2.1
Identification and Isolation of Extracellular Poly(HA)-Degrading Microorganisms

Poly(HA) granules in vivo are covered by a surface layer (see Sect. 3.2 and Steinbüchel and Hein in this volume), which is damaged after lysis of the bacteria or is removed during isolation of the polymer. As a consequence, poly(HA) granules become coalescent, and crystallization of the previously amorphous polymer begins [22]. Amorphous (intracellular) poly(HA) granules with an intact surface layer and crystalline (extracellular) poly(HA) without a surface layer or with a damaged surface layer will be referred to as "native" and "denatured poly(HA)," respectively, according to the nomenclature introduced by Merrick and Doudoroff [19].

Poly(HA)-degrading microorganisms can be enriched from soil or liquid samples collected from various ecosystems after inoculation of mineral salt solutions which contain poly(HA) as a sole source of carbon and energy. This procedure generally results in enrichment of only one or a few microbial species which exhibit the fastest growth under the chosen laboratory conditions. These organisms are not necessarily the most efficient poly(HA)-degrading strains. Contaminating microorganisms, which do not degrade poly(HA) but utilize the primary degradation products (oligomers of poly(HA)), might grow even faster. Such contaminants can be difficult to remove. A more suitable method for the assessment of the distribution of microbial poly(HA) degraders in a particular sample is to plate (soil) suspensions of the desired ecosystem on solid agar media which contain the polymer as a sole source of carbon in an opaque overlay prepared from denatured poly(HA) granules. Only true poly(HA)-degrading microorganisms secrete specific poly(HA) depolymerases, which hydrolyze the polymer extracellularly to water-soluble products, and produce transparent clearing zones around the depolymerase-secreting colonies (clear zone technique).

Unfortunately, only a few short-chain-length poly(HA)s [poly(HA_{SCL})], namely poly(3HB), its copolymers with 3-hydroxyvalerate, and poly(3-hydroxyvaleric acid) [poly(3HV)], can be prepared as a milky suspension of denatured granules. Most other poly(HA), including poly(4-hydroxybutyrate) and all medium-chain-length poly(HA) [poly(HA_{MCL})], form large rubber-like aggregations which cannot be used for the clear zone technique directly. Mineral agar plates containing ultrathin solution-cast films of such poly(HA), which have been stained with a dye (e.g., Sudan-red), have been successfully applied for the isolation of poly(HA_{MCL})-degrading bacteria [23]. However, microorganisms with only a low activity of poly(HA)-degrading enzymes can possibly be missed by applying this method. A significant improvement was achieved by the development of poly(HA) emulsions (latexes) [24, 25]. These are stable and heat-resistant emulsions which are prepared by adding acetone-dissolved poly(HA) to cold water and evaporating the solvent afterwards. The resulting emulsions have a milky appearance and can be used for the clear zone method. This technique, with slight modifications for different poly(HA), was successfully applied for the isolation of a large variety of poly(HA_{MCL})-degrading bacteria [26]. Another similar method for latex preparation was developed by Horowitz and Sanders [27, 28]. They prepared emulsions of poly(HA) by dissolving the polymer in chloroform, adding a surfactant and emulsifying with water by sonication. The solvent was removed by dialysis or evaporation, and an opaque, stable suspension of granules coated by the surfactant was obtained. Since poly(HA) generally are soluble in chloroform, this method can be used to prepare artificial granules from any kind of poly(HA). However, the surfactant might inhibit bacterial growth and poly(HA) depolymerase activity. In addition, artificial poly(HA) granules remain amorphous and thus resemble native poly(HA). This might prevent the isolation of poly(HA)-degrading bacteria with poly(HA) depolymerases specific for denatured poly(HA). Recently, an alternative method for preparing poly(3HB) latexes has been developed by heating aqueous suspensions of crystalline poly(3HB) granules above the melting temperature (≈ 180 °C) and subsequent rapid cooling to room temperature. The melted poly(3HB) granules remained in a metastable amorphous state for several weeks [29]. It is likely that this method can also be applied to poly(HA) other than poly(3HB).

Using the methods described above, poly(HA)-degrading microorganisms can be easily identified and isolated. Bacteria with fast growth and high poly(HA) hydrolysis rates can be distinguished from those with only low polymer-hydrolyzing abilities by measuring the diameter of the colonies and of the clearing zones. However, isolates with large clearing zones do not necessarily represent those strains which are most important for the degradation of poly(HA) in situ. Such organisms can be enriched by incubation of (thin) pieces of the polymer or polymer films in the ecosystem to be analyzed. After incubation for several weeks to months the organisms attached to the polymer can be investigated.

2.2
Characterization of Poly(HA)-Degrading Microorganisms

The ability to degrade extracellular poly(HA) is widely distributed among bacteria and fungi and depends on the secretion or surface-display of specific poly(HA) depolymerases, which hydrolyze the polymer by surface erosion to water-soluble monomers and/or oligomers. For a survey of SEM-analysis of surface-eroded poly(HA_{SCL}) and poly(HA_{MCL}) see Molitoris et al. [30]. Aerobic and anaerobic poly(HA)-degrading bacteria were isolated from various ecosystems such as soil, compost, aerobic and anaerobic sewage sludge, fresh and marine water, estuarine sediment, and air (for references see Table 1). Apparently, poly(HA)-degrading microorganisms, in particular poly(3HB)-degrading bacteria, are present in nearly all terrestrial and aquatic ecosystems. Ecological and taxonomical studies on the abundance and diversity of poly(HA)-degrading bacteria have been performed only with poly(3HB) and its copolymers with 3-hydroxyvaleric acid [31–40]. All of the poly(HA)-degrading bacteria tested so far are also able to accumulate poly(HA) if they are cultured under appropriate conditions. Poly(HA)-degrading bacteria differ with respect to the type of polyester they can degrade. Most of the characterized bacteria are specific for either poly(3HB) and other similar poly(HA_{SCL}) or for poly(HA_{MCL}) such as poly(3-hydroxyoktanoate) [poly(3HO)]. However, some bacteria revealed a rather broad polyester specificity and are able to utilize a large variety of polymers including poly(HA_{SCL}) and poly(HA_{MCL}) [16].

The ability to degrade PHA is not restricted to bacteria, and many poly(HA)-degrading fungi have been identified [41–46]. In a comparative study 95 genera of fungi [*Ascomycetes* (18 genera), *Basidiomycetes* (46 genera), *Deuteromycetes* (26 genera), *Mastigiomycetes* (1 genus), *Myxomycetes* (2 genera), *Zygomycetes* (2 genera)] have been identified which include at least one species of a poly(HA_{SCL})- or poly($H_{MCL}A$)-degrading fungus [47].

2.3 Biochemical Properties of Poly(HA) Depolymerases

The poly(HA) depolymerases of the bacteria *Alcaligenes faecalis* (strains AE122 and T1), *Comamonas acidovorans, Comamonas testosteroni, Comamonas* sp., *Pseudomonas fluorescens, Pseudomonas lemoignei, Pseudomonas stutzeri, Ralstonia pickettii, Streptomyces exfoliatus*, and of the fungi *Paecilomyces lilacinus, Penicillium funiculosum*, and *Penicillium pinophilum* have been purified and characterized (for details see Table 1). Poly(HA) depolymerases share several characteristics:

1. High stability at a wide range of pH, temperature, ionic strength etc.
2. A relatively small molecular weight (<70 kDa) with most depolymerases consisting of only one polypeptide.
3. Alkaline pH optimum (7.5–9.8) – only the depolymerases of *R. pickettii* and of the fungi *P. funiculosum* and *P. lilacinus* have pH optima between 5.5 and 7.0.
4. Most poly(HA) depolymerases are inhibited by reducing agents, e.g., dithioerythritol (DTT), which indicates the presence of essential disulfide bonds, and by serine hydrolase inhibitors such as diisopropyl-fluoryl phosphate (DFP) or acylsulfonyl derivates. The latter compounds covalently bind to the active site serine of serine hydrolases and irreversibly inhibit enzyme activity [48].

Table 1. Overview on poly(HA)-degrading microorganisms and biochemical characterization of purified poly(HA)-depolymerases

Strain analyzed	Depol. purified	Binding to DEAE	*Mr* (SDS PAGE)	Dependence on Ca^{2+}	T_{opt} (°C)	pH_{opt}	Carbohydrate content
Poly(HA_{SCL}) depolymerases of Gram negative aerobic bacteria							
Alcaligenes faecalis T1	Poly(3HB) depol.		50	no[b]		8.6	
A. faecalis T1	Poly(3HB) depol.						
A. faecalis T1	Poly(3HB) depol.						
A. faecalis T1	Recombinant hybrid poly(3HB) depol. domain mutants						
A. faecalis AE122	Poly(3HB) depol.						
A. faecalis AE122	PhaZ *Afa*122-*rec*[a]		62.5				
Aureobacterium saperdae	Poly(3HB) depol.		43	no	45	8.0	no
Comamonas sp.	Poly(3HB) depol.	no	44	no	29–35	9.4	no
C. acidovorans YM1609	Poly(3HB) depol.		45 (48.6)	no	≈ 37	9.0	
C. testosteroni ATSU	Poly(3HB) depol.		49		70	8.5	
C. testosteroni YM1004	Poly(3HB) depol.		50			9.5 – 10	
C. testosteroni YM1004	Glutathione-*S*-transferase- poly(3HB) depol. binding domain fusion						
Pseudomonas P1	Poly(3HB) depol.	no			45	9.8	
P. lemoignei	partial	no		yes			
P. lemoignei	Poly(3HB) d.ep A, B	no		yes			
P. lemoignei	A1, A2	no	54	yes		8.0	
	B1, B2	no	58	yes		8.0	
P. lemoignei	PhaZ1*Ple-rec*[a]	no	44	yes			
P. lemoignei	Poly(3HB) depol. A	no	59	yes	≈ 65		
	Poly(3HB) depol. B	no	67	yes			
	Poly(3HV) depol.	no	54	yes	55	8.0	
P. lemoignei	Poly(3HB) depol. A	no	55				yes
	Poly(3HB) depol. B	no	67				yes
	Poly(3HB) depol. C	no					
	Poly(3HV) depol.	no					yes
P. lemoignei	Poly(3HB) depol. A						yes
	Poly(3HB) depol. B						yes
	Poly(3HB) depol. C						yes
	Poly(3HV) depol.						yes
	PhaZ1*Ple-rec*[a]	no	44	yes	61.5		no
	PhaZ2*Ple-rec*[a]	no	46.5	yes	54.5		no
	PhaZ4*Ple-rec*[a]	no	65.5	yes	50		no
	PhaZ5*Ple-rec*[a]	no	49	yes	51		no
P. lemoignei	PhaZ4*Ple-rec*[a] (truncated)		51				
P. lemoignei	no						
P. lemoignei A62	no						
P. lemoignei	Poly(3HB) depol.	yes					
P. lemoignei	Poly(3HB) depol. A						
P. lemoignei							
P. pickettii	Poly(3HB) depol.		40		40	5.5	
P. pickettii	Poly(3HB) depol.		49			6.0	
isolate Z925	Poly(3HB) depol.		45			≈ 9.5	
isolate T107	Poly(3HB) depol.		45			≈ 9.5	
isolate S2	Poly(3HB) depol.		43			≈ 9.5	
isolate A1	Poly(3HB) depol.		49			≈ 9.5	

Table 1 (continued)

Remarks	Ref.
	182
Indication for substrate-binding site	70
Identification of the catalytic site serine by site-directed mutagenesis	54
Type and number of linker domains do not affect activity	55
Isolate from seawater, depolymerase with unusual high apparent *Mr*	183
Two poly(3HB) binding domains	64
Soil isolate	184
Monomers are the only hydrolysis end products	81, 185
Monomers are the end product of hydrolysis at high enzyme concentration	60
Depolymerase with the highest temperature opt., however unstable at 70 °C	186
Isolate from seawater	187
Function of C-terminal domain	61
Strains are not preserved	83
Detailed microbiological and physiological study	80
Thermal inactivation at 60 °C	84
	69
	82
	85
Poly(HA)$_{SCL}$ depolymerase with high poly(3HV) depolymerase activity	
Survey of poly(HA) depolymerases	51
Contains glucose and *N*-acetylglucosamine,	52
Survey of *P. lemoignei* poly(3HB) depolymerases	
Contains glucose and *N*-acetylglucosamine	
Contains glucose and *N*-acetylglucosamine	
Contains glucose and *N*-acetylglucosamine	
Glycosylation not essential for activity	
Glycosylation not essential for activity	
Glycosylation not essential for activity	
Glycosylation not essential for activity	
Identification of a C-terminal polyester-binding domain	56
Specific isolation procedure for *P. lemoignei*, no plasmid in type strain;	36
Strain A62 contains 200 kbp megaplasmid	
Identification of the catalytic site serine by site-directed mutagenesis	54
Simplified purification procedure for poly(3HB) depolymerase A	86
Relationship between poly(3HB) depolymerase synthesis and uptake of succinate	88
Depolymerase with acidic pH optimum	183
	75

Table 1 (continued)

Strain analyzed	Depol. purified	Binding to DEAE	*Mr* (SDS PAGE)	Dependence on Ca^{2+}	T_{opt} (°C)	pH_{opt}	Carbohydrate content
Poly(HA_{SCL}) depolymerases of Gram-negative aerobic bacteria continued							
P. stutzeri YM1414	Poly(3HB) depol.		48	yes	55	9.5	
P. stutzeri YM1006	Poly(3HB) depol.	yes	60			7–7.5	
P. stutzeri YM1006	PhaZ*Pst-rec*[a]		57.5				
6 *Pseudomonas* sp.	no						
several species	6 lipases and 9 poly(HA) depolymerases						
	no						
	no						
Poly(HA_{SCL}) depolymerases of Gram-positive aerobic bacteria							
Streptomyces exfoliatus K10	poly(3HB) depol.		49	no	40	8.5–9.0	
Poly(HA_{SCL}) depolymerases of anaerobic bacteria							
Ilyobacter delafieldii	no						
anaerobic consortium Clostridia-like isolates	no						
Poly(HA_{SCL}) depolymerases of fungi							
Paecilomyces lilacinus	Poly(3HB) depol.		48		45	7.0	
Penicillium funiculosum	Poly(3HB) depol.		37			6.0	yes
Poly(HA_{MCL}) depolymerases of Gram-negative aerobic bacteria							
P. fluorescens GK13	Poly(3HO) depol.		25	no	45	8.5	no
P. fluorescens GK13	Poly(3HO) depol.						
P. fluorescens GK13	PhaZ*Pfl-rec*[a]	no	28				
P. fluorescens GK13	PhaZ*Pfl-rec*[a]	no	28				
P. maculicola	no						
several species	6 lipases and 9 poly(HA) depolymerases						
Other microbial esterases related to poly(HA) depolymerases							
A. faecalis	PCL depolymerase/lipase					40	7.0
Pseudomonas sp. A1	Oligomer-hydrolase OhaZ*Psp-rec*		70–73				
Physarum polycephalum	PMA hydrolase		68	no		3.5	yes
Comamonas acidovorans	PMA hydrolase		43		40	8.1	yes
Comamonas acidovorans strain TB-35	Polyester-polyurethane hydrolase		62		45	6.5	

[a] enzyme purified from recombinant *E. coli*.

[b] addition of Ca^{2+} or Mg^{2+} not required for activity but influence of EDTA unknown; abbreviation: depolymerase (depol.), empty space in a column indicates that the value has not been determined.

Table 1 (continued)

Remarks	Ref.
	189
Poly(3HB) depolymerase from a marine isolate	190
Poly(3HB) depolymerase with cadherin-like linker domain and two poly(3HB) binding domains	57
Development of a method for preparing stable Poly(HA_{MCL}) emulsions	24
Lipases hydrolyze polymers of ω-hydroxyalkonoic acids	110
Ecological study	32
Taxonomical study	33–35, 37
K10 hydrolyzes poly(3HB) and poly(3HO)	49
Anaerobic poly(3HB) degrading bacterium that ferments poly(3HB) to acetate, butyrate and H_2	191
	192
	193
	194
	45
Contains mannose, galactose and glucose	42
Poly(3HO) depolymerase that does not hydrolyze poly(3HB)	23
Degradation of unsaturated and cross-linked poly(3HO) derivatives	72
In rec. *E. coli* processing of poly(3HO) depolymerase differs from wild type Site-directed mutagenesis of catalytic triad serine	67
Isolation of bacteria able to hydrolyse $poly(HA)_{SCL}$ and $poly(HA)_{MCL}$	26
Lipases hydrolyze polymers of ω-hydroxyalkanoic acids	110
	118
3HB-oligomer hydrolyse with no activity towards poly(3HB) No true lipase box within the deduced amino acid sequence	76a
Extracellular, *exo*-type poly(β-L-malic acid) hydrolase	124
Membrane-bound *exo*-type poly(β-L-malic acid) hydrolase	126
Cell-surface bound hydrolase	135
	136

5. Poly(HA) depolymerases – as far as it has been tested – do not bind to anion exchangers such as DEAE (at neutral pH) but have a pronounced affinity to hydrophobic materials. Therefore, many purification protocols include hydrophobic interaction chromatography.

Most poly(HA)-degrading bacteria apparently contain only one poly(HA) depolymerase. Interestingly, *P. lemoignei* has been shown to encode six poly(HA_{SCL}) depolymerases which differ slightly in their biochemical properties (Tables 1 and 2), and there is evidence for the presence of yet another one (unpublished results). At present, the selection advantage for having as many as seven poly(HA) depolymerases is not known. A cooperative effect on poly(3HB) hydrolysis in the presence of two or more poly(3HB) depolymerases might exist. It is likely that other bacteria will also prove to have more than one poly(HA) depolymerase (see below). The presence of several isoenzymes of a respective polymer hydrolase within a single degrading strain seems to be a common theme for the degradation of natural polymers such as cellulose or chitin.

All purified poly(HA) depolymerases are specific for either poly(HA_{SCL}) or poly(HA_{MCL}). Even a poly(3HB) depolymerase of *S. exfoliatus* K10, a strain that degrades both poly(3HB) and poly(3HO), is specific for poly(HA_{SCL}) [49]) indicating at least one additional depolymerase with specificity for poly(HA_{MCL}) in *S. exfoliatus.* Experiments with copolymers consisting of 3-hydroxybutyrate and 3-hydroxyhexanoate and *A. faecalis* T1 poly(3HB) depolymerase are in agreement with the results obtained with poly(HA_{SCL}) and poly(HA_{MCL}): the depolymerase was not able to hydrolyze ester bonds between two $3HA_{MCL}$ monomers and between 3-hydroxybutyrate and 3-hydroxyhexanoate [50].

In contrast to most other bacterial poly(HA) depolymerases the depolymerases of *P. lemoignei* and those of fungi are glycosylated. However, glycosylation appears not to be essential for activity of *P. lemoignei* depolymerases but may improve the resistance of the exoenzyme to elevated temperature and/or hydrolytic cleavage by proteases of competing microorganisms [51, 52].

2.4
Molecular Biology and Functional Analysis of Poly(HA_{SCL}) Depolymerases

Sixteen extracellular bacterial poly(HA) depolymerase genes (*phaZ*) have been cloned and analyzed since 1989 (Table 2). Except for the poly(3HB) depolymerase gene of *S. exfoliatus* all other genes were cloned from the Gram-negative bacteria *Acidovorax* sp., *A. faecalis* (strains AE122 and T1), *C. acidovorans, C. testosteroni, Comamonas* sp., *P. fluorescens, P. lemoignei* (six genes), *P. stutzeri*, and *R. pickettii.* Fifteen genes code for depolymerases with specificity for poly(HA_{SCL}), especially poly(3HB). Some of these depolymerase proteins have significant activity with poly(3HV) homopolyester (PhaZ6 > PhaZ1 > PhaZ4, all from *P. lemoignei*). None of the poly(HA_{SCL}) depolymerases has significant activity on poly(HA_{MCL}). The *P. fluorescens* poly(3HO) depolymerase is specific for poly(3HO) and related poly(HA_{SCL}). The poly(3HO) depolymerase is completely inactive with poly(3HB) but hydrolyses *p*-nitrophenylesters of fatty

Table 2. Characteristics of bacterial poly(3HA) depolymerase genes

Strain	Gene	Length (bp)	Protein	*Mr* (mature protein)	Lipase-box	Type of linker region[a]	No and type of polymer binding domain[b]	Reference Accession No
Extracellular poly(3H$_{SCL}$) depolymerases								
A. faecalis T1	*phaZ$_{Afa}$*	1464	Poly(3HB) depol.	47.0	GLSSG	Fn3[c]	1A	14
A. faecalis AE122	*phaZ$_{AflA22}$*	1905	Poly(3HB) depol.	62.5	GLSSG	Fn3	1A+1B	64
Acidovorax sp.	*phaZ$_{Asp}$*	1476	Poly(3HB) depol.	48.5	GLSSG	Fn3	1A	AB015309
Comamonas sp.	*phaZ$_{Csp}$*	1542	Poly(3HB) depol.	50.1	GLSSG	Fn3	1A	185
C. acidovorans	*phaZ$_{Cac}$*	1485	Poly(3HB) depol.	48.6	GLSSG	Fn3	1A	60
C. testosteroni	*phaZ$_{Cte}$*	1539	Poly(3HB) depol.	50	GLSSG	Fn3	1A	61
P. lemoignei	*phaZ1$_{Ple}$*	1242	Poly(3HB) depol. C	39.5	GLSAG	Thr[d]	1B	82
	phaZ2$_{Ple}$	1299	Poly(3HB) depol. B	41.8	GLSAG	Thr	1A	51
	phaZ3$_{Ple}$	1257	Poly(3HB) depol. D	41.2	GLSAG	Thr	1A	51
	phaZ4$_{Ple}$	1701	Poly(3HB) depol. E	57.5	GLSAG	Fn3	1B	52
	phaZ5$_{Ple}$	1299	Poly(3HB) depol. A	42.2	GLSSG	Thr	1A	52
	phaZ6$_{Ple}$	1254	PHV depolymerase	42.8	GLSSG	Thr	1A	195
R. pickettii	*phaZ$_{Ppi}$*	1476	Poly(3HB) depol.	47.5	GLSSG	Fn3	1A	196
P. stutzeri	*phaZPst*	1728	Poly(3HB) depol.	57.5	GLSSG	Cad[e]	2A	57
Streptomyces exfoliatus	*pha$_{Sex}$*	1467	Poly(3HB) depol.	47.9	GLSSG	Fn3	1A	49
Extracellular poly(3HA$_{MCL}$) depolymerases								
P. fluorescens GK13	*phaZ$_{Pfl}$*	837	Poly(3HO) depol.	26.6	GISSG			67
Extracellular 3HB-oligomer hydrolases								
Pseudomonas sp. strain A1	*ohaZ$_{Ppi}$*	2112	3HB-oligomer hydrolase	72.9	ATSSG and AISTG[f]			76a
Putative intracellular 3HB-oligomer hydrolases								
R. eutropha	*ohaZ$_{Aeu}$*	2274	3HB-oligomer hydrolase	78.4				AB003701
Putative intracellular poly(3HA$_{SCL}$) depolymerase								
R. eutropha	i-*phaZ$_{Reu}$*	1257	i-PHB depolymerase	47.3				AB017612

Table 2 (continued)

Strain	Gene	Length (bp)	Protein	*Mr* (mature protein)	Lipase-box	Type of linker region[a]	No and type of polymer binding domain[b]	Reference Accession No
Putative intracellular poly(3HA$_{MCL}$) depolymerases								
P. oleovorans	i-*phaZ*$_{Pol}$	852	putative i-poly(3HO) depol.	31.4	GVSWG			180
P. aeruginosa	i-*phaZ*$_{Pae}$	858	putative i-poly(3HO) depol.	31.5	GVSWG			181
P. resinovorans	i-*phaZ*$_{Pre}$	858	putative i-poly(3HO) depol.	31.5	GVSWG			AAD26366
Pseudomonas sp.	i-*phaZ*$_{Psp}$	858	putative i-poly(3HO) depol.	31.7	GVSWG			BAA36201

[a] region of the depolymerase protein between catalytic domain and substrate-binding domain.
[b] two types of poly(3HB) binding domain namely A and B were differentiated by sequence homology (see Fig. 2).
[c] fibronectin type III fingerprint.
[d] threonin-rich region.
[e] cadherin-like domain.
[f] instead of a true lipase box (G-X1-S-X2-G) two similar sequences (ATSSG and AISTG) were found. Empty space indicates that the parameter is not present or has not been identified.

acids with six or more carbon atoms. In conclusion, poly(HA) depolymerases are highly specific with respect to the length of the carbon side chain of the poly(HA) substrate.

All poly(HA_{SCL}) depolymerase proteins have a composite domain structure (Fig. 2) and consist of:

1. A 25–38 amino acid long signal peptide, which is characteristic for extracellular enzymes and is cleaved off during the passage across the cytoplasma membrane.
2. A large catalytic domain at the *N*-terminus.
3. A C-terminal substrate-binding domain.
4. A linking domain, which links the catalytic domain with the substrate-binding domain.

Three strictly conserved amino acids constitute the active center of the catalytic domain (catalytic triad). These amino acids are: serine, aspartate, and histidine (Fig. 3). The serine is part of a lipase-box pentapeptide Gly-Xaa_1-Ser-Xaa_2-Gly which is present in serine hydrolases such as lipases, esterases, and serine-proteases (for a review on lipases see Jaeger et al. [53, 53a]): the oxygen atom of the serine side chain is the nucleophile that attacks the ester bond. The acidity of the hydroxyl-function and thus the reactivity of the oxygen atom is enhanced by the imidazole ring of the histidine. The resulting positive charge of the imidazole ring is stabilized by the carboxyl group of the aspartate. Site-directed mutagenesis of the putative catalytic triad amino acids from three poly(HA)-depolymerases (*A. faecalis* and *P. lemoignei* poly(3HB) depolymerases and *P. fluorescens* poly(3HO) depolymerases) resulted in inactive proteins and confirmed the requirement of the correct side chain for activity [26, 54, 55]. In contrast to most bacterial lipases, in which Xaa_1 of the lipase-box is a histidine, a leucine is found in all poly(HA_{SCL}) depolymerases, and an isoleucine is present in the *P. fluorescens* poly(3HO) depolymerase. It is likely that the hydrophobic side-chains of leucine or isoleucine are more compatible with the hydrophobic polymers than the hydrophilic imidazole ring. A second histidine residue is also conserved in all poly(3HB) depolymerases (Fig. 3). The region near this histidine resembles the oxyanion hole known from lipases [53]. In lipases, the oxyanion amino acid stabilizes the transient state of the hydrolysis reaction by allowing the formation of a hydrogen bridge to the negatively charged oxygen atom (oxyanion) of the active site serine. It is assumed that the second strictly conserved histidine has a similar function in poly(3HB) hydrolysis.

Poly(HA_{SCL}) depolymerases are able to bind to poly(3HB)-granules. This ability is specific because poly(3HB) depolymerases do not bind to chitin or to (crystalline) cellulose [56, 57]. The poly(3HB)-binding ability is lost in truncated proteins which lack the C-terminal domain of about 60 amino acids, and these modified enzymes do not hydrolyze poly(3HB). However, the catalytic domain is unaffected since the activity with water-soluble oligomers of 3-hydroxybutyrate or with artificial water-soluble substrates such as *p*-nitrophenylesters is unaffected [55, 56, 58, 59]. Obviously, the C-terminal domain of poly(3HB) depolymerases is responsible and sufficient for poly(3HB)-binding [poly(3HB)-binding domain]. These results are in agreement:

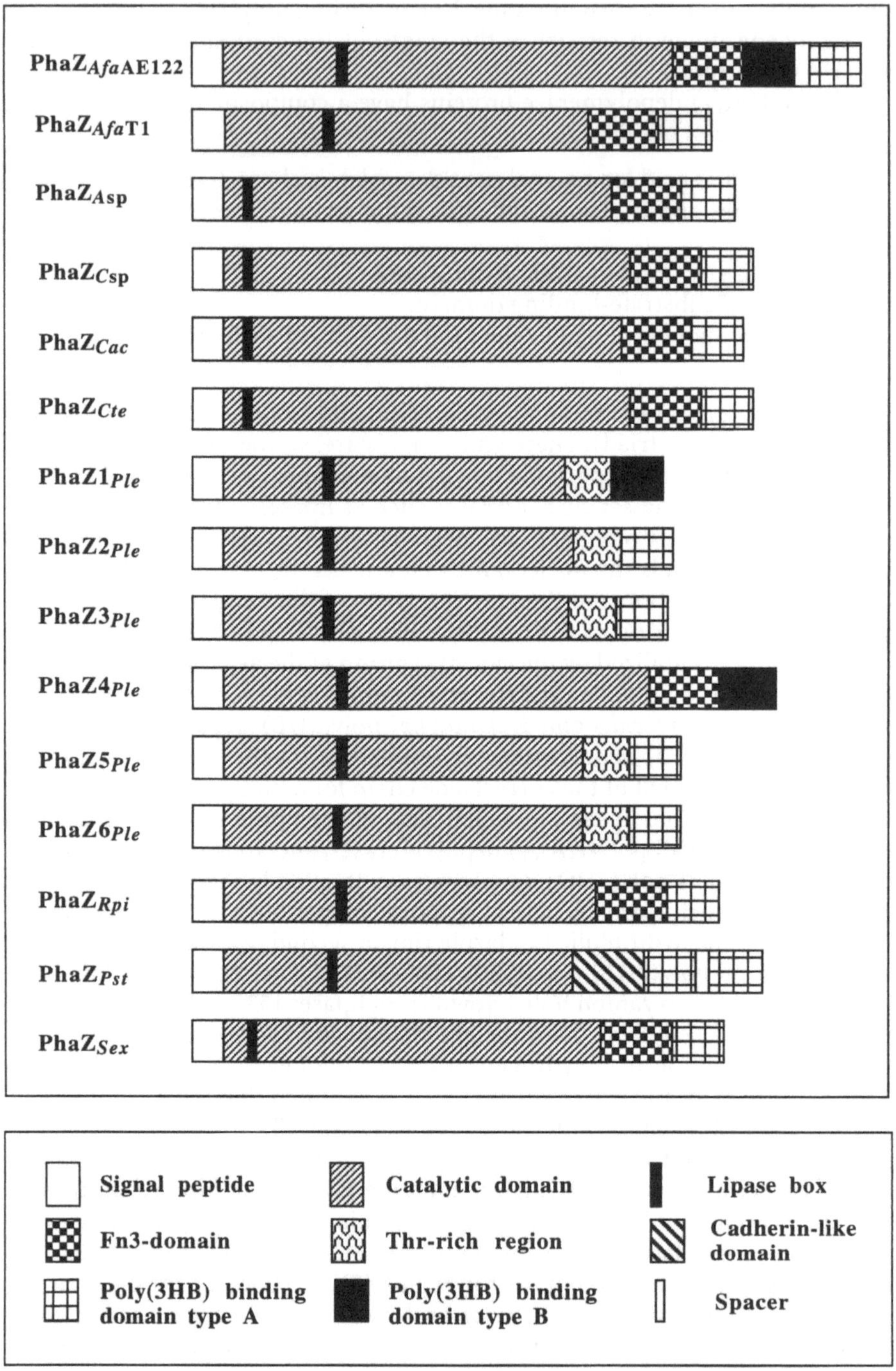

Fig. 2. Domain structure bacterial poly(HA) depolymerases. An interpretation of the amino acid sequences of microbial poly(HA) depolymerases is shown. For references see Table 2

Peptide	pos. (S)	Serine	pos. (D)	Aspartate	pos. (H)	Histidine	pos. (H)	Oxyanion hole	Reference
Extracellular poly(HA$_{SCL}$) depolymerases									
PhaZ$_{AfaT1}$	139	IDPNQVYVAGL**S**SGGGMT	214	VW-GTS**D**YTV	273	GMA**H**AWPAGTG	49	LVLVL**H**GCVQTAS	14
PhaZ$_{AfaAE122}$	119	IDPNQVYIAGL**S**SGASFA	191	SIHGDA**D**TTV	248	LWF**H**GLDHSWS	37	LLIVL**H**GCTQSID	64
PhaZ$_{Asp}$	20	IDKTQTTVSGL**S**SGGFMA	107	LFSGTL**D**SVV	141	ASE**H**AMVTDDY	233	LHVVL**H**GCKQNVN	AB015309
PhaZ$_{Sex}$	25	NATDQISVSGL**S**SGGFMA	111	LFHGTN**D**STV	143	ASG**H**AWVSPLG	237	LMVTL**H**GCYQYFG	49
PhaZ$_{Csp}$	20	IATDQISVSGL**S**SGGFMA	105	IFTGTS**D**YTV	140	GAA**H**VLPTDFD	258	LHVVL**H**GCQQSTD	185
PhaZ$_{Cac}$	20	IDKTQTTVSGL**S**SGGFMA	107	LFSGTL**D**SVV	141	AAE**H**AMVTDDY	233	LHVVL**H**GCQQSTD	60
PhaZ$_{Cte}$	20	IATDQISVSGL**S**SGGFMA	105	IFTGTS**D**YTVG	140	GAA**H**VLPTDFD	233	LHVVL**H**GCQQSTD	61
PhaZ1$_{Ple}$	117	IDPSRVYVTGL**S**AGAFMT	195	IWHGSS**D**YTV	253	GMG**H**GTPVDPG	35	LVVAL**H**GCDQTAA	82
PhaZ2$_{Ple}$	117	IDTNRVYVTGL**S**AGGYMV	195	IWHGDA**D**YTV	253	GMG**H**GTPVDPG	35	LVIAM**H**GCTQSAS	51
PhaZ3$_{Ple}$	117	IDANRVYVTGL**S**AGAFMT	195	VWQGSS**D**TTV	253	GMA**H**GTPVDPG	35	LIVAM**H**GCTQSAS	51
PhaZ4$_{Ple}$	137	IDPNQIYVSGL**S**AGAGET	212	VY-GDK**D**YLV	270	GMS**H**AWPSGPG	47	LMLSL**H**GCGQTAS	52
PhaZ5$_{Ple}$	138	IDPNQVYVTGL**S**SGGGET	213	VW-GTS**D**FTV	272	GMS**H**AWPAGTG	49	LVLVL**H**GCAQTAS	52
PhaZ6$_{Ple}$	136	IDPNQIYISGL**S**SGAAET	211	VY-GSN**D**AIV	269	GMS**H**AWPAGSG	46	LMLSL**H**GCGQTAS	195
PhaZ$_{Rpi}$	140	IDPNQVYVAGL**S**SGGGMT	215	VW-GTS**D**YTV	274	GMA**H**AWPAGTG	50	LVLVL**H**GCLQTAS	196
PhaZ$_{Pst}$	125	IDANRRFITGL**S**SGGAMA	202	ILQNRN**D**CTV	251	GCR**H**IAYTQDG	40	MVMAL**H**GCRQTND	57
CONSENSUS		id--q*-v-GL**S**sG+---		**-g--**D***tV		g--**H**-*p---g		l****H**GC-Q---	
CONSENSUS LIPASES		-V-**Gh**S**-G+---		------**D**-*v		---**H***------		--*****H**G*-----	

Fig. 3. Alignment of poly(HA) depolymerase catalytic active amino acids. Amino acids of the catalytic triad and the conserved histidine residue of the putative oxyanion are shown in *bold letters*. The positions (pos.) of respective amino acids of mature depolymerase proteins are indicated. Amino acids conserved in all poly(HA$_{SCL}$) depolymerases are indicated in *capital letters*, those which have been conserved in ten or more proteins are marked by *low letters* in the consensus sequence. * indicates amino acids with hydrophobic side chains; + indicates amino acids with a small side chain

Peptide	pos. (S)	Serine	pos. (D)	Aspartate	pos. (H)	Histidine	pos. (H)	Oxyanion hole	Reference
Extracellular poly(3HA$_{MCL}$) depolymerases									
PhaZ$_{Pfl}$	139	LNAQRQYATGI**S**SGGYNT	193	FLHGFV**D**AVV	227	LGG**H**EWFAASP	78	QNLLD**H**GYAVIAP	67
Cutinase$_{Fso}$	120	CPEATLIAGGY**S**QGAALA	175	VFCNTG**D**LVC	188	AAP**H**LAYGPDA			
Putative intracellular poly(3HA$_{MCL}$) depolymerases									
PhaZ*Pol*	102	LDYGQVNVIGV**S**WGGALA	221	VLAGDD**D**PLI	248	DDG**H**LFLITRA			
PhaZ*Pae*	102	LDYGQVNAIGV**S**WGGALA	221	VLAGDD**D**PII	248	DDG**H**LFLVTRA			
PhaZ*Pre*	102	LDYGQVSVIGV**S**WGGALA	221	VLAGDD**D**PII	248	DDG**H**LFLVTRA			
PhaZ*Psp*	102	LDYGQVNVVGV**S**WGGALA	221	VLAGDD**D**PLI	248	DDG**H**LFLITRA			

Fig. 3 (continued)

1. With data obtained by the analysis of fusion proteins consisting of a domain unrelated to poly(3HB) metabolism (e.g., maltose binding protein MalE or glutathione-*S*-transferase and the poly(3HB) depolymerase binding domain [57, 59–61].
2. With findings of Doi and coworkers who proposed a two-step reaction for enzymatic polyester hydrolysis, namely adsorption of the depolymerase to the polymer and subsequent hydrolysis [62, 63].

Recently, two new poly(3HB) depolymerase sequences from *A. faecalis* AE122 and from *P. stutzeri* were published which contain two instead of only one poly(3HB)-binding domain [57, 64]. Two types of poly(3HB) binding domains can be differentiated by amino acid alignment (types A and B in Fig. 4). Several amino acids are strictly conserved in both types of binding domains. It is not known whether these conserved amino acids are necessary to constitute a particular three-dimensional structure or whether these amino acids are directly involved in the interaction with the polymer chain.

The function of the linking region between the catalytic and the substrate-binding domain in poly(HA_{SCL}) depolymerases is unknown. In five poly(3HB) depolymerases, namely the poly(HA) depolymerases PhaZ1, PhaZ2, PhaZ3, PhaZ5, and PhaZ6 of *P. lemoignei*, this region consists of about 40 amino acids with 4–6 threonine residues clustered in 4 repetitions (threonine-rich region). This threonine-rich region is replaced by a fibronectin type III (Fn3) fingerprint in PhaZ4 of *P. lemoignei* and in most other sequenced poly(HA_{SCL}) depolymerases. Fn3 sequences have been described for many eukaryotic extracellular matrix proteins and in several prokaryotic polymer hydrolyzing proteins such as chitinases, cellulases and several glucoamylases [65]. The Fn3 sequence of the *Bacillus circulans* chitinase is not essential for activity [66]. In contrast, the Fn3 domain of the *A. faecalis* poly(3HB) depolymerase was essential for activity because deletion of the Fn3 domain resulted in a protein which had lost poly(3HB) depolymerase activity but not the dimer hydrolase activity [55]. Interestingly, the Fn3 domain of the *A. faecalis* depolymerase could be functionally replaced by a threonine-rich region of the poly(3HB) depolymerase A (PhaZ5) of *P. lemoignei.* These findings support the hypothesis that Fn3-sequences and/or the threonine-rich regions are necessary to provide a proper distance (linker) between the substrate-binding site and the catalytic center of the protein, but the nature of this linker appears to be less important. Recently, a third type of linker region has been described for the poly(3HB) depolymerase of *P. stutzeri* [57]. The *P. stutzeri* depolymerase contains a cadherin-like sequence about 100 amino acids long between the catalytic and the binding domain. Cadherin-like sequences have also been found in other polymer hydrolases such as the chitinases of *Vibrio harveyi* and *Clostridium paraputrificum.*

2.5 Molecular Biology and Functional Analysis of Poly(HA_{MCL}) Depolymerases

The poly(3HO) depolymerase of *P. fluorescens* GK13 is the only known bacterial poly(HA_{MCL}) depolymerase that has been purified and studied in detail [23].

The poly(3HO) depolymerase differs from poly(HA_{SCL}) depolymerases in several of its biochemical properties: it is specific for poly(HA_{MCL}) and for artificial esters such as *p*-nitrophenylacyl esters with six or more carbon atoms in the fatty acid moiety. Poly(3HB) and other poly(HA_{SCL}) are not hydrolyzed. The enzyme is not inhibited by dithioerythritol or by EDTA and therefore apparently does not contain essential disulfide bonds. It is also not dependent on Ca^{2+} or other divalent cations.

The deduced amino acid sequence of the cloned poly(3HO) depolymerase gene does not exhibit significant homology to poly(HA_{SCL}) depolymerases except for small regions in the neighborhood of a lipase-box (Gly-Xaa_1-Ser172-Xaa_2-Gly), an aspartate (Asp228), and a histidine (His260) residue. Site-directed mutagenesis of these amino acids confirmed their involvement in catalysis [26, 67]. Therefore, the poly(3HO) depolymerase apparently also belongs to the group of serine-hydrolases with a catalytic triad in the active center. The three amino acids of the putative active center (Ser172, Asp228, His260) are located in the C-terminal region of the protein which is in contrast to all known poly(HA_{SCL}) depolymerases. It is assumed that the *N*-terminal part of the protein constitutes a polymer-binding site. This assumption has been supported by PCR-induced mutagenesis of the poly(3HO) depolymerase gene and selection of mutants which were impaired in poly(3HO) depolymerase activity but still have almost normal levels of esterase activity with *p*-nitrophenyloctanoate. DNA-sequencing of 30 mutants revealed that most of the mutations were located in the *N*-terminal region of the mature depolymerase (amino acid 1–65) [68]. In particular amino acids near Leu15, Phe50, and Phe63 represented hot spots because these amino acids or neighboring amino acids were mutated in 16 independently isolated mutants (one mutant with Leu15Pro mutation, three Pro17Ser mutants, one Val20Glu mutant, two Ile26Thr mutants, four Phe50Leu mutants, three Phe63Leu mutants, one Phe63Ser mutant, and one Gly64Ser mutant). Since the esterase activity of these mutant poly(3HO) depolymerase proteins was not significantly changed, it is assumed that Leu15, Phe50, and Phe63 and neighboring amino acids are involved in the interaction of the enzyme with its polymeric substrate.

Although there is evidence that all poly(HA) depolymerases cleave the polyesters by the same mechanism (catalytic triad), the poly(3HO) depolymerase differs considerably from poly(HA_{SCL}) depolymerases in terms of primary sequence and polymer-binding. This might be due to different approaches of these enzymes to get access to the polymers reflecting the distinctive physicochemical properties of poly(HA_{SCL}) and poly($H_{MCL}A$) rather than coevolution.

2.6 Mechanism of Poly(HA) Hydrolysis by Poly(HA) Depolymerases

Hydrolysis of end-labeled 3-hydroxybutyrate oligomers by purified *A. faecalis* T_1 poly(3HB) depolymerase showed that the enzyme mainly cleaved the second and third ester linkage from the hydroxyl terminus [69]. However, since the enzyme also hydrolyzes cyclic oligomers, the *A. faecalis* depolymerase has *endo*-hydrolase activity in addition to *exo*-hydrolase activity [18, 70]. Results of

A

-60 -50 -40 -30 -20 -10

PhaZ6*Ple* VAATCYSA**SNYAH**VQ**AGRA**HLVISTGRAAANGSN**Q**DMGLYNTFQMTTLKMTGSNYYVI------GT-**C**--**P**
PhaZ5*Ple* VAATCYTS**SNYAH**VT**AGRA**HNSSGY--ALANGSN**Q**NMGLNNTFYTSTLKQTSPGYYVI------GT-**C**--**P**
PhaZ2*Ple* TAGACYNS**SNYAH**VT**AGRA**HDTGGY--AYTNGSN**Q**KMGLNNTFYTSKLRKTGTNYYVI-----DTT-**C**--**P**
PhaZ3*Ple* AAGACYNA**SNYAH**VT**AGRA**VNSMGY--AKAKGSN**Q**NMGLYNTFTTSKLREAPAGYFTI-----DST-**C**--**P**
PhaZ*Afa*T1 FTCTATTA**SNYAH**VQ**AGRA**HDSGGI--AYANGSN**Q**SMGLDNLFYTSTLAQTAAGYYIV------GN-**C**--P
PhaZ*Afa*AE122 DECQTQSTY**NY**Y**H**KV**AGRA**YSTGNYY--APDYFA**Q**GSDEPMAGSTWGLTALYSEDGGNVWRLGE---**C**--P
PhaZ*Pst*I FTCQEWNA**SN**LS**H**VM**AGRA**YYGYYTV-----GGN**Q**YLGSLSGLSTWVRETAQGHFQA-------GR-**C**SN
PhaZ*Pst*II ETCQVVSA**S**PS**AH**IG**AGRA**YA-GGTSNLRAYAKDDGVDIGGSFDTWSNVPLYEGEPGRWYAQRPAA-**C**G
PhaZ*Rpi* FACTATTA**SNYAH**VQ**AGRA**HDSGG---AYANGSN**Q**SMGLDNVFYTNTLAQTAAGYYVI------GN-**C**--P
PhaZ*Asp* GTGTCTTA**SNYAH**TM**AGRA**YVYGGY--TFALGSN**Q**SMGLWNIFVNKTLKQTSPNYYVI------GT-**C**--P
PhaZ*Csp* PAGTCTTA**SNYAH**VQ**A**N**RA**YQQGGY--AYATASG**Q**NMGLWNVFYTTTLKQTGSNYYVI------GT-**C**--**P**
PhaZ*Cac* GTATCTTA**SNYAH**TL**AGRA**YAAGGY--TYALGSN**Q**NMGLWNVFVTNTLKQTSTNYYVI------GT-**C**--**P**
PhaZ*Cte* PAGTCTTA**SNYAH**VQ**A**N**RA**YQQGGY--AYANGSG**Q**NMGLWNVFYTTTLKQTGSNYYVI------GT-**C**--**P**
PhaZ*Sex* GAAVCVTA**SNYAH**TQ**AGRA**HQSGGY--TYANGSN**Q**NLGLWNVLASSTIKEETAPGYWV-------T-**C**

consensus A: --------**SNYAH**--**AGRA**---------------**Q**------------------------------------**C**--**P**

B

PhaZ1*Ple* GYTQTTSATVTN**H**YV**AGR**------------INVT**Q**YNVLGAR-YGYVTTIPLYYCPSLSGWTDKAN-**C**S-**P**I
PhaZ4*Ple* TYTEKVTATVTG**H**YS**AGR**------------INVN**Q**YLQLGAK-YGYNASLTLYKCEGV--WTNSSS-**C**G-**P**LQ
PhaZ*Afa*AE122 -------ATLDQ**H**IS**AGR**------------LDYTNYADCYLEYGADAFKLTEATVSGQCQWQDDDAS**C**AG**P**V

Fig. 4. Alignment of poly(HA_{SCL}) depolymerase C-terminal substrate binding domains. Two types of poly(HA_{SCL}) depolymerase C-terminal substrate binding domains (*A* and *B*) can be distinguished by amino acid alignment. Amino acids strictly conserved in all depolymerase proteins are indicated by *bold letters*

Brandl et al. [71] using culture fluid of *Acidovorax delafieldii* and cyclic 3HB oligomers were in agreement with the presence of *endo*-hydrolase activity of poly(3HB) depolymerases. Similar results were obtained by de Koning et al. [72] who demonstrated that covalently cross-linked poly(HA_{MCL}) was hydrolyzed completely by *P. fluorescens*. It is assumed that most – if not all – extracellular poly(HA) depolymerases have *endo*- and *exo*-hydrolase activity. Depending on the depolymerase the hydrolysis products are only monomers, monomers and dimers, or a mixture of oligomers (mono- to trimers).

All poly(HA) depolymerases analyzed so far are specific for polymers consisting of monomers in the (*R*)-configuration. Poly[(*S*)-3HB] is not degraded by poly(3HB) depolymerases [73]. However, poly[(*R*,*S*)-3HB] with isotactic diad fractions between 0.68–0.92 showed increased erosion when compared with biologically produced poly[(*R*) 3HB]. In that case a larger fraction of oligomers (dimers, trimers, and tetramers) was observed. The degradation rate of atactic poly[(*R*,*S*)-3HB] was moderately slower [73] or drastically slower [74] than that of poly[(*R*)-3HB]. Apparently, bacterial poly(HA) depolymerases are not able to hydrolyze ester bonds between monomers of the (*S*)-configuration. This conclusion was confirmed recently by Bachmann and Seebach [18]. They synthesized defined linear 3HB-oligomers including 3HB-octamers with different stereoregulatories and demonstrated that (*R*)-3HB-oligomers are cleaved to the dimer and monomer as end products. The 3HB-trimer accumulated as an intermediate hydrolysis product but was finally cleaved to equimolar amounts of mono- and dimer. The hydrolysis rate was significantly lower for the trimer than for the tetramer or the higher oligomers. Oligomers of only (*S*)-3HB-units or any other oligomer without (*R*)-3HB-(*R*)-3HB-linkages such as $[(S)\text{-3-HB-}(R)\text{-3-HB}]_4$ were not hydrolyzed. 3-HB-Octamers, which contained one to several adjacent (*S*)-3HB-units [1 × (*S*)-7 × (*R*); 2 × (*S*)-6 × (*R*), 3 × (*S*)-5 × (*R*) etc.], were hydrolyzed to the same end products as all-(*R*)-octamers (monomer + dimer), but one additional product with one 3-HB-unit more than the number of *S*-3HB-units of the original octamer appeared. In conclusion, the *A. faecalis* poly(3HB) depolymerase needs four 3HB-units of the poly(3HB) chain for maximal hydrolysis rates, two (adjacent) of which must be present in the (*R*)-configuration [18]. The enzyme hydrolyses only (*R*-*R*)-linkages, and adjacent 3-HB units – either (*R*)-units or (*S*)-3HB-units – contribute to the higher activity of the enzyme for oligomers with four or more 3-HB-units compared to the trimer.

2.7 3-Hydroxybutyrate Dimer-Hydrolases and 3-Hydroxybutyrate Oligomer-Hydrolases

The first products of enzymatic hydrolysis of poly(3HB) by purified poly(3HB) depolymerases are a mixture of monomeric and/or oligomeric 3-hydroxybutyrate esters. Some enzymes are able to hydrolyze oligomers and dimers to monomeric 3-hydroxybutyrate after prolonged time of hydrolysis in the presence of an excess of the appropriate depolymerase. These poly(3HB) depolymerases have high endogenous dimer-hydrolase activities (e.g., the poly(3HB) depolymerases of *Comamonas* strains, *P. stutzeri*, *S. exfoliatus*, and the depolymerases of the isolates T107 and Z925 [75]. Other bacteria, which have poly(3HB) depolymerases without or with only low dimer-hydrolase activities, have additional enzymes that are necessary for efficient hydrolysis of the oligomers/dimers to monomeric 3-hydroxybutyrate. These oligomer- or dimer-hydrolases can be located extracellularly (e.g., in *A. faecalis* T1, *Pseudomonas* sp. [70, 76, 76a]), or intracellularly (e.g., in *P. lemoignei, Rhodospirillum rubrum*, and *Zoogoea ramigera* [77–79]). Bacteria, which have intracellular located oligomer- or dimer-hydrolases, must have carrier systems for the uptake of the oligomers/dimers. Recently, an extracellular 3-hydroxybutyrate oligomer hydrolase of *Pseudomonas* sp. including its structural gene has been described [76a]. The purified protein (70 kDa) hydrolyzed the dimeric and trimeric ester of 3-hydroxybutyrate at comparably high rates. Methyl-(*R*)-3-hydroxybutyrate, *p*-nitrophenyl acetate, and *p*-nitrophenylbutyrate, but not *p*-nitrophenyloctanoate could also serve as substrates. Poly(3HB) or olive oil were not cleaved by the purified oligomer hydrolase. The DNA-deduced amino acid sequence encoded a protein of 72.9 kDa and contained a potential *N*-terminal signal-peptide. Surprisingly, neither a typical lipase-box fingerprint nor any homologies to poly(3HB) depolymerases or other proteins of databases could be identified. The authors hypothesized that one of two lipase-box-related sequences of the oligomer hydrolase, in which the first glycine of the lipase-box was replaced by an alanine, could be involved in catalysis [76a].

2.8 Regulation of Poly(HA) Depolymerase Synthesis

The synthesis of poly(HA) depolymerases is highly regulated in most poly (HA)-degrading bacteria, with the expression being generally repressed in the presence of utilizable soluble carbon sources such as glucose or organic acids. This can be shown by culturing the bacteria on solid opaque media which contain poly(3HB) and an additional soluble carbon source. The reduction or inhibition of clearing zone formation in comparison to a control [poly(3HB) without additional carbon source] indicates the degree of depolymerase repression. By this simple method a high number of strains and carbon sources can be analyzed simultaneously. Alternatively, extracellular poly(3HB) depolymerase activity of microorganisms grown in liquid media can be assayed quantitatively in cell-free culture supernatants photometrically [80] or semiquantitatively by a drop test on opaque pure poly(3HB) agar [81, 82]. Most poly(3HB)-degrading bacteria repress poly(3HB) depolymerase synthesis in the presence of a soluble carbon source that permits high growth rates (e.g., glucose or organic acids). However, after exhaustion of the nutrients, the synthesis of poly(HA) depolymerases is derepressed in many strains and halo formation begins [81]. At least in some bacterial strains poly(3HB) depolymerase is expressed even in the absence of the polymer after cessation of growth in liquid cultures. Therefore, an induction mechanism by the polymer itself, as proposed by Chowdhury [83], is not necessary.

In contrast to all other known poly(3HB)-degrading bacteria poly(3HB) depolymerase production by *P. lemoignei* is maximal during growth on succinate in batch culture, and the isolation of poly(3HB) depolymerase from *P. lemoignei* is usually performed from succinate-grown cells [69, 80, 84–86]. At least four poly(3HB) depolymerase isoenzymes have been detected in the supernatant of succinate-grown cells (poly(3HB) depolymerases A, B, C, and a novel yet unpurified poly(3HB) depolymerase). It was found that synthesis of poly(3HB) depolymerase on succinate was pH-dependent and occurred only above pH 7 [87]. Recently, it was shown that transport of succinate is pH-dependent and does not work well above pH 7 in *P. lemoignei* [88]. As a consequence, the bacteria starve even in the presence of residual succinate, and poly(3HB) depolymerase synthesis is derepressed. Analysis of the succinate transport system of *P. lemoignei* revealed that the succinate carrier is energy-dependent and utilizes only the monocarboxylate form of succinate (H-succinate^{-}) but is not able to take up the dicarboxylate (succinate^{2-}). The pH of the culture fluid increases during growth of *P. lemoignei* due to the uptake of succinic acid. As a consequence, the *H*-succinate^{1-} concentration ($pK_A2=5.6$) decreases. The initiation of poly(3HB) depolymerase synthesis above pH 7 can be considered as carbon starvation induced because of insufficient uptake of succinate at high pH [88].

P. lemoignei has at least six poly(HA_{SCL}) depolymerase genes (*phaZ*, Table 2). The depolymerase genes are not clustered on the chromosome except for *phaZ5* [poly(3HB) depolymerase A] and *phaZ2* [poly(3HB) depolymerase B] which are located on the same DNA strand on a 4-kbp DNA-fragment of the chromosome. Both genes are separated by one open reading frame (*phaR*) which is transcribed in the opposite direction (unpublished result). Since the deduced amino acid sequence of *phaR* contains a helix-turn-helix motif, it was hypothesized that *phaR* encodes a DNA-binding protein which regulates the expression of one or several poly(3HB) depolymerases. Based on analysis of *phaZ: lacZ* transcriptional fusions in *P. lemoignei* wild type and in *phaR* null mutants we were able to demonstrate that PhaR is a negative regulator of *phaZ2* (poly(3HB) depolymerase B) but does not affect the expression of *phaZ5* [poly(3HB) depolymerase A] (unpublished result).

Regulation of poly(HA_{MCL}) depolymerase synthesis apparently is similar to that of poly(3HB) depolymerases: high levels of poly(3HO) depolymerase activity were found during growth of *P. fluorescens* GK13 on poly(HA_{MCL}) and on low concentrations of MCL-monomers. The presence of sugars or fatty acids repressed synthesis of poly(3HO) depolymerase [23]. Similar results have been reported for a poly(3HO) depolymerase of *P. maculicola* [89] and a polycaprolactone depolymerase of *F. solani* [44]. Expression of the poly(3HO) depolymerase structural gene of *P. fluorescens* (*phaZ*$_{Pfl}$) is regulated at the level of transcription by a promoter similar in sequence to σ [70] promoters of *E. coli* [67].

2.9 Influence of Physico-Chemical Properties of the Polymer on its Biodegradability

As it has been pointed out above, poly(HA)-degrading microorganisms, in particular the poly(HA)-hydrolyzing enzymes, differ highly in their substrate specificities for various poly(HA). Furthermore the physico-chemistry of the polymer itself also has a strong impact on its biodegradability. The most important factors are: (i) stereoregularity of the polymer, (ii) crystallinity of the polymer, (iii) composition of poly(HA), and (iv) accessibility of the poly(HA) surface. With regard to stereoregularity of the polymer, only ester linkages of monomers in the (*R*)-configuration are hydrolyzed by the depolymerases [18, 73, 74, 90–92] (see above). Regarding crystallinity of the polymer, the degradability of a polyester decreases as the overall crystallinity or its crystallinity phase perfection increases [38, 86, 93]. The amorphous phase of poly(3HB) and copolymers of 3-hydroxybutyrate and 3-hydroxyvalerate is preferentially degraded which has been demonstrated by ^{1}H-NMR imaging [94]. However, even single crystals of poly(3HB) or of copolymers of 3-hydroxybutyrate and 3-hydroxyvalerate can be hydrolyzed by poly(3HB) depolymerases. The splintering of poly(HA_{SCL}) single crystals into fragments with a needle-like morphology suggested that enzymatic hydrolysis proceeds by an edge-attack mechanism [95–100]. Syndiotactic or atactic poly(3HB), which are completely amorphous, are not biodegradable by extracellular poly(3HB) depolymerases [101–103]. However, atactic poly(3HB) is biodegradable by poly(3HB) depolymerase A of *P. lemoignei* if a crystalline support is provided by blending atactic (synthetic) poly(3HB) with a (natural) copolymer of 3-hydroxybutyrate and 3-hydroxyvalerate or by using block polymers of atactic poly(3HB) and pivalolactone [102–104]. This is astonishing because neither atactic poly(3HB) nor poly(pivalolactone) are hydrolyzed by poly(3HB) depolymerase A. Similar results were obtained with blends of atactic poly(3HB) with polycaprolactone and blends of atactic poly(3HB) with poly(L-lactide). Although each of the homopolyesters was resistant to hydrolysis, the blends could be hydrolyzed by poly(3HB) depolymerase A [105]. Recently, it was shown that atactic poly(3HB) oligomers can be biodegraded and utilized as a carbon source by selected bacteria [106]. Therefore, blends of atactic poly(3HB) with polymers that support a crystalline phase can be considered as completely biodegradable. Interestingly, the ability to utilize atactic poly(3HB) oligomers was not restricted to poly(3HB)-degrading bacteria such as *Comamonas* sp. and *A. faecalis* T1. *Ralstonia eutropha* H16 (formerly *Alcaligenes eutrophus*, formerly *Hydrogenomonas eutropha*), which does not produce any extracellular poly(3HB) depolymerase activity, was also able to utilize a significant portion of atactic poly(3HB) oligomers. Therefore, *R. eutropha* must have an enzyme with 3-hydroxybutyrate oligomer hydrolase activity. Recently, the DNA sequence of a 3-hydroxybutyrate oligomer hydrolase of *R. eutropha* has been published (accession no: ABO003701). This enzyme might be responsible for the (partial) utilization of atactic poly(3HB) oligomers. Regarding composition of poly(HA), as an example the rate of degradation by purified *A. faecalis* poly(3HB) depolymerase was slower for poly(3HB) than for copolymers of 3-hydroxybutyrate and 3-hydroxyvalerate [107]. However, similar experiments with other copolymers (41% 3-hydroxyvalerate-content) and results obtained with two of the poly(HA) depolymerases of *P. lemoignei* as well as in situ studies with compost soils showed the reverse order [32, 34, 108]. Apparently, degradation of poly(HA) in complex ecosystems can not be predicted from laboratory experiments using pure cultures and/or purified enzymes only. In addition, not only the length of the side chain but also the position of the hydroxyfunction strongly affects the rate of degradation: polymers of ω-hydroxy fatty acids, which do not have side chains, are good substrates for many lipases and thus are likely to be more susceptible to biodegradation in complex ecosystems [62, 109, 110]. It should be mentioned that the monomeric composition determines the physical properties of a given poly(HA), e.g., changing the monomeric composition will also change the crystallinity of the polymer. With regard to accessibility of poly(HA) surface, maximal rates of enzymatic poly(HA) hydrolysis require the surface to be accessible for all depolymerase molecules. The surface of a piece of a typical poly(3HB) sample used in an in vitro degradation experiment is much lower in comparison to the surface of the same amount of a poly(3HB) granule suspension. Therefore, the depolymerase concentration above which the available polymer surface becomes rate-limiting is lower for polymer films as compared to polymer granule suspensions. For a detailed study see Scandola et al. [111].

2.10
Extracellular Degradation of Polymers Related to Bacterial Poly(HA)

Beside poly(HA) other biodegradable polyesters have been described:

1. Poly(caprolactone) [poly(6-hydroxyhexanoate) (PCL)] is a synthetic unbranched polyester and has been used by man for a several decades. The biodegradability of PCL is well-known, and many PCL-degrading microorganisms have been described [26, 38, 112–115].
2 Cutin is the major component of the cuticle of plant leaves and consists of a mixture of long-chain-length ω-hydroxy fatty acids. Cutin can be hydrolyzed by cutinases which are secreted by many phytopathogenic fungi such as *F. solani*. The cutinase of this fungus has a catalytic triad in its active center and belongs to the family of serine esterases [116]. However, it shares no significant sequence homologies to poly(HA) depolymerases [117]. It has been shown that PCL depolymerases of two *Fusarium* species hydrolyzed cutin in addition to PCL. Since PCL is a chemically produced polymer, which apparently does not occur in nature, and since PCL depolymerase expression was induced by cutin monomers, it is assumed that at least some PCL depolymerases are actually cutinases [44]. Several lipases have PCL depolymerase activity [110, 118]. However, the activity of the enzymes with cutin has not been determined. Cutinases are described in detail in the chapter by Kolzzukudy in this book.
3. Several novel biodegradable polyesters of different compositions have been developed during the last decade, e.g., an aliphatic copolymer of various glycols and dicarboxylic acids (Bionolle) or copolymers of aliphatic diols and aromatic monomers [119–121].
4. Poly(L-malate) [poly(malic acid) (PMA)], is a water-soluble polyanion produced by slime molds and some yeasts such as *Physarum polycephalum* or *Aureobasidium pullulans*, respectively. Its function and metabolism has been studied during the last few years [122–125]. Recently, several PMA-degrading bacteria have been isolated, and a cytoplasmic membrane-bound PMA hydrolase was purified from *Comamonas acidovans* strain 7789 [126] that

differed in many respects from the PMA hydrolase purified from *P. polycephalum* [124]. PMA hydrolases and poly(HA) depolymerases appear to be unrelated.

5. Other polyesters of 2-hydroxyacids are polylactides and copolymers of lactic acid and glycolic acid. They are produced by chemical synthesis and are applied as biocompatible and bioresorbable materials for medical applications. Polylactides of high molecular weight apparently are hardly biodegradable but can be hydrolyzed chemically in the presence of water at slow rates, and low molecular hydrolysis products can be utilized by some microorganisms [127]. Recently, a polylactide-degrading *Amycolatopsis* strain has been described which was able to produce clearing zones on polylactide-containing opaque agar [128].
6. Polyesteramides and polyester-polyurethanes represent two additional classes of chemically produced but biodegradable polymers. The polyesteramide BAK 1095 is a terpolymer of butanediol, aminocaproic acid, and adipinic acid that has been commercialized as a biodegradable packing material by the Bayer AG [129]. Several microorganisms, mainly spore-forming bacteria, have been isolated which produce clearing zones on BAK 1095-containing agar and utilize the degradation products for growth [130]. The microbial degradation of polyester-polyurethanes (PUR) was investigated [131–134]. A cell-associated PUR esterase was purified from *Comamonas acidovorans* TB35 [135]. Beside PUR, low molecular weight poly(lactide), tributyrin (but not triolein), and *p*-nitrophenylacetate were hydrolyzed by the purified enzyme. Poly(3HB) and poly(HA_{SCL}) as well as high molecular weight poly(lactides) did not serve as a substrate. Diethylene glycol and adipinic acid were identified as degradation products. The DNA-deduced amino acid sequence of the PUR esterase contained a lipase-box fingerprint and showed homology to an acetylcholinesterase [136]. Interestingly, the protein contained three putative polymer-binding domains one of which was related to a poly(3HB)-binding domain.

3 Intracellular Degradation of Poly(HA)

3.1 Mobilization of Accumulated Poly(HA) by Bacteria

The mechanism of the intracellular degradation of poly(HA) by bacteria, i.e., the mobilization of a previously accumulated polyester, is poorly understood (see also the chapter by Babel et al. in this book). Most of the research on intracellular poly(3HB) mobilization was done more than 30 years ago. Lemoigne observed in 1925 that 3-hydroxybutyrate was the main product of anaerobic breakdown of poly(3HB) in *Bacillus* "M" [12, 137]. Macrae and Wilkinson [138, 139] noticed a reduction of the poly(3HB) content of *Bacillus megaterium* upon aerobic incubation of poly(3HB)-rich cells in phosphate buffer. The authors found that autolysis of poly(3HB)-rich cells occurred later and to a minor extent compared to poly(3HB)-poor cells and proposed that poly(3HB) might function as a storage compound. Hayward et al. [140] observed that the intracellular poly(3HB) content of *Rhizobium*, *Spirillum*, and *Pseudomonas* species had a maximum followed by a decrease in the stationary growth phase. Similar reports have been published for *Micrococcus halodenitrificans* and *R. eutropha* H16 [21, 141–144]. The authors found that survival of bacteria in the absence of exogeneous carbon sources was dependent on the intracellular poly(3HB) content.

The investigation of intracellular degradation of poly(3HB) was resumed during the last decade by several research groups: Doi and coworkers [145] provided evidence on the cyclic nature of poly(HA) metabolism: resting cells of *R. eutropha* with accumulated poly(3HB) homopolyester incorporated 3-hydroxyvalerate monomer units into the polymer upon incubation in nitrogen-free medium supplemented with pentanoic acid whereas the poly(3HB) content decreased simultaneously. Correspondingly, resting cells with accumulated copolymer of 3-hydroxybutyrate and 3-hydroxyvalerate synthesized poly(3HB) homopolyester upon incubation in nitrogen-free medium supplemented with butyric acid, but the copolymer content decreased simultaneously. Apparently, synthesis and degradation of poly(HA) can occur simultaneously and might depend on the intracellular concentrations of poly(3HB)-related metabolites and cofactors (e.g., NAD, NADH, CoA) [146]. Further evidence for cyclic poly(HA) metabolism was obtained from experiments in which a pulse of ^{14}C-labeled glucose was applied to poly(3HB) accumulating cells of *R. eutropha* [147]. The bacteria incorporated the label into poly(3HB) at a high rate even after the end of net poly(3HB) accumulation. This can only be explained by the assumption of simultaneous accumulation and degradation of poly(3HB). However, the physiological role of the energy-consuming cycle of degradation and synthesis of poly(HA) remains unknown (see also the chapter by Babel et al. in this book). Recently, it was shown that *R. eutropha* H16 could even grow poorly in the absence of any exogeneous carbon source by utilizing previously accumulated poly(3HB) [68, 68a]. Rapid intracellular mobilization of poly(3HB) was also for *Legionella pneumophila* and *Hydrogenophaga pseudoflava* in the absence of an exogeneous carbon source [148, 149].

3.2
Hydrolysis of Native Poly(HA) Granules In Vitro

The mobilization and hydrolysis of native poly(HA) granules by intracellular poly(HA) depolymerases requires the presence of amorphous poly(HA) granules which have retained an intact surface structure. Many efforts have been made to elucidate the fine structure of poly(3HB) granules and its in vivo surface structure over the last four decades. It is generally accepted that intracellular poly(3HB) granules are amorphous [150–152] and are covered by a surface layer ["native" poly(3HB) granules] that is highly sensitive to chemical and physical stress [19, 20, 153]. Analysis of purified native poly(3HB) granules revealed that they contained about 2% of protein and significant traces of phospholipids, especially phosphatidic acid, in addition to around 98% poly(3HB) [154]. Poly(3HB) granules with removed or damaged surface structure are partially crystalline and have lost their catalytic properties. Such granules are referred to as "denatured" poly(3HB) granules. The surface layer of poly(3HB) granules can be seen in thin sections or freeze-fractures of *Rhodospirillum rubrum* [155], *Bacillus cereus, Bacillus megaterium* [156–159], and *Ferrobaccillus ferrooxidans* [160]. Boatman [155] clearly demonstrated that the surface layer of poly(3HB) granules in *R. rubrum* appeared as one electron-dense borderline after staining with osmium tetroxide and uranyl acetate (e.g., Fig. 7 in [155]). The thickness of the surface layer was estimated to be about 4 nm provided that the poly(3HB) granule had been cut in the middle. In contrast, a cytoplasmic unit membrane consists of two separate electron-dense layers with a thickness of about 8.5 nm (Fig. 15 in [155]). The molecular architecture of the poly(3HB) granule surface layer is still controversial. For more details on the architecture, composition and function of poly(HA) granules see [161–170] and the chapter by Babel et al. in this book.

The first detailed biochemical study on intracellular poly(3HB) degradation was done by Merrick and Doudoroff [19]. They analyzed the hydrolysis of native poly(3HB) granules from *B. megaterium* in vitro. The granules themselves had hardly any self-hydrolyzing activity but were rapidly hydrolyzed by crude extracts of *Rhodospirillum rubrum* to mainly hydroxybutyrate and oligomeric esters. Remaining dimers and oligomers were hydrolyzed to monomers by a soluble 3-hydroxybutyrate dimer-hydrolase [78] and metabolized to acetoacetate by NAD-dependent 3-hydroxybutyrate dehydrogenase [171]. Interestingly, crude extracts of *R. rubrum* contain two soluble components which are both necessary for hydrolysis of the polymer. One compound is heat-sensitive and is thought to be the intracellular poly(3HB) depolymerase itself (i-poly(3HB) depolymerase). The second component is heat stable and is called "activator" [19]. Since the effect of the activator on native poly(3HB) granules could be mimicked by trypsin or mild alkali treatment [172], the activator could be a protease that removes proteins from the surface layer of native granules, thus making the polyester chain accessible for the i-poly(3HB) depolymerase. This assumption is supported by the finding that artificial poly(3HB) granules prepared by the Horowitz procedure or by related protocols [24, 25, 27, 28] do not require activator or trypsin for hydrolysis by i-poly(3HB) depolymerase [173, 174]. Artificial granules resemble native poly(3HB) granules but contain only detergents such as SDS or phospholipids as a surface layer and do not contain any proteins. However, and in contrast to the activation of native poly(3HB) granules by trypsin, the activator could not be inhibited by serine protease inhibitors, metallo protease inhibitors, or cysteine protease inhibitors. This indicates that the mechanism of activation by trypsin could be different from activation by the activator. Interestingly, isolated native poly(3HB) granules purified from *R. eutropha* could be hydrolyzed by extracellular poly(3HB) depolymerase B (PhaZ2) from *P. lemoignei* without the presence of trypsin or activator. The same result was obtained with extracellular poly(3HB) depolymerases of most other origins and with extracellular poly(3HB) depolymerase B purified from recombinant *E. coli*. In contrast, hydrolysis of native poly(3HB) granules by purified extracellular poly(3HB) depolymerase A (PhaZ5) of *P. lemoignei* strictly required the presence of trypsin, and trypsin could not be replaced by the activator fraction (unpublished results). The reason for these differences is unknown.

Isolated native poly(3HB) granules of *R. eutropha* have a very low rate of self-hydrolysis which is about two orders of magnitude lower as compared to the hydrolysis rates obtained by *R. rubrum* extracts. However, this endogenous activity can be enhanced about threefold if the poly(3HB)-rich bacteria have been exposed to carbon starvation for some hours before the cells are harvested [68]. The pH optimum of this endogeneous i-poly(3HB) depolymerase activity was at pH 7. A second pH optimum around pH 9 and a soluble i-poly(3HB) depolymerase activity has been published for *R. eutropha* earlier [175]. Self-hydrolysis of native poly(3HB) granules has also been described for native poly(3HB) granules purified from *Zoogloea ramigera* [176] and for native poly(3HO) granules of *P. oleovorans* [177–179]. The pH optimum was in the alkaline range in all cases. Inhibition of self-hydrolysis by serine esterase inhibitors indicated that the active center of i-poly(HA) depolymerases might be related to that of extracellular depolymerases and other serine esterases. This assumption is supported by the presence of potential catalytic triad amino acids (Ser, Asp, His) including a typical lipase box fingerprint in the DNA-deduced amino acid sequences of an open reading frame that has been identified between the two poly(HA) synthase genes in genomes of *P. oleovorans* [180], *Pseudomonas aeruginosa* [181], *Pseudomonas resinovorans* (accession no. AAD26366), and *Pseudomonas* sp. (accession no. BAA36201). These open reading frames are assumed to be the structural genes of the intracellular poly(3HO) depolymerases.

3.3
Properties of the i-Poly(3HB) Depolymerase of *R. rubrum*

The i-poly(3HB) depolymerase of *R. rubrum* is the only i-poly(3HB) depolymerase that has been purified [174]. The enzyme consists of one polypeptide of 30–32 kDa and has a pH and temperature optimum of pH 9 and 55 °C, respectively. A specific activity of 4 mmol released 3-hydroxybutyrate/min × mg protein was determined (at 45 °C). The purified enzyme was inactive with denatured poly(3HB) and had no lipase-, protease-, or esterase activity with *p*-nitrophenyl fatty acid esters (2–8 carbon atoms). Native poly(3HO) granules were not hydrolyzed by i-poly(3HB) depolymerase, indicating a high substrate specificity similar to extracellular poly(3HB) depolymerases. Recently, the DNA sequence of the i-poly(3HB) depolymerase of *R. eutropha* was published (AB07612). Surprisingly, the DNA-deduced amino acid sequence (47.3 kDa) did not contain a lipase box fingerprint. A more detailed investigation of the structure and function of bacterial i-poly(HA) depolymerases will be necessary in future.

Acknowledgments. This work was supported by the Deutsche Forschungsgemeinschaft, the Graduiertenkolleg "Chemische Aktivitäten von Mikroorganismen," and the "Fonds der Chemischen Industrie." I thank Andreas Schirmer for critically reading the manuscript.

References

1. Anderson AJ, Dawes EA (1990) Microbiol Rev 54:450-472
2. Doi Y (1990) VHC, New York
3. Steinbüchel A (1991) In: Biomaterials Byrom B (ed) Macmillan, Basingstoke, pp 122-213
4. Müller HM, Seebach D (1993) Angew Chem Int Ed Engl 32:477-502
5. Seebach D, Brunner A, Bachmann BM, Hoffmann T, Kühnle FNM, Lengweiler UD (1995) Ernst Schering Research Foundation, Berlin
6. Sharma R, Ray AR (1995) J Mat Sci: Rev Macromol Phys C35:327-359
7. Marchessault RH (1996) TRIP 4:163-168
8. Sasikala C, Ramana CV (1996) Adv Appl Miccrobiol 42:97-218
9. Steinbüchel A, Füchtenbusch B (1998) Tibtech 16:419-426
10. Braunegg G, Lefebvre G, Genser KF (1998) J Biotechnol 65:127-161
11. Madison LL, Huisman GW (1999) Microbiol Mol Biol Rev 63:21-53
12. Lemoigne M (1925) Ann Inst Pasteur 39:144-155
13. Steinbüchel A, Valentin HE (1995) FEMS Microbiol Lett 128:219-228
14. Saito T, Suzuki K, Yamamoto J, Fukui T, Miwa K, Tomita K, Nakanishi S, Odani S, Suzuki JI, Ishikawa K (1989) J Bacteriol 171:184-189
15. Brandl H, Bachhofen R, Mayer J, Wintermantel E (1995) Can J Microbiol 41[Suppl 1]:143-153
16. Jendrossek D, Schirmer A, Schlegel HG (1996) Appl Microbiol Biotechnol 46:451-463
17. Jendrossek D (1998) Polym Degrad Stab 59:317-325
18. Bachmann BM, Seebach D (1999) Macromolecules 32:1777-1784
19. Merrick JM, Doudoroff M (1964) J Bacteriol 88:60-71
20 Merrick JM, Lundgren DG, Pfister RM (1965) J Bacteriol 89:234-239
21. Hippe H, Schlegel HG (1967) Arch Mikrobiol 56:278-299
22. de Koning GJM, Lemstra PJ (1992) Polymer 33:3304-3306
23. Schirmer A, Jendrossek D, Schlegel HG (1993) Appl Environm Microbiol 59:1220-1227
24. Ramsay BA, Saracovan I, Ramsay JA, Marchessault RH (1994) J Environ Polym Deg 2:1-7
25. Marchessault RH, Morin FG, Wong S, Saracovan I (1995) Can J Microbiol 41[Suppl 1]:138-142
26. Schirmer A, Matz C, Jendrossek D (1995) Can J Microbiol 41[Suppl 1]:170-179
27. Horowitz DM, Sanders JKM (1994) J Am Chem Soc 116:2695-2702
28. Horowitz DM, Sanders JKM (1995) Can J Microbiol 41[Suppl 1]:115-123
29. Horowitz DM, Brennan EM, Koon JJ, Gerngross TU (1999) Macromolecules 32:3347-3352
30 Molitoris KP, Moss ST, de Koning G, Jendrossek D (1996) Appl Microbiol Biotechnol 46:570-579
31. Brandl H, Püchner P (1992) Biodegradation 2:237-243
32. Briese B-H, Jendrossek D, Schlegel HG (1994) FEMS Microbiol Lett 117:107-112
33. Mergaert J, Webb A, Anderson C, Wouters A, Swings J (1993) Appl Environ Microbiol 59: 3233-3238
34. Mergaert J, Webb A, Anderson C, Wouters A, Swings J (1994) J Environ Polym Deg 2: 177-183
35. Mergaert J, Wouters A, Anderson C, Swings J (1995) Can J Microbiol 41[Suppl 1]:154-159
36. Mergaert J, Schirmer A, Hauben L, Mau M, Jendrossek D, Swings J (1996) Int J Syst Bacteriol 46:769-773
37. Mergaert J, Swings J (1996) J Ind Microbiol 17:463-469
38. Nishida H, Tokiwa Y (1993) J Environm Polym Degrad 1:65-80
39. Kasuya K, Takagi K, Ishiwatari S, Yoshida Y, Doi Y (1998) Polym Degrad Stab 59:327-332
40. Ohura T, Aoyagi Y, Takagi K, Yoshida Y, Kasuya K, Doi Y (1999) Polym Degrad Stab 63:23-29
41. McLellan DW, Halling PJ (1988) FEMS Microbiol Lett 52:215-218
42. Brucato CL, Wong SS (1991) Arch Biochem Biophys 290:497-502
43. Matavulj M, Molitoris HP (1992) FEMS Microbiol Rev 103:323-332
44. Murphey CA, Cameron JA, Huang SJ, Vinopal RT (1996) Appl Environm Microbiol 62:456-460

45. Oda Y, Osaka H, Urakami T, Tonomura K (1997) Curr Microbiol 34:230-232
46. Han J-S, Son Y-J, Chang C-S, Kim M-N (1998) J. Microbiol 36:67-73
47. Neumeier S (1994) Diploma thesis, Universität Regensburg
48. Gruchulski P, Bouthillier F, Kazlauskas RJ, Serreqi AN, Schrag JD, Ziomek E, Cygler M (1994) Biochemistry 33:3494-3500
49. Klingbeil B, Kroppenstedt R, Jendrossek D (1996) FEMS Microbiol Lett 142:215-221
50 Doi Y, Kitamura S, Abe H (1995) Macromolecules 28:4822-4828
51. Briese B-H, Schmidt B, Jendrossek D (1994) J Environm Polym Degrad 2:75-87
52. Jendrossek D, Frisse A, Behrends A, Andermann M, Kratzin HD, Stanislawski T, Schlegel HG (1995) J Bacteriol 177:596-607
53. Jaeger K-E, Ransac S, Dijkstra BW, Colson D, van Heuvel M, Misset O (1994) FEMS Microbiol Rev 15:29-63
53a. Jaeger K-E, Dijkstra BW, Reetz MT (1999) Annu Rev Microbiol 53:315-51
54. Shinohe T, Nojiri M, Saito T, Stanislawski T, Jendrossek D (1996) FEMS Microbiol Lett 141:103-109
55. Nojiri M, Saito T (1997) J. Bacteriol 179:6965-6970
56. Behrends A, Klingbeil B, Jendrossek D (1996) FEMS Microbiol Lett 143:191-194
57. Ohura T, Kasuya K, Doi Y (1999) Appl Environm Microbiol 65:189-197
58. Fukui T, Narikawa T, Miwa K, Shirakura Y, Saito T, Tomita K (1988) Biochim Biophys Acta 952:164-171
59. Briese B-H, Jendrossek D (1998) Macromol Sympos 130:205-216
60. Kasuya K, Inoune Y, Tanaka T, Akehata T, Iwata T, Fukui T, Doi Y (1997) Appl Environm Microbiol 63:4844-4852
61. Shinomiya M, Iwata T, Kasuya K, Doi Y(1997) FEMS Microbiol Lett 154:89-94
62. Doi Y, Mukai K, Kasuya K, Yamada K (1994) In: Doi Y, Fukuda K (eds) Biodegradable plastics and polymers. Elsevier, Amsterdam, pp 39-51
63. Kasuya K, Inoune Y, Yamada K, Doi Y (1995) Polym Degrad Stab 48:167-174
64. Kita K, Mashiba S, Nagita M, Ishimaru K, Okamoto K, Yanase H, Kato N (1997) Biochim Biophys Acta 1352:113-122
65. Little E, Bork P, Doolittle RF (1994) J Mol Evol 39:631-643
66. Watanabe T, Ito Y, Yamada T, Hashimoto M, Sekine S, Tanaka H (1994) J Bacteriol 176:4465-4472
67. Schirmer A, Jendrossek D (1994) J Bacteriol 176:7065-7073
68. Jendrossek D, Schirmer A, Handrick R (1997) In: Eggink G, Steinbüchel A, Poirier Y, Witholt B (eds) (1996) International Symposium on bacterial polyhydroxyalkanoates. NRC Research Press, Ottawa, Canada, pp 89-101
68a. Handrik R, Jendrossek D (2000) J Bacteriol 182:5916-5918
69. Nakayama K, Saito T, Fukui T, Shirakura Y, Tomita K (1985). Biochim Biophys Acta 827:63-72
70. Shirakura Y, Fukui T, Saito T, Okamoto Y, Narikawa T, Koide K, Tomita K, Takemasa T, Masamune S (1986) Biochim Biophys Acta 880:46-53
71. Brandl H, Aeberli B, Bachofen R, Schwegler I, Müller H-M, Bürger MH, Hoffmann T, Lengweiler UD, Seebach D (1995) Can J Microbiol 41[Suppl 1]:180-186
72. de Koning GJM, van Bilsen HMM, Lemstra PJ, Hazenberg W, Witholt B, Preusting H, van der Galien JG, Schirmer A, Jendrossek D (1994) Polymer 35:2090-2097
73. Doi Y, Kumagai Y, Tanahashi N, Mukai K (1992) In: Vert M (ed) Biodegradable polymers and plastics. Royal Society of Chemistry, London, pp 139-148
74. Abe H, Matsubara I, Doi Y, Hori Y, Yamaguchi A (1994) Macromolecules 27:6018-6025
75. Shiraki M, Shimada T, Tatsumichi M, Saito T (1995) J Environm Polym Deg 3:13-21
76. Shirakura Y, Fukui T, Tanio T, Nakayama K, Matsuno R, Tomita K (1983) Biochim Biophys Acta 748:331-339
76a. Zhang K, Shiraki M, Saito T (1997) J Bacteriol 179:72-77
77. Delafield FP, Cooksey KE, Doudoroff M (1965) J Biol Chem 240:4023-4028
78. Merrick JM, Yu CI (1966) Biochemistry 5:3563-3568
79. Tanaka Y, Saito T, Fukui T, Tanio T, Tomita K (1981) Eur J Biochem 118:177-182

80. Delafield FP, Doudoroff M, Palleroni NJ, Lusty CJ, Contopoulos R (1965) J Bacteriol 90:1455–1466
81. Jendrossek D, Knoke I, Habibian RB, Steinbüchel A, Schlegel HG (1993) J Environ Polym Deg 1:53–63
82. Jendrossek D, Müller B, Schlegel HG (1993) Eur J Biochem 218:701–710
83. Chowdhury AA (1963) Arch Mikrobiol 47:167–200
84. Lusty CJ, Doudoroff M (1966) Proc Natl Acad Sci USA 56:960–965
85. Müller B, Jendrossek D (1993) I Appl Microbiol Biotechnol 38:487–492
86. Tomasi G, Scandola M, Briese B-H, Jendrossek D (1996) Macromolecules 29:507–513
87. Stinson MW, Merrick JM (1974). J Bacteriol 119:152–161
88. Terpe K, Kerkhoff K, Pluta E, Jendrossek D (1999) Appl Environm Microbiol 65: 1703–1709
89. Foster LJR, Zervas SJ, Lenz RW, Fuller RC (1995) Biodegradation 6:67–73
90. Kemnitzer JE, McCarthy SP, Gross RA (1992) Macromolecules 25:5927–5934
91. Abe H, Matsubara I, Doi Y (1995) Macromolecules 28:844–853
92. Hocking PJ, Timmins MR, Scherer TM, Fuller RC, Lenz RW, Marchessault RH (1995) J M S Pure Appl Chem A32:889–894
93. Kumagai Y, Kanasawa Y, Doi Y (1992) Makromol Chem 193:53–57
94. Spyros A, Kimmich R, Briese BH, Jendrossek D (1997) ^{1}H NMR imaging study of enzymatic degradation in poly(3-hydroxybutyrate) and poly(3-hydroxybutyrate-*co*-3-hydroxyvalerate). Macromolecules 30:8218–8225
95. Hocking PJ, Marchessault RH, Timmins MR, Lenz RW, Fuller RC (1996) Macromolecules 29:2472–2478
96. Sykes KE, Barham PJ, Hobbs JK, Fuljames C, Debouzy I, Baker AA, McMaster TJ, Miles MJ, Jackson R, Farrell R, Gross RA (1997) In: Eggink G, Steinbüchel A, Poirier Y, Witholt B (eds) 1996 International symposium on bacterial polyhydroxyalkanoates, NRC research press, Ottawa, Canada, pp 66–77
97. Nobes GAR, Marchessault RH, Chanzy H, Briese BH, Jendrossek D (1996) Macromolecules 29:8330–8333
98. Nobes GAR, Marchessault RH, Briese BH, Jendrossek D (1998) J Environm Polym Degrad 5:99–107
99. Iwata T, Doi Y, Kasuya K, Inoue Y (1997) Macromolecules 30:833–839
100. Iwata T, Doi Y, Tanaka T, Akehata T, Shiromo M, Teramachi S (1997) Macromolecules 30:5290–5296
101. Abe H, Doi Y (1996) Macromolecules 29:8683–8688
102. Scandola M, Focarete ML, Adamus G, Sikorska W, Baranowska I, Swierczek S, Gnatowski M, Kowalczuk M, Jedlinski Z (1997) Macromolecules 30:2568–2574
103. Scandola M, Focarete ML, Gazzano M, Matuszowicz A, Sikorska W, Adamus G, Kurcok P, Kowalczuk M, Jedlinski Z (1997) Macromolecules 30:7743–7748
104. Scandola M (1997) In: Eggink G, Steinbüchel A, Poirier Y, Witholt B (eds) 1996 International symposium on bacterial polyhydroxyalkanoates, NRC Research Press, Ottawa, Canada, pp 20–27
105. Focarete ML, Ceccorulli G, Scandola M, Kowalczuk M (1998) Macromolecules 31:8485–8492
106. Focarete ML, Scandola M, Jendrossek D, Adamus G, Sikorska W, Kowalczuk M (1999) Macromolecules 32:4814–4818
107. Doi Y, Kanesawa Y, Kunioka M, Saito T (1990) Macromolecules 23:26–31
108. Kanewasa Y, Tanahashi N, Doi Y, Saito T (1994) Polym Degrad Stab 45:179–185
109. Mukai K, Doi Y, Sema Y, Tomita K (1993) Biotechnol Lett 15:601–604
110. Jaeger K-E, Steinbüchel A, Jendrossek D (1995) Appl Environm Microbiol 61:311–3118
111. Scandola M, Focarete ML, Frisoni G (1998) Macromolecules 31:3846–3851
112. Fields RD, Rodriguez F, Finn RK (1974) J Appl Polym Sci 18:3571–3579
113. Kavelman R, Kendrick B (1978) Mycologia 70:87–103
114. Benetict CV, Cook WJ, Jarrett P, Cameron JA, Huang SJ, Bell JP (1983) J Appl Polym Sci 28:327–334

115. Goldberg D (1995) J Environm Polym Degrad 3:61-67
116. Martinez C, de Geus P, Lauwereys M, Matthyssens G, Cambillau C (1992) Nature 356:615-618
117. Soliday CL, Flurkey WH, Okita TW, Kolatturkudy PE (1984). Proc Natl Acad Sci USA 81:3939-3943
118. Oda Y, Oida N, Urakami T, Tonomura K (1997) FEMS Microbiol Lett 152:339-343
119. Nishioka M, Tuzuki T, Wanajyo Y, Oonami H, Horiuchi T (1994). In: Doi Y, Fukuda K (eds) Biodegradable plastics and polymers. Elsevier Science B V, pp 584-590
120. Witt U, Müller R-J, Deckwer W-D (1995) J Environ Polym Deg 3:215-223
121. Witt U, Müller R-J, Deckwer W-D (1997) J Environm Polym Degrad 5:81-89
122. Fischer H, Erdmann S, Holler E (1989). Biochemistry 28:5219-5226
123. Angerer B, Holler E (1995) Biochemistry 34:14,741-14,751
124. Korherr C, Roth M, Holler E (1995) Can J Microbiol 41 [Suppl 1]:192-199
125. Liu S, Steinbüchel A (1996) Appl Microbiol Biotechnol 46:273-278
126. Gödde C, Liebergesell M, Steinbüchel A (1999) FEMS Microbiol Lett 173:365-372
127. Torres A, Li SM, Roussos S, Vert M (1996) Appl Environm Microbiol 62:23,393-23,397
128. Pranamuta H, Tokiwa Y, Tanaka H (1997) Appl Environm Microbiol 63:1637-1640
129. Bayer AG (1995) Biodegradable thermoplastic BAK1095*. KU-marketing, Informations-systeme, KU 48.872
130. Kohlhaussen S, Steffen M, Steger R, Koch R (1999) Poster shown on the annual meeting of the Vereinigung für Allgemeine und Angewandte Mikrobiologie, 7-10 March 1999, Göttingen, Germany
131. Nakajima-Kambe T, Onuma F, Kimpara N, Nakahara T (1995) FEMS Microbiol Lett 129:39-42
132. Nakajima-Kambe T, Onuma F, Akutso Y, Nakahara T (1997) J Ferm Biotechnol 83:456-460
133. Nakajima-Kambe T, Shigeno-Akutsu Y, Nomura N, Onuma F, Nakahara T (1999) Appl Environm Microbiol 65:134-140
134. Hiltunen K, Seppälä JV, Itävaara M, Härkönen M (1997) J Environm Polym Degrad 3:167-173
135. Akutsu T, Nakajima-Kambe T, Nomura N, Nakahara T (1998) Appl Environm Microbiol 64:62-67
136. Nomura N, Shigeno-Akutsu Y, Nakajima-Kambe T, Nakahara T (1998) J Ferm Bioeng 86:339-345
137. Lemoigne M (1925) Ann Inst Pasteur 41:148-159
138. Macrae RM, Wilkinson JF (1958) J Gen Microbiol 19:210-222
139. Macrae RM, Wilkinson JF (1958) Proc R Phys Soc Edinburgh 27:73-82
140. Hayward AC, Forsyth WGC, Roberts JB (1959) J Gen Microbiol 20:510-518
141. Bartha R (1962) Arch Mikrobiol 41:313-350
142. Hippe H (1967) Arch Mikrobiol 56:248-277
143. Sierra G, Gibbons NE (1962) Can J Microbiol 8:255-269
144. Braunegg G, Kornetti L (1984) Biotechnol Lett 6:825-829
145. Doi Y, Segawa A, Kawaguchi Y, Kunioka M (1990) FEMS Microbiol Lett 67:165-170
146. Senior PJ, Dawes EA (1973) Biochem J 134:225-238
147. Taidi B, Mansfield DA, Anderson A (1995) FEMS Microbiol Lett 129:201-206
148. James BW, Mauchline WS, Dennis PJ, Keevil CW, Wait R (1999) Appl Environm Microbiol 65:822-827
149. Choi MH, Yoon SC, Lenz RW (1999) Appl Environm Microbiol 65:1570-1577
150. Barnard, GN, Sanders JKM (1988) FEBS Let 231:16-18
151. Barnard, GN, Sanders JKM (1989) Macromolecules 24:4583-4588
152. Amor SR, Rayment T, Sanders JKM (1991) Macromolecules 24:4583-4588
153. Merrick JM (1965) J Bacteriol 90:965-969
154. Griebel R, Smith Z, Merrick JM (1968) Biochem 7:3676-3681
155. Boatman ES (1964) J Cell Biol 20:297-311
156. Lundgren DG, Pfister RM, Merrick JM (1964) J Gen Microbiol 34:441-446

157. Pfister RM, Lundgren DG (1964) J Bacteriol 88:1119–1129
158. Ellar D, Lundgren DG, Okamura K, Marchessault RH (1968) J Mol Biol 35:489–502
159. Dunlop W, Robards AW (1973) J Bacteriol 114:1271–1280
160. Wang WS, Lundgren DG (1969) J Bacteriol 97:947–950
161. Pieper-Fürst U, Madkour MH, Mayer F, Steinbüchel A (1994) J Bacteriol 176:4328–4337
162. Pieper-Fürst U, Madkour MH, Mayer F, Steinbüchel A (1995) J Bacteriol 177:2513–2523
163. Wieczoreck R, Pries A, Steinbüchel A, Mayer F (1995) J Bacteriol 177:2425–2435
164. Wieczoreck R, Steinbüchel A, Schmidt B (1996) FEMS Microbiol Lett 135:23–30
165. Valentin HE, Stuart ES, Fuller RC, Lenz RW, Dennis D (1998) J Biotechnol 64:145–157
166. McCool GJ, Cannon MC (1999) J Bacteriol 181:585–592
167. Steinbüchel A, Aerts K, Babel W, Föllner C, Liebergesell M, Madkour MH, Mayer F, Pieper-Fürst U, Pries A, Valentin HE, Wieczorek R (1995) Can J Microbiol 41[Suppl 1]:94–105
168. Gerngross TU, Reilly P, Stubbe J, Sinskey AJ, Peoples OP (1993) J Bacteriol 175: 5289–5293
169. Stuart ES, Lenz RW, Fuller RC (1995)Can J Microbiol 41[Suppl 1]:84–93
170. Mayer F, Madkour MH, Pieper-Fürst U, Wieczorek R, Liebergesell M, Steinbüchel A (1996) J Gen Appl Microbiol 42:445–455
171. Shuster CW, Doudoroff M (1962) J Bacteriol 108:782–789
172. Griebel RJ, Merrick JM (1971) J Bacteriol 108:782–789
173. Merrick JM (1998) Lecture on occasion of the international symposium on biological polyhydroxyalkanoates 1998, Tokyo, 9–11 Sept
174. Handrick R, Jendrosssek D (1998) In: Steinbüchel A (ed) Biochemical principles and mechanisms of biodegradation and biodegradation of polymers. Wiley-VCH, Weinheim, pp 57–67
175. Saito T, Takizawa K, Saegusa H (1995) Can J Microbiol 41[Suppl 1]:187–191
176. Saito T, Saegusa H, Miyata Y, Fukui T (1992). FEMS Microbiol Rev 103:333–338
177. Stuart ES, Foster LJR, Lenz RW, Fuller RC (1996) Int J Biol Macromol 19:171–176
178. Foster LJR, Lenz RW, Fuller RC (1994) FEMS Microbiol Lett 118:279–282
179. Foster LJR, Stuart ES, Tehrani A, Lenz RW, Fuller RC (1996) Int J Biol Macromol 19:177–183
180. Huisman GW, Wonink E, Meima R, Terpstra B, Witholt B (1991) J Biol Chem 266:2191–2198
181. Timm A, Steinbüchel A (1992) Eur J Biochem 209:15–30
182. Tanio T, Fukui T, Shirakura Y, Saito T, Tomita K, Kaiho T (1982) Eur J Biochem 124:71–77
183. Kita K, Ishimaru K, Teraoka M, Yanase H, Kato N (1995) Appl Environm Microbiol 61:1727–1730
184. Sadocco P, Nocerino S, Dubini-Paglia E, Seves A, Elegir G (1997) J Environm Polym Degrad 5:57–65
185. Jendrossek D, Backhaus M, Andermann M (1995) Can J Microbiol 41[Suppl 1]:160–169
186. Kasuya K, Doi Y, Yao T (1994) Polym Degrad Stab 45:379–386
187. Mukai K, Yamada K, Doi Y (1993). Polym Degrad Stab 41:85–91
188. Yamada K, Mukai K, Doi Y (1993) Int J Biol Macromol 15:215–220
189. Mukai K, Yamada K, Doi Y (1994. Polym Degrad Stab 43:319–327
190. Uefuji M, Kasuya K, Doi Y (1997) Polym Degrad Stab 58:275–281
191. Janssen PH, Harfoot CG (1990) Arch Microbiol 154:253–259
192. Janssen PH, Schink B (1993) Biodegradation 4:179–185
193. Budwill K, Fedorak PM, Page WJ (1992) Appl Environ Microbiol 58:1398–1401
194. Mergaert J, Glorieux G, Hauben L, Storms V, Mau M, Swings J (1996) Syst Appl Microbiol 19:407–413
195. Schöber U, Thiel C, Jendrossek D (2000) Appl Environm Microbiol 66:1385–1392
196. Kurusu Y, Kohama K, Uchida Y, Saito T, Yukawa H (1994) In: Doi Y, Fukuda K (eds) Biodegradable plastics and polymers. Elsevier Science B V, Amsterdam, pp 357–361

Received: December 1999

Author Index Volume 1–71

Author Index Volumes 1–50 see Volume 50

Subject Index

MIX
Papier aus verantwortungsvollen Quellen
Paper from responsible sources
FSC® C105338

If you have any concerns about our products,
you can contact us on
ProductSafety@springernature.com

In case Publisher is established outside the EU,
the EU authorized representative is:
Springer Nature Customer Service Center GmbH
Europaplatz 3, 69115 Heidelberg, Germany

Printed by Libri Plureos GmbH
in Hamburg, Germany